Spectroscopy and Optical Diagnostics for Gases

Ronald K. Hanson • R. Mitchell Spearrin
Christopher S. Goldenstein

Spectroscopy and Optical Diagnostics for Gases

Springer

Ronald K. Hanson
Department of Mechanical Engineering
Stanford University
Stanford, CA, USA

R. Mitchell Spearrin
Mechanical and Aerospace Engineering Department
University of California, Los Angeles (UCLA)
Los Angeles, CA, USA

Christopher S. Goldenstein
School of Mechanical Engineering
Purdue University
West Lafayette, IN, USA

ISBN 978-3-319-23251-5 ISBN 978-3-319-23252-2 (eBook)
DOI 10.1007/978-3-319-23252-2

Library of Congress Control Number: 2015949092

Springer Cham Heidelberg New York Dordrecht London

Cover illustration: Toluene planar laser-induced fluorescence of nitrogen shock bifurcation (courtesy J. Yoo, D.F. Davidson, 2010, High Temperature Gasdynamics Laboratory, Stanford University)

Printed on acid-free paper

Springer International Publishing AG Switzerland is part of Springer Science+Business Media (www.springer.com)

Preface

This text provides an introduction to the science that governs the interaction of light and matter (in the gas phase). It provides readers with the basic knowledge to exploit light-matter interactions to develop quantitative tools for gas analysis (i.e. optical diagnostics) and understand and interpret the results of spectroscopic measurements. The text is organized to cover three sub-topics of gas-phase spectroscopy: (1) spectral line positions, (2) spectral line strengths, and (3) spectral lineshapes by way of absorption, emission, and scattering interactions. Greater focus is dedicated to absorption and emission interactions. The latter part of the book describes optical measurement techniques and equipment for practical applications. The text is written for graduate students, advanced undergraduate students, and practitioners across a range of applied sciences including mechanical, aerospace, and chemical engineering.

The text grew out of a course, *Introduction to Spectroscopy and Laser Diagnostics for Gases (ME364)*, first offered in 1977, in response to the growing use of spectroscopic diagnostics in the research conducted by graduate students in the High Temperature Gasdynamics Laboratory (HTGL) at Stanford. At the time, the field of spectroscopy was undergoing a revolution owing to the development and application of lasers, and many of the standard texts on laser physics and spectroscopy dealt primarily with theory, e.g., quantum mechanics and optics, rather than engineering applications. There was thus no single textbook that seemed suitable for students (or for the professor!) with traditional mechanical engineering backgrounds, nor was there a suitable text that focused on applied measurements in high temperature gases. As a result, I initially used various introductory texts, written for undergraduates and master's level students in physics and chemistry, and I supplemented these materials with my own notes for topics not treated in existing books but critical to the diagnostics employed in the HTGL.

Over time, my own notes became more complete, until finally in 2001 an energetic student, Michael Webber, helped put the notes into electronic form for use as a course reader at Stanford. The material continued undergoing expansion and refinement until two recent doctoral students, Mitchell Spearrin and Christopher Goldenstein, agreed to help convert my reader into a textbook. They are now my co-authors, having made new contributions to the latter half of the book.

I am deeply indebted to the many graduate students who have passed through their doctoral studies in my group at Stanford and contributed in many ways to the evolution of this text. Though too many to mention all by name, I must acknowledge one student, Xing Chao, who heroically converted my notes into powerpoint lectures and created many improved figures for the lecture slides and the text.

Finally, I want to acknowledge the pleasure I have enjoyed in working to develop and apply laser-based spectroscopic diagnostics to engineering problems. It has been a very rewarding experience, particularly in watching Stanford's mechanical engineering graduates become leaders in the field of applied spectroscopy.

Stanford, CA, USA
June 2015

Ronald K. Hanson

Contents

Preface ... v

1 Introduction ... 1
- 1.1 Role of Quantum Mechanics ... 1
- 1.2 Emission and Absorption Spectra ... 1
- 1.3 Planck's Law ... 3
- 1.4 Wavelength, Frequency, and Other Units and Conversions ... 4
- 1.5 Spectral Regions ... 5
- 1.6 Basic Elements of Spectroscopy ... 6
  - 1.6.1 Positions, Strengths, and Shapes of Lines ... 6
- 1.7 Typical Absorption Spectroscopy Setup ... 6
- 1.8 Beer's Law of Absorption ... 7
- 1.9 Spectral Absorption Coefficient ... 8
- 1.10 Boltzmann Distribution ... 8
- Reference ... 8

2 Diatomic Molecular Spectra ... 9
- 2.1 Interaction Mechanism for EM Radiation with Molecules ... 9
  - 2.1.1 Microwave Region: Rotation ... 10
  - 2.1.2 Infrared Region: Vibration ... 11
  - 2.1.3 Ultraviolet and Visible Regions: Electronic ... 11
  - 2.1.4 Summary of Background ... 11
- 2.2 Rotational Spectra: Simple Model ... 12
  - 2.2.1 Rigid Rotor (RR) ... 12
  - 2.2.2 Classical Mechanics ... 13
  - 2.2.3 Quantum Mechanics ... 13
  - 2.2.4 Rotational Energy ... 14
  - 2.2.5 Absorption Spectrum ... 15
  - 2.2.6 Usefulness of Rotational Line Spacing ... 16
  - 2.2.7 Rotational Partition Function ... 17
  - 2.2.8 Rotational Temperature ... 17
  - 2.2.9 Intensities of Spectral Lines ... 19
- 2.3 Vibrational Spectra: Simple Model ... 19
  - 2.3.1 Simple Harmonic Oscillator ... 19
  - 2.3.2 Classical Mechanics ... 19

2.3.3 Quantum Mechanics 21
2.3.4 Vibrational Partition Function 21
2.3.5 Vibrational Temperature 22
2.4 Improved Models of Rotation and Vibration 23
2.4.1 Non-rigid Rotation 23
2.4.2 Anharmonic Oscillator 24
2.4.3 Typical Correction Magnitudes 25
2.5 Rovibrational Spectra: Simple Model 27
2.5.1 Born–Oppenheimer Approximation 27
2.5.2 Spectral Branches 28
2.6 Rovibrational Spectra: Improved Model 30
2.6.1 Breakdown of Born–Oppenheimer Approximation 30
2.6.2 Spectral Branches 31
2.6.3 Rotational Constant 32
2.6.4 Bandhead 32
2.6.5 Finding Key Parameters: B_e, α_e, ω_e, x_e 33
2.6.6 Effects of Isotopic Substitution 35
2.6.7 Hot Bands 36
2.7 Electronic Spectra of Diatomic Molecules 37
2.7.1 Potential Energy Wells 37
2.7.2 Types of Spectra 39
2.7.3 Rotational Analysis 40
2.7.4 Vibrational Analysis 44
2.8 Summary 46
2.9 Exercises 48
References 49

3 Bond Dissociation Energies 51
3.1 Birge–Sponer Method 51
3.2 Thermochemical Approach 53
3.3 Predissociation 53
3.3.1 HNO 54
3.3.2 N_2O 55
3.4 Exercises 56
Reference 57

4 Polyatomic Molecular Spectra 59
4.1 Rotational Spectra of Polyatomic Molecules 59
4.1.1 Linear Molecules 60
4.1.2 Symmetric Top 61
4.1.3 Spherical Top 64
4.1.4 Asymmetric Rotor 64
4.1.5 Rotational Partition Function 64
4.2 Vibrational Bands of Polyatomic Molecules 66
4.2.1 Number of Vibrational Modes 66
4.2.2 Parallel and Perpendicular Modes 66

4.2.3 Types of Bands ... 69
4.2.4 Relative Strengths ... 70
4.2.5 Vibrational Partition Function ... 71
4.3 Rovibrational Spectra of Polyatomic Molecules ... 71
4.3.1 Linear Polyatomic Molecules ... 71
4.3.2 Symmetric Top Molecules ... 73
4.4 Exercises ... 76
References ... 78

5 Effects of Nuclear Spin: Rotational Partition Function and Degeneracies ... 79
5.1 Introduction ... 79
5.2 Nuclear Spin and Symmetry ... 80
5.3 Case I: Linear Molecules ... 82
5.3.1 Asymmetric (e.g., CO and N_2O) ... 82
5.3.2 Symmetric (e.g., O_2, CO_2, and C_2H_2) ... 83
5.4 Case II: Nonlinear Molecules ... 87
5.4.1 Asymmetric Rotor (e.g., CHFClBr and N_2H_4) ... 87
5.4.2 Symmetric Top ... 87
5.4.3 Others (e.g., C_6H_6, CH_4, and P_4) ... 90
5.5 Exercises ... 90
References ... 90

6 Rayleigh and Raman Spectra ... 91
6.1 Light Scattering ... 91
6.1.1 Cross-Sections ... 92
6.2 Quantum Model ... 95
6.3 Classical Theory ... 95
6.4 Rotational Raman Spectra ... 96
6.4.1 Linear Molecules ... 96
6.4.2 Symmetric Top Molecules ... 98
6.5 Vibrational Raman Spectra ... 99
6.5.1 Polarization ... 100
6.5.2 Selection Rules ... 101
6.5.3 Diatomics ... 101
6.5.4 Temperature ... 102
6.5.5 Typical Raman Shift ... 103
6.6 Summary of Rayleigh and Raman Scattering ... 103
6.7 Exercises ... 103
References ... 105

7 Quantitative Emission and Absorption ... 107
7.1 Spectral Absorption Coefficient ... 107
7.2 Equation of Radiative Transfer: Classical Approach ... 109
7.2.1 Case 1: Emission Experiments ($I_\nu^0 = 0$) ... 110
7.2.2 Case 2: Absorption Experiments ($I_\nu^0 \gg I_\nu^{bb}$) ... 111

7.3 Einstein Theory of Radiation ... 112
7.3.1 Einstein Coefficients ... 112
7.3.2 Equilibrium ... 113
7.3.3 What is k_ν? ... 114
7.4 Revised Treatment of Einstein Theory (with Lineshape) ... 116
7.5 Radiative Lifetime ... 122
7.6 Alternate Forms ... 124
7.6.1 Line Strengths ... 124
7.6.2 Beer's Law ... 125
7.7 Temperature-Dependent Linestrengths ... 126
7.8 Concept of Band Strength ... 127
7.9 Exercises ... 129
References ... 129

8 Spectral Lineshapes ... 131
8.1 Lineshape Introduction ... 131
8.2 Line Broadening Mechanisms ... 132
8.2.1 Natural Broadening ... 133
8.2.2 Collisional Broadening (Pressure Broadening) ... 134
8.2.3 Doppler Broadening ... 137
8.2.4 Stark Broadening ... 139
8.2.5 Artifactual/Instrument Broadening ... 139
8.3 Voigt Profile ... 140
8.3.1 Analytical Expressions ... 140
8.3.2 Numerical Approximations ... 141
8.4 Line Shifting Mechanisms ... 141
8.4.1 Pressure Shift of Absorption Lines ... 141
8.4.2 Doppler Shift Measurements of Velocity ... 142
8.5 Lineshapes Beyond the Voigt Profile ... 142
8.5.1 Line Narrowing Mechanisms ... 142
8.5.2 Rautian and Galatry Profiles ... 143
8.5.3 Speed Dependent Voigt Profile ... 143
8.6 Quantitative Lineshape Measurements ... 144
8.6.1 Species Concentration and Pressure ... 144
8.6.2 Temperature ... 144
8.6.3 Examples ... 145
8.7 Exercises ... 147
References ... 148

9 Electronic Spectra of Atoms ... 149
9.1 Electron Quantum Numbers ... 149
9.2 Electronic Angular Momentum ... 149
9.3 Single Electron Atoms ... 150
9.3.1 Orbital Angular Momentum ... 150
9.3.2 Spin Angular Momentum ... 150
9.3.3 Total Electronic Angular Momentum ... 150

9.3.4 Term Symbols 151
9.3.5 Example: Hydrogen Atom 151
9.4 Multi-Electron Atoms 152
9.4.1 Building-Up Principle 152
9.4.2 Examples 153
9.4.3 Hydrogen-Like Species 156
9.4.4 Zeeman Effect 157
9.4.5 Nuclear Spin 157
9.5 Exercises 158
References 159

10 Electronic Spectra of Diatomic Molecules: Improved Treatment 161
10.1 Term Symbols for Diatomic Molecules 161
10.2 Common Molecular Models for Diatomics 162
10.2.1 Rigid Rotor ($^1\Sigma$) 162
10.2.2 Symmetric Top 164
10.2.3 Interaction of Λ and Σ 168
10.2.4 Hund's Case ***a*** 169
10.2.5 Hund's Case ***b*** 169
10.2.6 Λ-Doubling 171
10.2.7 Comment on Models 171
10.3 Quantitative Absorption 171
10.3.1 Boltzmann Fraction 172
10.3.2 Oscillator Strength 172
10.4 Exercises 174
Reference 175

11 Laser-Induced Fluorescence 177
11.1 Introduction 177
11.1.1 Background 177
11.1.2 Typical Experimental Set-Up 179
11.1.3 Measurement Volume 179
11.1.4 Signal Level (Two-Level System) 180
11.2 Two-Level Model 181
11.2.1 Weak Excitation Limit 183
11.2.2 Saturation Limit 184
11.2.3 Intermediate Result 184
11.2.4 Typical Values for A and Q in Electronic Transitions 185
11.2.5 Typical Values for A and Q: Vibrational Transitions (IR) 187
11.3 Detection Limits (Pulsed Laser) 187
11.3.1 Weak Excitation Limit 187
11.3.2 Saturation Limit 189
11.4 Characteristic Times 190
11.5 Modifications of the Two-Level Model 191
11.5.1 Hole-Burning Effects 191

11.5.2 Multi-Level Effects ... 191
11.5.3 Predissociation ... 193
11.6 Example: Acetone LIF ... 194
11.6.1 Background ... 194
11.6.2 Acetone Photophysics ... 194
11.7 Applications of LIF ... 195
11.7.1 Species Density ... 195
11.7.2 Species Mole Fraction ... 195
11.7.3 Temperature ... 195
11.7.4 Velocimetry ... 197
11.8 Exercises ... 198
Reference ... 199

12 Diagnostic Techniques for Gaseous Flows ... 201
12.1 Absorption Techniques ... 201
12.1.1 Frequency-Modulation Spectroscopy (FMS) ... 201
12.1.2 Wavelength-Modulation Spectroscopy (WMS) ... 202
12.1.3 Cavity-Ringdown Spectroscopy (CRDS) ... 203
12.1.4 Off-Axis Integrated Cavity Output Spectroscopy ... 204
12.2 Fluorescence Techniques ... 205
12.2.1 Planar Laser-Induced Fluorescence (PLIF) ... 205
12.2.2 Narrow Linewidth LIF ... 205
12.2.3 Laser-Induced Breakdown Spectroscopy (LIBS) ... 207
12.3 Photothermal Techniques ... 208
12.3.1 Photoacoustic Spectroscopy (PAS) ... 208
12.3.2 Photothermal Deflection (PTD) ... 209
12.4 Scattering Techniques ... 210
12.4.1 Spontaneous Raman Scattering ... 210
12.4.2 Coherent Anti-Stokes Raman Spectroscopy ... 211
12.5 Laser Ionization Spectroscopy ... 212
References ... 213

13 Spectroscopy Equipment ... 217
13.1 Sources ... 217
13.1.1 The Helium–Neon Laser ... 217
13.1.2 The Nd:YAG Laser ... 218
13.1.3 The Excimer Laser ... 219
13.1.4 The CO_2 Laser ... 219
13.1.5 Semiconductor Lasers ... 219
13.1.6 Dye Lasers ... 220
13.1.7 Non-Laser Sources ... 220
13.2 Detectors ... 220
13.2.1 The Photomultiplier ... 221
13.2.2 Photoconductive Detectors ... 221

13.2.3 Photodiode Detectors 222
13.2.4 Selecting a Detector 222
References 224

14 Case Studies: Molecular Spectroscopy 227
14.1 Ultraviolet OH Spectroscopy: The $A^2\Sigma^+ - X^2\Pi$ System 227
14.1.1 OH Energy Levels 227
14.1.2 Allowed Radiative Transitions 232
14.1.3 Absorption Coefficient 235
14.2 Visible/Near-Infrared O_2 Spectroscopy 244
14.3 Near-Infrared H_2O Spectroscopy: Lineshapes 245
14.3.1 Experimental Setup 246
14.3.2 Experimental Results 247
14.4 Mid-Infrared Spectroscopy of Hydrocarbons and Other Organic Compounds 249
References 252

A Glossary 257

B Voigt Tables 259

C Matlab Voigt Fitting Program 265

D HITRAN Database 267
D.1 Example Calculation of H_2O Absorbance Spectra using the HITRAN Database 270
D.1.1 Calculation of Linestrength at T 270
D.1.2 Calculation of Lineshape Function 271
D.1.3 Calculation of Absorbance 272

E Center of Symmetry 275

F Fluorescence Yield: Multi-Level Models 277
References 279

List of Figures

Fig. 1.1 Absorption and emission energy transitions in an atom/molecule ... 2
Fig. 1.2 Calculated infrared absorption spectra for HBr. Here frequency (cm^{-1}) is simply 1/wavelength (cm) ... 2
Fig. 1.3 Typical emission spectra of high-temperature air between 560 and 610 nm. The seven large features are part of the $N_2(1^+)$ system of transitions ... 3
Fig. 1.4 Electromagnetic spectrum ... 4
Fig. 1.5 Representative absorption spectra. The *left-hand side* shows a single line, while the *right-hand side* shows a band ... 6
Fig. 1.6 Laser absorption experimental schematic ... 7

Fig. 2.1 (**a**) Electric dipole oscillation for a heteronuclear diatomic molecule rotating at frequency $1/\tau_{rot}$; and (**b**) E-field for an incident wave of frequency ν, shown here as resonant and in phase with the dipole oscillation in (**a**); (**c**) oscillation of the vertical component of the electric dipole ... 10
Fig. 2.2 Stretching vibrational mode for carbon monoxide ... 11
Fig. 2.3 Electronic structure of carbon monoxide ... 12
Fig. 2.4 Diatomic molecule with rigid rotor approximation ... 13
Fig. 2.5 Rotational energy level spacing ... 16
Fig. 2.6 Energy level spacing and introduction of "first difference" in energy ... 16
Fig. 2.7 Absorption spectrum spacing for heteronuclear rotation ... 17
Fig. 2.8 Oscillation of a linear diatomic molecule; r_{min} corresponds to molecule at distance of greatest compression ... 20
Fig. 2.9 Potential energy and vibrational levels for diatomic SHO and THO molecules ... 21
Fig. 2.10 Potential energy and vibrational levels for a diatomic molecule ... 25
Fig. 2.11 Energy level diagram denoting P and R absorption transitions from a ground vibrational state for a heteronuclear diatomic molecule ... 28

Fig. 2.12 Simulated absorption spectrum of the P and R branches of a ground state rovibrational transition of a heteronuclear diatomic molecule 29
Fig. 2.13 Unequal line spacing due to non-rigid rotation leads to a bandhead in the R branch for diatomic molecules 33
Fig. 2.14 Energy level diagram for the method of common upper states 34
Fig. 2.15 Sample potential wells for the X and A electronic energy states 37
Fig. 2.16 Sample potential wells for discrete electronic spectra 39
Fig. 2.17 Samples of potential wells (*top*) that result in continuous spectra (*bottom*). The parameter k_ν is an absorption coefficient, i.e. a measure of absorption strength 40
Fig. 2.18 Fortrat parabola for the case with $B' < B''$ 43
Fig. 2.19 Second difference rotational analysis for $a < 0$ 44
Fig. 2.20 Rotational spectrum in the 0–0 band of $^{35}Cl_2$ 44
Fig. 2.21 Common upper states 45
Fig. 2.22 Absorption and emission between two potential wells 45
Fig. 2.23 Deslandres table with row and column analysis 46
Fig. 2.24 Example potential wells and corresponding absorption spectrum for upper state properties
Typical analyses for absorption include:

1. using band origin data to give $G(v')$ and hence $G(v' = 0)$.
2. using measured $\nu_0 = T_e + G(v' = 0) - G(v'' = 0)$ to find T_e.
3. using measured Δ to give D_e' via $\Delta + G(v'' = 0) = T_e + D_e'$ 47

Fig. 2.25 Potential curves and emission spectrum for lower-state properties
Typical analyses for emission include:

1. using band origin data (Deslandres table) from fixed v' to find $G(v'')$.
2. using measured Δ and known T_e and $G(v')$ to find D_e'' via $D_e'' + \Delta = T_e + G(v')$ 47

Fig. 3.1 Illustration of Birge–Sponer method for finding the dissociative vibrational level, v_D, and the dissociation energy, D_e 52
Fig. 3.2 HNO structure 54
Fig. 3.3 Curve-crossing method for HNO dissociation 54
Fig. 3.4 Allowed absorption spectrum for HNO 55
Fig. 3.5 N_2O structure 55

Fig. 3.6 Potential energy wells for N_2O 56

Fig. 4.1 Orthogonal axes and moments of inertia for ammonia, a polyatomic molecule 60
Fig. 4.2 Model of the linear polyatomic molecule OCS 61
Fig. 4.3 Symmetric top structure. The A-axis passes through C–F for CH_3F; for BCl_3, it is perpendicular to the plane formed by the atoms and passes through B 62
Fig. 4.4 Molecular structure of CH_4 64
Fig. 4.5 Molecular structure of H_2O 65
Fig. 4.6 Structure, symmetry, and vibrational modes for H_2O 67
Fig. 4.7 Symmetric stretching of carbon dioxide (Fig. 1.6, Banwell) 68
Fig. 4.8 Asymmetric stretching vibrational mode for carbon dioxide 68
Fig. 4.9 Structure, symmetry, and vibrational modes for CO_2 69
Fig. 4.10 Energy levels and absorption spectrum for the P, Q, and R branches of a linear polyatomic perpendicular band 72
Fig. 4.11 The resolved components of a parallel band showing the contributions from each of the K levels of the $v = 0$ state. The small but discernible splitting evident in the superposed P- and R-branch spectra (*bottom row* of this figure) is due to a difference in the magnitude of $(A - B)$ in the upper and lower vibrational levels; see Eq. (4.20). The discernible splitting in the superposed Q-branch is due mostly to the difference in B in the upper and lower vibrational levels ($B' < B''$, so spectra degrade to lower frequencies) 74
Fig. 4.12 The energy levels of a symmetric top molecule showing the transitions that are allowed for a perpendicular band. Figure from Barrow [2, p. 151] 75
Fig. 4.13 The components of a perpendicular band of a symmetric top molecule. Note that the lines with $\Delta J = \Delta K$ have greater intensity than those with $\Delta J = -\Delta K$, i.e., R-branch lines with $\Delta J = \Delta K = +1$ are stronger than the P-branch lines of $\Delta J = -1$, when $\Delta K = +1$. See Herzberg [1, pp. 424–426], for the selection rules that characterize this effect. Figure from Barrow [2, p. 152] 76

Fig. 6.1 Schematic of a light scattering experiment 92
Fig. 6.2 Photon scattering due to interaction with a molecule 95
Fig. 6.3 Sample energy levels for rotational Raman transitions 97
Fig. 6.4 Two branches of Raman spectra. Note that the signal strength is proportional to the population of molecules in the initial state ($N_{J\text{-initial}}$) 98
Fig. 6.5 Sample energy levels for symmetric top rotational Raman transitions 99

Fig. 6.6 S, Q, and O branches for Stokes and anti-Stokes vibrational and rotational Raman spectra. Note that the signal strength is dependent on the population of molecules in the initial state ($N_{v\text{-initial}}$) 102

Fig. 7.1 Radiation energy balance across a slab of gas 109
Fig. 7.2 Radiation energy across a slab of gas of width L 110
Fig. 7.3 Transition probabilities between states 1 and 2 112
Fig. 7.4 T_ν and k_ν versus frequency for a structureless absorption line of width $\delta\nu$ 115
Fig. 7.5 Transmission of laser intensity across a gas slab of depth dx 115
Fig. 7.6 T_ν, k_ν, and ϕ versus frequency for an absorption line with typical structure 117
Fig. 7.7 Transition probabilities per second per molecule in level 2 or 1 118
Fig. 7.8 Energy/power balance on an incremental gas slab 119

Fig. 8.1 Sample lineshape as a function of frequency 132
Fig. 8.2 Comparison of Gaussian and Lorentzian lineshapes with the same FWHM 139

Fig. 9.1 Motions for orbital and spin angular momentum of a single electron atom 150
Fig. 9.2 Some of the lower energy levels for the hydrogen atom 152
Fig. 9.3 States for a carbon atom 155
Fig. 9.4 States for a nitrogen atom 155
Fig. 9.5 Some of the lower energy levels for the lithium atom and transitions "allowed" by the selection rules 156

Fig. 10.1 Rigid rotor model for molecular motion 163
Fig. 10.2 Symmetric top model for molecular motion 164
Fig. 10.3 Symmetric top model for molecular motion, $^1\Delta \leftarrow ^1\Delta$ case 166
Fig. 10.4 Coupling of Σ and Λ 168
Fig. 10.5 Spin–orbit splitting 168
Fig. 10.6 Molecular model for Hund's ***b*** 170
Fig. 10.7 Lambda-doubling results in two different energy levels 171

Fig. 11.1 Simple model for laser-induced fluorescence 178
Fig. 11.2 Typical experimental setup for LIF systems 179
Fig. 11.3 Measurement volume for LIF 179
Fig. 11.4 Spontaneous emission corresponds to an atomic (or molecular) transition from level 2 to 1, with release of a photon 180
Fig. 11.5 Different transitions that are included in the two-level model 181

Fig. 11.6 Pulsed lasers are often spectrally broad compared to absorption lines ... 182
Fig. 11.7 Intermediate fluorescence signal levels ... 185
Fig. 11.8 Characteristic laser pulse times can be 5–20 ns ... 190
Fig. 11.9 Characteristic steady-state and decay times for the population in state 2 ... 190
Fig. 11.10 Upper and lower manifold states ... 192
Fig. 11.11 Potential curves for ground, excited, and predissociative states of OH ... 193
Fig. 11.12 Molecular levels for acetone ... 194
Fig. 11.13 OH line pair for temperature measurements ... 196
Fig. 11.14 The Doppler shift from a flowing gas can be used to measure the gas velocity along the direction of the laser beam ... 197
Fig. 11.15 The Doppler shift from a flowing gas can be used to measure a component of velocity ... 198

Fig. 12.1 Schematic for laser-absorption measurements using FMS ... 202
Fig. 12.2 Schematic for laser-absorption measurements using WMS ... 203
Fig. 12.3 Schematic for laser-absorption measurements using CRDS ... 204
Fig. 12.4 Schematic for laser-absorption measurements using OA-ICOS .. 205
Fig. 12.5 Schematic of a PLIF experiment to measure species concentration ... 206
Fig. 12.6 A typical experimental schematic for narrow-linewidth LIF ... 206
Fig. 12.7 Experimental schematic for LIBS analysis of a condensed-phase specimen ... 207
Fig. 12.8 Experimental configuration for optoacoustic measurements ... 209
Fig. 12.9 Schematic and data traces for photothermal deflection ... 209
Fig. 12.10 Schematic velocity measurements using photothermal deflection ... 210
Fig. 12.11 Energy levels for spontaneous Raman scattering ... 210
Fig. 12.12 Typical experimental setup for CARS (4-wave mixing) ... 211
Fig. 12.13 Energy levels for CARS spectroscopy with 4-wave mixing ... 211
Fig. 12.14 Conservation of momentum for 4-wave mixing ... 212
Fig. 12.15 CARS spectra for scanning and broadband sources ... 212
Fig. 12.16 Different energy levels for laser ionization spectroscopy. ... 213
Fig. 12.17 Experimental schematic for laser ionization spectroscopy ... 213

Fig. 13.1 Wavelengths of operation for many common lasers ... 218

Fig. 13.2 Wavelength range for common detectors. *Black boxes* indicate photovoltaic detectors (photodiodes). *Gradient from left to right* indicates a photoconductor and *gradient from top to bottom* indicates a photomultiplier 223
Fig. 13.3 Typical bandwidth for common detectors. *Black boxes* indicate photovoltaic detectors (photodiodes). *Gradient from left to right* indicates a photoconductor and *gradient from top to bottom* indicates a photomultiplier 223

Fig. 14.1 Energy level progression for $A^2\Sigma^+$; Hund's Case (b). The symbol "*p*" denotes parity (+ or −) 230
Fig. 14.2 Energy level progression for OH $X^2\Pi$ (inverted) 233
Fig. 14.3 Energy level progression for a regular $X^2\Pi$ configuration 233
Fig. 14.4 Allowed rotational transitions from $N'' = 13$ in the $A^2\Sigma^+ \leftarrow X^2\Pi$ system 234
Fig. 14.5 Allowed rotational transitions from $N'' = 13$ in the $A^2\Sigma^+ \leftarrow X^2\Sigma^+$ system 235
Fig. 14.6 Plot of OH line strengths for a selected region of the $A^2\Sigma^+ \leftarrow X^2\Pi(0,0)$ band at 2000 K.............................. 243
Fig. 14.7 Potential energy diagram for diatomic molecular oxygen. 244
Fig. 14.8 The *A*-Band of O_2.. 246
Fig. 14.9 Experimental setup used to measure H_2O lineshapes............ 246
Fig. 14.10 Measured I_o and I_t for single scans across H_2O transitions near 1345 nm at various pressures. $I_t = I_o$ at non-resonant wavelengths 247
Fig. 14.11 Measured absorbance spectra and best-fit lineshape residuals for H_2O transitions near 7413 cm^{-1} with $J'' = 8$ (*left*) and 6919.9 cm^{-1} with $J'' = 14$ (*right*). The Voigt profile is significantly less accurate for the transition near 6919.9 cm^{-1} 248
Fig. 14.12 Measured γ_{N_2}, γ_{2,N_2}, and β_{N_2} for the H_2O transitions near 7413.02 cm^{-1} with $J'' = 8$ (*left*) and 7471.6 cm^{-1} with $J'' = 16$ (*right*). *Dotted lines* correspond to best-fit power-law model 249
Fig. 14.13 Measured $\gamma_{N_2,\mathrm{qSDVP}}$ and corresponding best-fit power-law model for all transitions studied. Error bars are too small to be seen 250
Fig. 14.14 FTIR measurements of octane and methane at 1 atm and 25 °C 251
Fig. 14.15 FTIR measurements of acetone and ethanol at 1 atm and 25 °C 252

Fig. D.1 Comparison of linestrength as a function of temperature calculated using Eq. D.6 and Eq. D.2 for Lines 1 and 2 271

Fig. D.2 Voigt lineshapes of Line 1 and 2 at 1000 K, 1 atm, with 10% H_2O in air 272
Fig. D.3 Total absorbance spectrum and individual contributions of each line at the conditions of interest 272

List of Tables

Table 2.1 Characteristic rotational temperatures for some diatomic species 18
Table 2.2 Characteristic vibrational temperatures for some diatomic species 22
Table 2.3 AHO frequencies and corrections 24
Table 2.4 Typical values for vibration (Banwell [5], p. 62, Table 3.1) 27
Table 2.5 Analysis techniques and their related fundamental quantities 46

Table 4.1 Symmetry factors for a few polyatomic molecules 65
Table 4.2 Active modes of CO_2 (*S* strong, *VS* very strong); *N.A.* not active 69
Table 4.3 Fundamental vibrations, frequencies, types, and descriptions for NH_3 69
Table 4.4 Band strength comparison: HCN 72

Table 5.1 Spin of several nuclei 81
Table 5.2 Key to addition or subtraction of nuclear degeneracies in Eqs. (5.15) and (5.16) for different rotational statistics and electronic manifold configurations 85
Table 5.3 Sample species and ground state configurations 85

Table 6.1 Rayleigh scattering parameters for selected molecules, taken from [1] and [3] 93
Table 6.2 Vibrational Raman cross-sections for common molecules at 532 nm (in units of 10^{-30} cm^2/molec/Sr) 94
Table 6.3 Raman and IR spectra information for N_2O 100

Table 7.1 Oscillator strengths of selected sodium transitions, abstracted from [1] 122
Table 7.2 Absorption oscillator strengths of selected vibrational and vibronic bands of a few molecules 123

Table 8.1 Some collisional broadening coefficients 2γ [cm^{-1}/atm] in Argon and Nitrogen at 300 K 137

Table 8.2 Some collisional broadening coefficients 2γ [cm^{-1}/atm] in Argon and Nitrogen at 2000 K 137

Table 9.1 Building-up principle for the first few quantum numbers 153
Table 9.2 Building-up principle for the first ten elements 153
Table 9.3 Some possible electronic configurations for the carbon atom 154
Table 9.4 Terms symbols for equivalent electrons 154

Table 14.1 OH term energy constants (in cm^{-1}) 230
Table 14.2 Band oscillator strengths for the OH $A^2\Sigma^+ - X^2\Pi$ System 238
Table 14.3 Hönl–London factors for selected OH transitions 239
Table 14.4 Spectroscopic constants for O_2 245
Table 14.5 Typical characteristic frequencies for hydrocarbon vibrations 250

Table D.1 Sample HITRAN2012 output for CO_2 for v_o from 2311.105 to 2311.12 cm^{-1} 268
Table D.2 Key to symbols in Table D.1 269
Table D.3 Sample HITRAN2012 output for the H_2O doublet of interest 271

About the authors

Ronald K. Hanson is the Woodard Professor of Mechanical Engineering at Stanford University. Prof. Hanson has been actively involved in teaching and applied spectroscopy research at the High Temperature Gasdynamics Laboratory at Stanford for over 40 years, resulting in over 95 Ph.Ds being awarded under his supervision. The Hanson research group has published over 1000 technical papers, contributing to many advances in optical diagnostics, and also shock wave physics, chemical kinetics, combustion science and advanced propulsion. Co-authors Dr. Mitchell Spearrin and Dr. Christopher Goldenstein are former students of Prof. Hanson's research group.

R. Mitchell Spearrin is an Assistant Professor of Mechanical and Aerospace Engineering at the University of California Los Angeles (UCLA). Prof. Spearrin's research focuses on spectroscopy and optical sensors with experimental application to dynamic flow fields in aerospace, energy, and biomedical systems.

Christopher S. Goldenstein is an Assistant Professor of Mechanical Engineering at Purdue University. Prof. Goldenstein's research focuses on the development and application of laser-based sensors for studying energetic materials, energy systems, and trace gases.

1 Introduction

1.1 Role of Quantum Mechanics

This text focuses on the application of spectroscopic diagnostics to gaseous flows. In order to keep the length and scope manageable, we will accept, prima facie, the results of quantum mechanics (explaining quantum mechanics in detail would require another complete text, of which many are already available). That is, we accept that an atom or molecule may exist only in specific quantum states (characterized by quantum "numbers"), with each quantum state having a discrete amount of energy and angular momentum. *Hence, molecular energy is quantized.* Furthermore, we will view the internal energy, which excludes kinetic energy, as the sum of the energy stored in three modes: (1) rotation, (2) vibration, and (3) electronic:

$$E_{\text{tot}} = E_{\text{rot}} + E_{\text{vib}} + E_{\text{elec}}. \tag{1.1}$$

Note that quantized internal energies lead to discrete differences in energy when molecules change quantum states. These transitions correspond directly with the energy of emitted or absorbed photons (and hence the emission or absorption wavelengths) in discrete spectra. Quantum mechanics also places restrictions on the allowable changes in quantum states during emission and absorption, described by *selection rules*. These selection rules (also simply accepted in this course) dictate which transitions are *allowed* and which are *forbidden*, and greatly simplify the resulting spectra.

1.2 Emission and Absorption Spectra

Emission results when a molecule or atom changes quantum states from higher to lower energy and releases a photon. *Absorption* occurs when a molecule or atom changes quantum states from lower to higher energy by absorbing a photon.

R.K. Hanson et al., *Spectroscopy and Optical Diagnostics for Gases*,
DOI 10.1007/978-3-319-23252-2_1

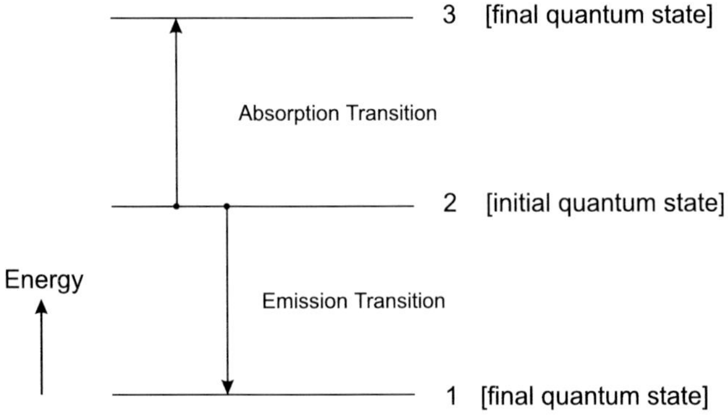

Fig. 1.1 Absorption and emission energy transitions in an atom/molecule

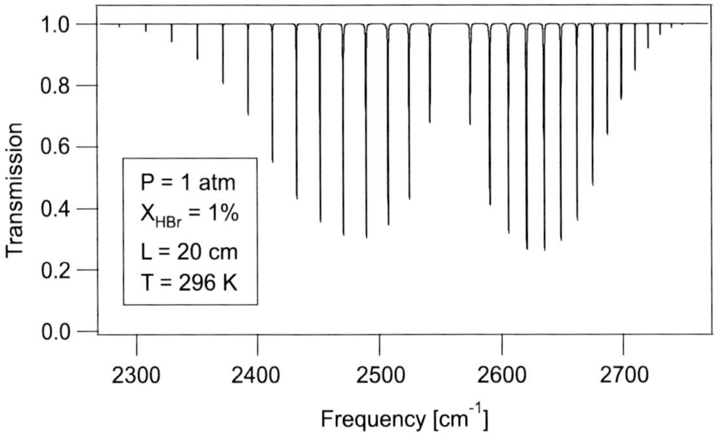

Fig. 1.2 Calculated infrared absorption spectra for HBr. Here frequency (cm^{-1}) is simply 1/wavelength (cm)

The energy of the photon emitted or absorbed is equal to the difference in energy of the two quantum states (by conservation of energy, of course!). Figure 1.1 shows how the energy of a molecule changes during these processes. Example absorption and emission spectra, obtainable by measuring the energies of the photons involved in these processes, are shown in Figs. 1.2 and 1.3.

Our goal: to be able to interpret and predict absorption and emission spectra.

Before being able to interpret spectra, however, we must first know a few basics about light.

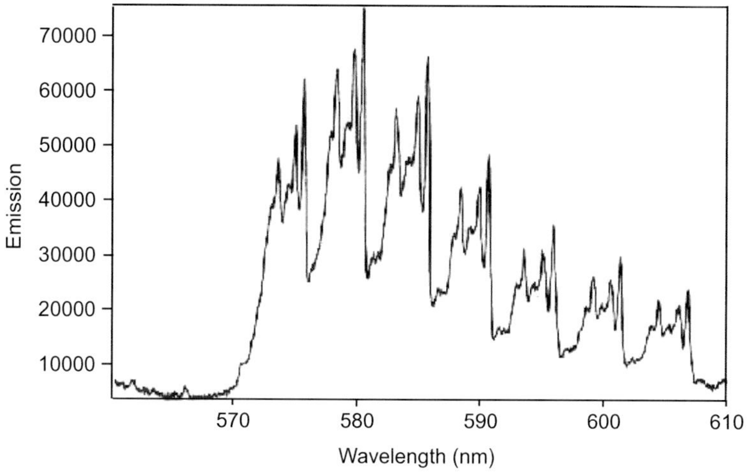

Fig. 1.3 Typical emission spectra of high-temperature air between 560 and 610 nm. The seven large features are part of the $N_2(1^+)$ system of transitions

1.3 Planck's Law

Planck's Law is a fundamental relation that links two ways of thinking about radiation, namely the *particle* and *wave* concepts:

$$\Delta E = E_{\text{upper}} - E_{\text{lower}} = h\nu \tag{1.2}$$

Here ΔE is the energy of the photon (i.e., the *particle*) associated with a molecular transition (emission or absorption) between two quantum states, while ν is the frequency of the corresponding electromagnetic *wave*, and h is Planck's constant:

$$h = 6.63 \times 10^{-34}\ \text{J s} \tag{1.3}$$

Extending Eq. (1.1), we may also express the change in energy of the molecule (i.e., the energy of the photon) as

$$\Delta E = \Delta E_{\text{rot}} + \Delta E_{\text{vib}} + \Delta E_{\text{elec}}. \tag{1.4}$$

That is, the total change in internal energy of a molecule/atom is the sum of the individual changes in rotational, vibrational, and electronic energy. Energy changes in each of these modes have different magnitudes (as the reader will see in Fig. 1.4, if he/she doesn't already know) and can be classified according to the following:

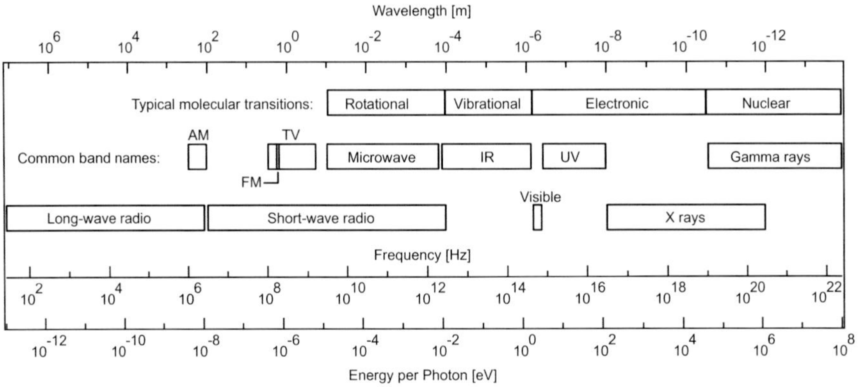

Fig. 1.4 Electromagnetic spectrum

$$\begin{aligned} \Delta E_{\text{rot}} &\rightarrow \text{microwave transitions (rotational)} \\ \Delta E_{\text{rot}} + \Delta E_{\text{vib}} &\rightarrow \text{IR transitions (rovibrational)} \\ \Delta E_{\text{rot}} + \Delta E_{\text{vib}} + \Delta E_{\text{elec}} &\rightarrow \text{UV/Vis transitions (rovibronic)} \end{aligned}$$

Note that changes in vibrational energy are generally accompanied by changes in rotational energy, and changes in electronic energy are accompanied by changes in vibrational and rotational energy (although ΔE_{rot} and ΔE_{vib} can sometimes be zero).

Thus, absorption or emission spectra generally consist of a number of "lines," corresponding to the discrete differences in energy between a molecule's states which are allowed (quantum mechanically) to undergo radiative transitions.

1.4 Wavelength, Frequency, and Other Units and Conversions

Spectroscopy utilizes several different units of length and energy. The most common usage varies with the spectral region in question.

Term	Symbol	Units
Wavelength	λ	Å, nm, μm (microns); 1 Å $= 10^{-8}$ cm $= 0.1$ nm
Frequency	ν	s^{-1}, Hz, MHz, GHz, THz
Wavenumber	$\bar{\nu}, \omega$	cm^{-1}, mK ($10^{-3}\,cm^{-1} = 1$ milliKayser) (a wavenumber is the number of wavelengths per cm)
Energy	$E, \Delta E$	J, cm^{-1}, ergs, eV (all per photon); kcal/mole

The relationship between wavelength and (temporal) frequency is given by

$$c = \lambda \nu \tag{1.5}$$

where λ is the wavelength of the radiation and c is the speed of light, nominally 3.0×10^8 m/s.

Wavelength and wavenumber (sometimes also called frequency, as in spatial frequency) are related by

$$\overline{\nu} = \omega = 1/\lambda \tag{1.6}$$

From Eqs. (1.5) and 1.6, Planck's Law can be written in several ways:

$$\boxed{\Delta E = h\nu = hc/\lambda = hc\overline{\nu} = hc\omega} \tag{1.7}$$

Hence, a transition is interchangeably expressed in terms of nanometers, angstroms or microns for wavelength and in Hertz or wavenumbers for frequency. In the infrared (IR), use of wavenumber units is quite common; however, in the visible or ultraviolet (UV), nanometer or angstrom units are the norm.

1.5 Spectral Regions

Figure 1.4 shows some of the different regions of the electromagnetic spectrum, along with regions where rotational, vibrational, and electronic transitions tend to occur. The boundaries are somewhat variable in common usage.

As can be seen from Fig. 1.4, units of energy can have many equivalent forms. Knowing the conversion for a few reference points and the scaling may facilitate fluency in every unit. For example,[1]

$$\begin{aligned} \lambda_{1\,\mathrm{eV}} &\approx 12{,}400\,\text{Å} \\ \lambda_{2\,\mathrm{eV}} &\approx 620\,\mathrm{nm} \\ \lambda_{4\,\mathrm{eV}} &\approx 310\,\mathrm{nm} \end{aligned}$$

and

$$\begin{aligned} \lambda\ [\mathrm{nm}] &= 10^7/\overline{\nu}\ [\mathrm{cm}^{-1}] \\ \rightarrow 500\,\mathrm{nm} &= 20{,}000\,\mathrm{cm}^{-1} \\ \rightarrow 1000\,\mathrm{nm} &= 10{,}000\,\mathrm{cm}^{-1} \end{aligned}$$

[1] $1\,\mathrm{eV} \approx 1.6 \times 10^{-19}$ J.

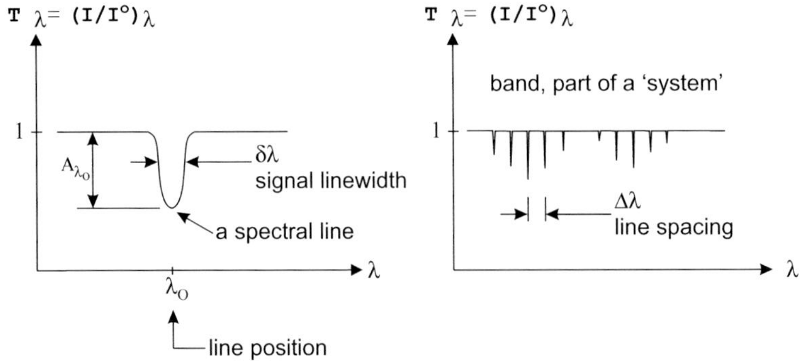

Fig. 1.5 Representative absorption spectra. The *left-hand side* shows a single line, while the *right-hand side* shows a band

1.6 Basic Elements of Spectroscopy

A *line* is the part of a spectrum that corresponds to a transition from one quantum state to another, e.g. a change in a molecule's rotational energy may show a rotational line in a spectrum as shown on the left of Fig. 1.5. Groups of individual lines, with common upper and lower vibrational quantum numbers, may comprise a *vibrational band*, shown on the right of Fig. 1.5. Furthermore, several vibrational bands may comprise an *electronic system*.

1.6.1 Positions, Strengths, and Shapes of Lines

It is convenient to think of spectroscopy as comprised of three primary elements: i.e., the positions, strengths, and shapes of lines. Line positions, which receive emphasis in classical spectroscopy, depend on molecular structure. The strengths and shapes of lines are of particular importance in diagnostics applications. All three elements will be addressed in this reader.

What is the relationship of these spectral features to gaseous properties?

1. Line positions and spacing (λ_0, $\Delta\lambda$ in Fig. 1.5) → molecular parameters (internuclear spacing, bond angles)
2. Absorbance and linewidth (A_λ, $\delta\lambda$ in Fig. 1.5) → composition, temperature, pressure

1.7 Typical Absorption Spectroscopy Setup

Most modern absorption experiments use a laser as the radiation source for several reasons. First, their high intensities allow absorption measurements in very hot gases (with high background emissions levels). Also, because they are

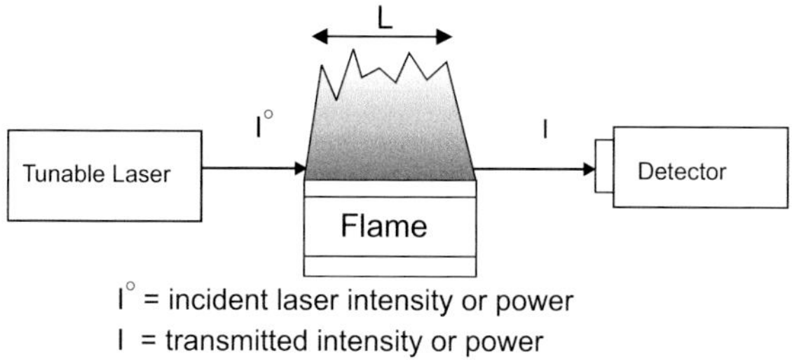

Fig. 1.6 Laser absorption experimental schematic

highly collimated, laser light can travel across substantial distances—up to several kilometers. Furthermore, lasers are often spectrally narrow and tunable in wavelength, allowing high-resolution measurements of absorption spectra. Alternatively, absorption spectra may be recorded with a spectrally broad light source (e.g., globar) and a tunable spectral filter (e.g., grating monochromator), but such spectra are *artificially broadened* owing to the (relatively) large spectral bandwidth of the filters currently available.

1.8 Beer's Law of Absorption

The governing law for absorption spectroscopy describes the relationship between the incident and transmitted laser intensities, shown in Fig. 1.6. By way of introduction, Beer's Law is simply listed below. In Chap. 7 it will be derived.

$$\textbf{Beer's Law} \qquad T_\nu = \left(\frac{I}{I^0}\right)_\nu = \exp(-k_\nu L) \tag{1.8}$$

T_ν is the fractional transmission at frequency ν, k_ν is the spectral absorption coefficient [cm^{-1}], and L is the absorption path length [cm]. The combined quantity $k_\nu L$ is known as the *spectral absorbance*. The spectral absorption coefficient for a single, spectrally isolated transition (typical of atoms and small molecules) is given by

$$k_\nu = S \times \phi(\nu \text{ or } \lambda) \times P_i \tag{1.9}$$

For Eq. (1.9), S is the "strength" of the transition (common units are cm^{-2} atm^{-1}), ϕ is the "lineshape function" (common unit is cm), and P_i is the partial pressure of the absorbing species (unit is atm). Beer's Law can also be expressed in terms of absorption:

$$A_\nu = 1 - T_\nu = \left(\frac{I^0 - I}{I^0}\right)_\nu = 1 - \exp(-k_\nu L) \tag{1.10}$$

1.9 Spectral Absorption Coefficient

Later in the course we will derive relations for the spectral absorption coefficient, k_ν, which appears in Beer's Law of absorption (Eq. (1.8)). We will find that k_ν is proportional to: (1) the population density n_i in the lower level i of an absorption transition, and (2) a parameter (e.g., f_{osc} or B_{12}) that characterizes the strength of the transition. Here f_{osc} is known as the oscillator strength and B_{12} is an Einstein coefficient for absorption between levels 1 and 2.

Thus, absorption from a level i will only occur at frequencies corresponding to quantum-mechanically allowed transitions, and the magnitude of the absorption coefficient depends on the population density in i and the oscillator strength of the transition.

1.10 Boltzmann Distribution

Boltzmann's equation for the fraction of molecules in energy level i is [1]

$$F_i = \frac{n_i}{n} = \frac{g_i \exp\left(-\frac{\epsilon_i}{kT}\right)}{Q}, \tag{1.11}$$

where the partition function, Q, is given by

$$Q = \sum_i g_i \exp\left(-\frac{\epsilon_i}{kT}\right) = Q_{rot} Q_{vib} Q_{elec} \tag{1.12}$$

The Boltzmann distribution function describes the equilibrium distribution of molecules (or atoms) of a single species over its allowed quantum states. Here g_i is the degeneracy of level i (i.e., the number of individual states with a common energy, ϵ_i) and Q is a specific, energy-weighted sum over all levels known as the partition function. There are some subtle differences between "states" and "levels" which will require careful attention later in this course. Note that the overall partition function Q may be written as a product of partition functions for the different types of internal energy, since these energies are taken to be additive. The parameter T is the temperature, and the species may be said to be in local thermodynamic equilibrium (LTE) if the populations in all the quantum levels obey Boltzmann's equation. In essence, this equation *defines temperature*.

Reference

1. W.G. Vincenti, C.H. Kruger, *Physical Gas Dynamics* (Krieger Publishing Company, Malabar, FL, 1965)

2 Diatomic Molecular Spectra

2.1 Interaction Mechanism for EM Radiation with Molecules

The primary interactions of light and matter take the form of emission, absorption, or scattering. There are multiple possibilities for the interaction, of which the most likely are:

- Electric dipole moment (emission/absorption)
- Induced polarization (Raman scattering)
- Elastic scattering (Rayleigh scattering)

There are other, rarer, mechanisms for electromagnetic (EM) interaction such as magnetic dipoles, electric quadrupoles, octopoles, etc., but we will limit most of our discussion to electric dipoles. Scattering processes will be discussed briefly in Chap. 6.

Heteronuclear diatomic molecules, which carry a permanent net positive charge on one end and a net negative charge on the other (e.g., HCl, NO), have a permanent electric dipole moment. The motion of this electric dipole moment, through rotation or vibration, gives rise to the possibility of emitting or absorbing (receiving) electromagnetic radiation, much like a miniature antenna. The strength or probability of emission or absorption is a function of the electric dipole moment and its variation with internuclear spacing. EM radiation can also interact with diatomics through the rearrangement of the electron distribution in the molecule's shells.

R.K. Hanson et al., *Spectroscopy and Optical Diagnostics for Gases*,
DOI 10.1007/978-3-319-23252-2_2

2.1.1 Microwave Region: Rotation

When heteronuclear diatomic molecules rotate, their dipole moments also rotate their orientation. Molecular motions, at characteristic frequencies, create opportunities for resonances with EM waves, leading to absorption or emission at these frequencies.

The electric dipole moment is specified by

$$\vec{\mu} = \sum_i q_i \vec{r}_i, \tag{2.1}$$

where i refers to particles in a molecule or system, q_i is the particle charge, and $\vec{r}_i$ is the vector specifying location. For carbon monoxide (CO), the C atom has a net positive charge, while the O atom has a net negative charge (see the left panel of Fig. 2.1). Thus, the dipole points upwards when the molecule is oriented along the vertical axis with the C atom above the O atom, as drawn in panel (a) of Fig. 2.1.

When $1/\nu = \tau_{\text{rot}}$ (see Fig. 2.1), resonance occurs, increasing the chance of "exchange" between the EM wave and molecule by absorption or stimulated emission. The frequency of molecular rotation is in the microwave region. All molecules with a permanent electric dipole can interact with light as a result of their rotation, and hence are considered "microwave active." Homonuclear diatomic molecules with no permanent electric dipole (N_2, Cl_2, etc.) are termed "microwave inactive."

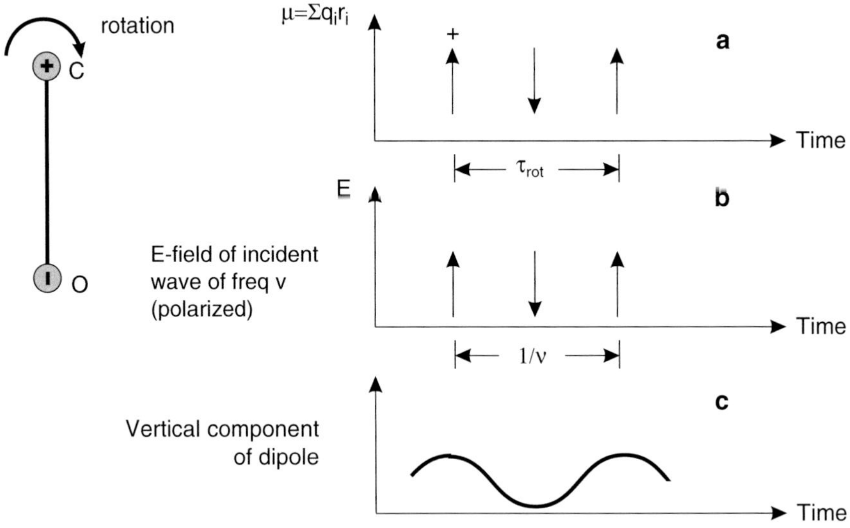

Fig. 2.1 (**a**) Electric dipole oscillation for a heteronuclear diatomic molecule rotating at frequency $1/\tau_{\text{rot}}$; and (**b**) E-field for an incident wave of frequency ν, shown here as resonant and in phase with the dipole oscillation in (**a**); (**c**) oscillation of the vertical component of the electric dipole

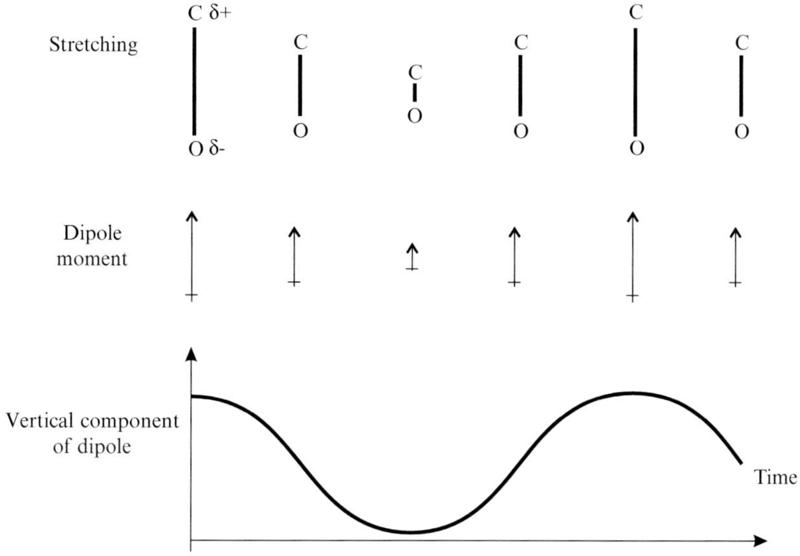

Fig. 2.2 Stretching vibrational mode for carbon monoxide

2.1.2 Infrared Region: Vibration

For the IR region, it is a heteronuclear molecule's vibration that leads to the changes in electric dipole moment and the possibility for interaction with light. Figure 2.2 depicts the change in CO's electric dipole moment with time due to the molecule's stretching motion.

2.1.3 Ultraviolet and Visible Regions: Electronic

For the ultraviolet (UV) and visible regions of the spectrum, allowed changes in a molecule's electronic structure (and hence electric dipole moment) introduce the possibility for interaction with light. For infrared and microwave spectra, energy transitions are related to the motions of the molecule. For electronic spectra, energy transitions are related to the distribution of electrons in the molecule's shells (Fig. 2.3).

2.1.4 Summary of Background

Quantum mechanics tells us that energy levels of most molecules (and atoms) are discrete and that optically allowed transitions (i.e., emission, absorption) may occur only in certain cases. The result is that absorption and emission spectra are

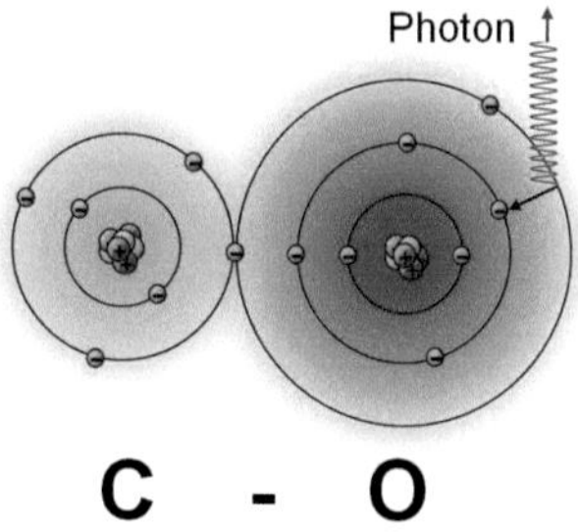

Fig. 2.3 Electronic structure of carbon monoxide

typically discrete. The molecular energies of interest are: rotational, vibrational, and electronic, with progressively larger energy spacings.

$$\text{Energy spacing}: \quad \Delta E_{\text{rot}} < \Delta E_{\text{vib}} < \Delta E_{\text{elec}} \tag{2.2}$$

Although our primary interest will be in vibrational and electronic spectra, rotational spectra are embedded. In order to understand and simulate actual spectra, we first begin with a discussion of rotational spectra and a physical model that helps us understand the processes involved. The simplest rotational model for the diatomic molecule is the Rigid Rotor (Sect. 2.2.1), while the simplest vibrational model is the Simple Harmonic Oscillator (SHO) (Sect. 2.3.1). Once we have introduced these simple models and have shown how they can describe each mode separately (with the help of the results of quantum theory), we will relax some of their assumptions to form improved models: the Non-rigid Rotor and the Anharmonic Oscillator (AHO) (Sect. 2.4.1). In essence, these more complex models require only minor corrections to the original, simpler models. Having introduced these models for each mode, we will then combine them and use them to understand *rovibrational* spectra, using at first, the simple models (Sect. 2.5), and then the improved models (Sect. 2.6). Finally, in Sect. 2.7, we will incorporate electronic transitions into the conceptual framework.

2.2 Rotational Spectra: Simple Model

2.2.1 Rigid Rotor (RR)

Our approach for the rigid rotor (RR) model is a blend of classical and quantum mechanics. For this model, we assume that the atoms are point masses ($d_{\text{nuc}} \approx 10^{-13}$ cm) with an equilibrium separation distance r_e that is constant or "rigid." That is, the rotating diatomic is analogous to a rotating dumbell that has a massless, inflexible rod connecting the weights at the end. Typical separation lengths are $r_e \approx 10^{-8}$ cm (Fig. 2.4). The "rigid" assumption for the bond length will be relaxed later.

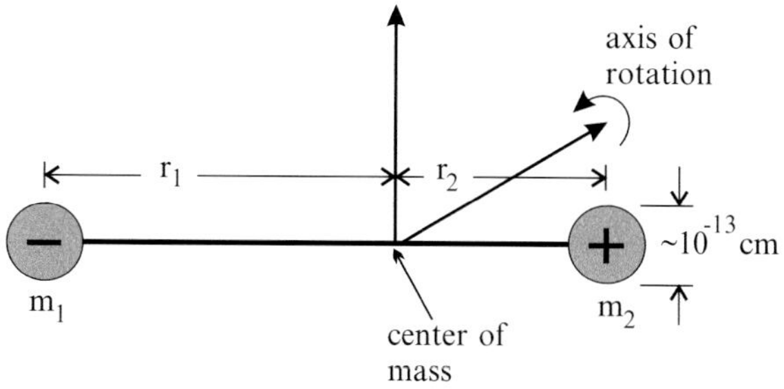

Fig. 2.4 Diatomic molecule with rigid rotor approximation

2.2.2 Classical Mechanics

Classical mechanics can be used to describe the moment of inertia about the center of mass for a diatomic molecule. The center of mass is the location along the internuclear axis at which $r_1m_1 = r_2m_2$. The moment of inertia I is given by:

$$I = \sum m_i r_i^2 = \mu r_e^2, \tag{2.3}$$

where μ, the reduced mass, is

$$\mu = \frac{m_1 m_2}{m_1 + m_2}. \tag{2.4}$$

So, the two-body problem is equivalent to the motion of a single-point mass, μ, rotating about the center of mass at a distance, r_e. The angular momentum of the molecule is then $I\omega_{\text{rot}}$ where ω_{rot} is the angular velocity.

2.2.3 Quantum Mechanics

Although angular momentum is a vector quantity, whose allowed values and directions are quantized, we often care only for its magnitude. Quantum theory gives the following relationship for the allowed magnitudes of angular momentum:

$$| I\omega_{\text{rot}} | = \sqrt{J(J+1)}\,\hbar, \quad J = 0, 1, 2, 3 \ldots, \tag{2.5}$$

where

$$\hbar = h/2\pi. \tag{2.6}$$

Here, "J" is an integer called a quantum number. There are several different quantum numbers needed to completely describe the state of a molecule. J is the one that characterizes the total angular momentum.

2.2.4 Rotational Energy

Classical mechanics can now be used to relate the rotational energy of a molecule to its moment of inertia, thereby yielding an expression for the allowed values of rotational energy as a function of rotational quantum number.

$$\begin{aligned} E_{\text{rot}} &= \frac{1}{2} I \omega_{\text{rot}}^2 && (2.7) \\ &= \frac{1}{2I} (I \omega_{\text{rot}})^2 && (2.8) \\ &= \frac{1}{2I} J(J+1)\hbar^2 && (2.9) \\ &= J(J+1)\frac{h^2}{8\pi^2 I} && (2.10) \\ &= E_J && (2.11) \end{aligned}$$

E_J, calculated in this manner, is usually in units of *Joules*. By convention, however, spectroscopists usually denote rotational energy by $F(J)$, in units of cm^{-1}. Referring to Eq. (1.7), the conversion is

$$\begin{aligned} F(J)\ \left[\text{cm}^{-1}\right] &= \frac{E_J\ [\text{J}]}{hc} && (2.12) \\ &= \left[\frac{h}{8\pi^2 Ic}\right] J(J+1). && (2.13) \end{aligned}$$

The "rotational constant," known as B, is

$$B\ \left[\text{cm}^{-1}\right] = \left[\frac{h}{8\pi^2 Ic}\right]. \tag{2.14}$$

Thus, Eq. (2.13) reduces to

$$\boxed{F(J) = BJ(J+1)}. \tag{2.15}$$

Note: So far we have only considered molecular rotation, so J, which commonly represents the total angular momentum, also represents the rotational angular momentum in this case. Later in the text, we will differentiate between angular momentum due to molecular rotation and angular momentum from electrons, and several new quantum numbers will be introduced that will contribute to J.

2.2.5 Absorption Spectrum

Schrödinger's wave equation is a key relation in quantum mechanics. Its basic form is as follows:

$$\frac{d^2\psi}{dx^2} + \frac{2m}{\hbar^2}[E - U(x)]\psi(x) = 0 \tag{2.16}$$

This is the time-independent form of the Schrödinger equation that describes a particle of mass m moving in a potential field described by $U(x)$. The wave function, ψ, is the solution to Schrödinger's differential wave equation, and $\psi\psi^*$ is proportional to the probability that the particle will occupy the portion of configuration space in $x \rightarrow x + dx$. The transition probability is directly related to the integral of the wave functions for the initial and final quantum states (m and n), and the permanent electric dipole moment, over all the configuration space elements, $d\tau$ [1].

$$\text{Transition probability} \propto \int \psi_m \mu \psi_n^* d\tau \rightarrow \Delta J = \pm 1 \tag{2.17}$$

where

$$\begin{aligned} \psi &\sim \text{Wave function} \\ \psi^* &\sim \text{Complex conjugate of the wave function} \\ \mu &\sim \text{Dipole moment} \end{aligned} \tag{2.18}$$

The quantum mechanical solution to Schrödinger's equation also yields "selection rules" for rotational transitions, namely that the change in rotational quantum number ($J_{\text{final}} - J_{\text{initial}}$) for a diatomic rigid rotor, can only be ± 1. For pure rotational transitions (meaning there are no changes in vibrational or electronic configuration), we can restrict the change in J to +1 if we define the change in J as:

$$\Delta J = \underset{\substack{\text{$'$ (upper)} \\ \downarrow}}{J'} - \underset{\substack{\text{$''$ (lower)} \\ \downarrow}}{J''} = +1$$

Here we have introduced commonly used notation in which the upper state is denoted with a single prime superscript and the lower state with a double prime. For example,

$$\begin{aligned} \overline{\nu}_{J'=1\leftarrow J''=0} &= F(J=1) - F(J=0) \\ &= 2B - 0 \\ &= 2B \end{aligned}$$

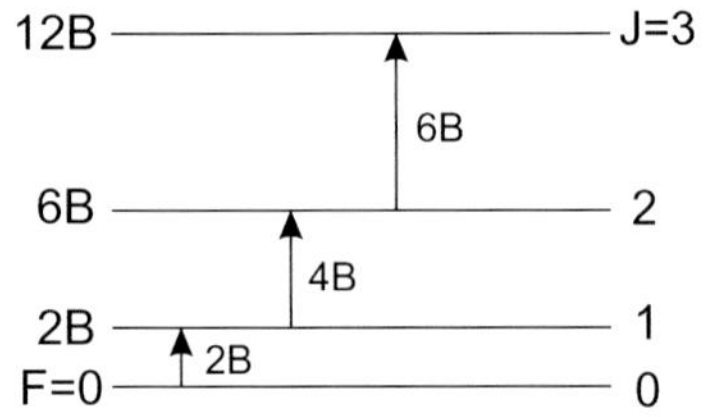

Fig. 2.5 Rotational energy level spacing

J	F		1st diff$=\overline{\nu}$
0	0		
		$\rangle$	2B
1	2B		
		$\rangle$	4B
2	6B		
		$\rangle$	6B
3	12B		
		$\rangle$	8B
4	20B		

Fig. 2.6 Energy level spacing and introduction of "first difference" in energy

Figure 2.5 shows each rotational state's energy level and the allowed transitions between them. Figure 2.6 provides the same information in tabular form.

In general, the rotational frequencies for transitions obeying the $\Delta J = 1$ selection rule are

$$\overline{\nu}_{J+1 \leftarrow J} = \overline{\nu}_{J' \leftarrow J''} = B(J'' + 1)(J'' + 2) - B(J'')(J'' + 1), \tag{2.19}$$

so, in terms of the lower state J''

$$\boxed{\overline{\nu}_{J' \leftarrow J''} = 2B(J'' + 1)}. \tag{2.20}$$

Let's look at the following rotational absorption spectrum for CO (Fig. 2.7).

2.2.6 Usefulness of Rotational Line Spacing

The line spacing of rotational absorption spectra can be used to deduce accurate physical characteristics of the molecule under investigation.

$$\text{Line spacing} \rightarrow B \rightarrow I \rightarrow r_e$$

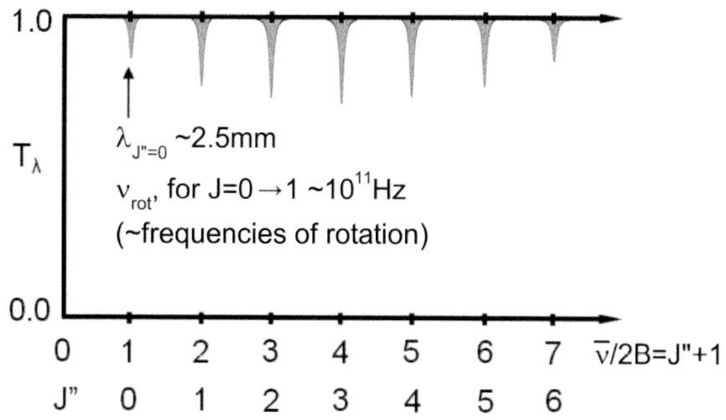

Fig. 2.7 Absorption spectrum spacing for heteronuclear rotation

Note:
1. lines have uniform spacing, making them easy to identify/interpret
2. $B_{CO} \approx 2\,cm^{-1} \rightarrow \lambda_{J''=0} = 1/\bar{\nu} = 1/4\,cm = 2.5$ mm (microwaves/mm waves)
3. $\nu_{rot} = c/\lambda = (3 \times 10^{10})/0.25 = 120\,GHz$ (μwave)

Consider carbon monoxide (CO) for an example. The rotational constant can be spectroscopically measured to six significant figures, leading to a highly precise determination of the CO internuclear spacing.

$$B = 1.92118\ cm^{-1} \rightarrow r_{CO} = 1.128227\,\text{Å}$$

The last digit, 7, is associated with an increment of length of only $7 \times 10^{-6}\,\text{Å} = 7 \times 10^{-16}\,m$!

2.2.7 Rotational Partition Function

The rotational partition function for diatomic molecules that are rigid rotors can be approximated by Vincenti and Kruger [2]

$$Q_{rot} = \frac{1}{\sigma}\frac{kT}{hcB}, \tag{2.21}$$

where σ is the symmetry number (the number of ways of rotating the molecule to achieve the same orientation of the molecule, treating identical atoms as indistinguishable). For heteronuclear molecules such as CO, $\sigma = 1$, while for homonuclear molecules such as N_2, $\sigma = 2$.

2.2.8 Rotational Temperature

Rotational excitation can be described in terms of a characteristic temperature. From statistical mechanics, we know that the Boltzmann fraction of molecules with rotational quantum number J is

$$\frac{N_J}{N} = \frac{(2J+1)\exp(-E_J/kT)}{Q_{\mathrm{rot}}} \tag{2.22}$$

$$= \frac{(2J+1)\exp(-\theta_r J(J+1)/T)}{T/\sigma\theta_r}. \tag{2.23}$$

It may be noted here that $2J + 1$ is the degeneracy for energy level E_J, a result from quantum mechanics associated with the number of possible directions (orientations) of the angular momentum vector with the same energy.

Since

$$\frac{E_J}{k} = \frac{hcF(J)}{k} \tag{2.24}$$

$$= \left(\frac{hc}{k}\right) BJ(J+1) \tag{2.25}$$

$$\equiv \theta_{\mathrm{rot}} J(J+1), \tag{2.26}$$

the rotational temperature θ_{rot} is

$$\theta_{\mathrm{rot}}\ [\mathrm{K}] = \left(\frac{hc}{k}\right) B. \tag{2.27}$$

Thus the "characteristic temperature," in this case θ_{rot}, is determined simply by multiplying a characteristic energy in cm^{-1} units by hc/k (Table 2.1). The relation

$$\boxed{\left(\frac{hc}{k}\right) = 1.44\,\mathrm{K/cm^{-1}}} \tag{2.28}$$

Table 2.1 Characteristic rotational temperatures for some diatomic species

Species	θ_{rot} [K]
O_2	2.1
N_2	2.9
NO	2.5
Cl_2	0.351

is worth memorization. We will use it often! Using the rotational temperature, the equation for the rotational partition function (Eq. (2.21)) can be simplified to

$$Q_{\text{rot}} = \frac{1}{\sigma}\frac{T}{\theta_r}. \tag{2.29}$$

2.2.9 Intensities of Spectral Lines

The absorption (or emission) probability, per molecule, is *approximately* independent of J; we shall call this crude approximation the principle of "equal probability." Therefore, the absorption/emission spectrum varies with J similar to the Boltzmann distribution for the population in J.

Using the Boltzmann fraction again, we have

$$\frac{N_J}{N} = \frac{(2J+1)\exp\left(-E_J/kT\right)}{Q_{\text{rot}}}. \tag{2.30}$$

The strongest peaks occur near where the population is at a local maximum (i.e., at the maxima of the Boltzmann distribution, Eq. (2.30)), i.e.,

$$\frac{d(N_J/N)}{dJ} = 0, \tag{2.31}$$

giving

$$J_{\max} = (T/2\theta_{\text{rot}})^{1/2} - 1/2. \tag{2.32}$$

2.3 Vibrational Spectra: Simple Model

2.3.1 Simple Harmonic Oscillator

The simplest model for diatomic vibration is the SHO. This model assumes that two masses, m_1 and m_2, have an equilibrium separation distance r_e. The bond length, or separation distance between the masses, r, oscillates about the equilibrium distance as if the bond were a spring (see Fig. 2.8). We will begin the investigation of the SHO model with classical mechanics.

2.3.2 Classical Mechanics

Hooke's law describing linear spring forces can be applied for the SHO:

$$\text{Force} = -k_s(r - r_e). \tag{2.33}$$

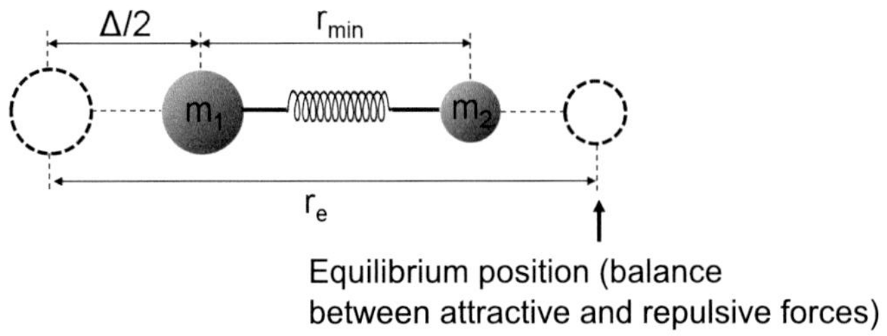

Fig. 2.8 Oscillation of a linear diatomic molecule; $r_{\min}$ corresponds to molecule at distance of greatest compression

In this equation, k_s is the spring constant (not to be confused with Boltzmann's constant, k), and the restoring force is linearly proportional to the extension (or compression) of the spring from its equilibrium length. Such systems have a fundamental resonant frequency of vibration, ν_{vib}, which depends on the stiffness of the bond (i.e., the spring constant) and the magnitudes of the masses at both ends of the bond,

$$\nu_{\text{vib}} = \frac{1}{2\pi}\sqrt{k_s/\mu}\ \left[\text{s}^{-1}\right]. \tag{2.34}$$

As before, the reduced mass, μ, is given by

$$\mu = \frac{m_1 m_2}{m_1 + m_2}. \tag{2.35}$$

Note that this vibration frequency does *not* depend on the amplitude of vibration. In wavenumber units the fundamental frequency is

$$\omega_e = \nu_{\text{vib}}/c\ \left[\text{cm}^{-1}\right]. \tag{2.36}$$

The potential energy, U, stored in the oscillator (owing to compression or extension of the spring), is

$$U = \frac{1}{2}k_s(r - r_e)^2. \tag{2.37}$$

Thus, according to Eq. (2.37), the potential energy for an SHO molecule varies as the square of the extension (or compression) of the internuclear spacing. A plot of U versus r is thus a parabola, with a minimum value ($U = 0$) at $r = r_e$ (see Fig. 2.9). Also shown in Fig. 2.9 is the truncated harmonic oscillator (THO).

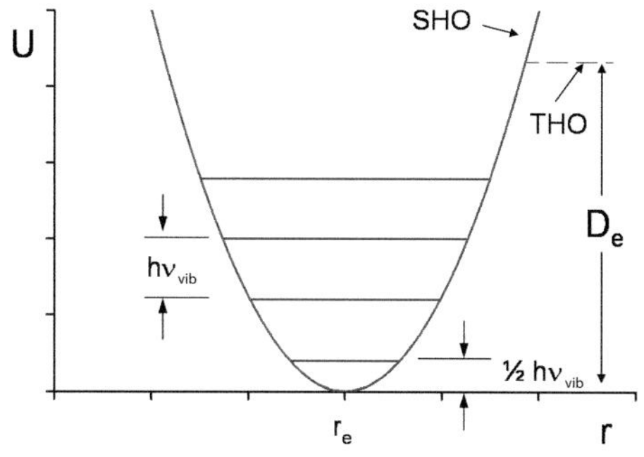

Fig. 2.9 Potential energy and vibrational levels for diatomic SHO and THO molecules

2.3.3 Quantum Mechanics

The results of quantum mechanics for an SHO lead to an expression for the energy of a vibrating diatomic molecule

$$G(v) = \omega_e(v + 1/2) \ [\text{cm}^{-1}] \tag{2.38}$$

where v is the vibrational quantum number:

$$v = 0, 1, 2, 3 \ldots .$$

Note that the SHO has *equal energy spacing* between adjacent quantum states, i.e. $G(v + 1) - G(v) = \omega_e$ independent of v. This independence is one of the attractive simplifications that result from the SHO model. Another virtue of the SHO model is that the quantum mechanics solution for absorption and emission of a heteronuclear diatomic molecule leads to a very simple selection rule, namely that the vibrational quantum number can change only by 1 [3].

$$\boxed{\Delta v = v' - v'' = +1} \tag{2.39}$$

2.3.4 Vibrational Partition Function

For diatomics whose vibrational potential energy can be approximated by the SHO model (Eq. (2.38)), the vibrational partition function, Q_{vib}, is [2]

$$Q_{\text{vib}} = \left[1 - \exp\left(\frac{-hc\omega_e}{kT}\right)\right]^{-1} \exp\left(\frac{-hc\omega_e}{2kT}\right). \tag{2.40}$$

It is common with the SHO model to choose an alternate reference (zero) energy at $v = 0$, so that

$$G(v) = \omega_e v \tag{2.41}$$

In this case,

$$Q_{\text{vib}} = \left[1 - \exp\left(\frac{-hc\omega_e}{kT}\right)\right]^{-1} \tag{2.42}$$

It is important to keep in mind that the magnitude of the vibrational partition function depends on the choice of the zero energy, and that the same zero must be used in specifying molecular energies E_i for any level i and in evaluating the associated partition function.

2.3.5 Vibrational Temperature

Just as rotations have a characteristic temperature, θ_{rot}, vibrations have their own temperatures, θ_{vib}, that are typically much higher (Table 2.2). If we define the vibrational temperature as

$$\theta_{\text{vib}} \quad [\text{K}] = \left(\frac{hc}{k}\right)\omega_e, \tag{2.43}$$

and employ Eq. (2.41) for $G(v)$, then the Boltzmann fraction for vibrational states is

$$\frac{N_{\text{vib}}}{N} = \frac{g_{\text{vib}}\exp(-v\theta_{\text{vib}}/T)}{Q_{\text{vib}}} \tag{2.44}$$

$$= \exp(-v\theta_{\text{vib}}/T)\left(1 - \exp(-\theta_{\text{vib}}/T)\right), \tag{2.45}$$

where $g_{\text{vib}} = 1$.

Table 2.2 Characteristic vibrational temperatures for some diatomic species

Species	θ_{vib} [K]
O_2	2270
N_2	3390
NO	2740
Cl_2	808

2.4 Improved Models of Rotation and Vibration

2.4.1 Non-rigid Rotation

The model for molecular rotation can be improved by relaxing the initial assumption of rigidity. There are two dominant effects that lead to non-rigid rotation and hence affect B and $F(J)$:

1. vibrational stretching causes the average spacing $\bar{r}$ to be a function of E_{vib}, i.e. $\bar{r}(E_{\text{vib}})$. The trends are:

$$E_{\text{vib}} \uparrow, \bar{r} \uparrow, I \uparrow, B \downarrow \tag{2.46}$$

   That is, as the vibrational energy E_{vib} increases, the average nuclear separation increases, thus increasing the moment of inertia, and the rotational constant decreases.
2. centrifugal distortion causes the average spacing $\bar{r}$ to be a function of J, i.e. $\bar{r}(J)$. The trends are:

$$J \uparrow, \bar{r} \uparrow, B \downarrow \tag{2.47}$$

   That is, as rotational energy (J) increases, the average nuclear separation increases, and the rotational constant decreases.

The effects of vibrational stretching are much larger than the effects of centrifugal distortion. The result of these non-rigidities is a new expression for the rotational energy, $F_{\upsilon}(J)$,

$$F_{\upsilon}(J) = B_{\upsilon}J(J+1) - D_{\upsilon}J^2(J+1)^2, \tag{2.48}$$

where D_{υ} is the centrifugal distortion constant (written with a subscript υ to denote its dependence on vibrational quantum number), and B_{υ} is the vibrationally dependent rotational constant. The rotational transition frequency for rotators in vibrational level υ, after accounting for distortion, becomes

$$\overline{\nu}_{J' \leftarrow J'', \upsilon} = 2B_{\upsilon}(J''+1) - 4D_{\upsilon}(J''+1)^3 \tag{2.49}$$

The distortion constant term is subtracted in Eq. (2.49) (compared with Eq. (2.20)), and thus the rotational spacings are *reduced* by non-rigid rotation. The vibrationally dependent constants for rotation, B_{υ}, and centrifugal distortion, D_{υ}, are given by:

$$B_{\upsilon} = B_e - \alpha_e(\upsilon + 1/2) \tag{2.50}$$

$$D_{\upsilon} = D_e + \beta_e(\upsilon + 1/2) \tag{2.51}$$

where both α_e and β_e have positive values. See Sect. 2.4.3 for typical correction values for vibrational and centrifugal distortion.

2.4.2 Anharmonic Oscillator

True diatomics do not adhere exactly to the idealized SHO model, but rather have anharmonicities that affect the shape of the potential well and the spacing between energy levels. Thus, the models for vibrational energy can be improved by accounting for the effects of anharmonic oscillation. The total energy for an oscillating diatomic, after correcting for higher order anharmonicities, is

$$G(v) = \omega_e(v + 1/2) - \underbrace{\omega_e x_e(v + 1/2)^2}_{\text{anharmonicity corr.}} + \cdots + H.O.T. \tag{2.52}$$

Correcting for anharmonicity *decreases* the energy spacing. In addition, the selection rule for allowed changes in vibrational quantum number is modified to permit the additional possibility of (relatively weak) transitions with $\Delta v = v' - v''$ other than 1 (Table 2.3). That is, relaxing the SHO model to allow for anharmonicity leads to finite probabilities for $\Delta v = 2, 3, \ldots$ and higher transitions, though these probabilities diminish rapidly with increasing magnitude of Δv.

Note: Transition probabilities for the first overtone of CO are about 100 times weaker than for the fundamental.

Potential Energy

The potential energy well for an anharmonic diatomic molecule can more accurately be described by the Morse function (see Fig. 2.10) than by Hooke's Law. The Morse function is

$$U = D_{\text{eq}} \left[1 - \exp(-\beta(r - r_e))\right]^2, \tag{2.53}$$

where U is the potential energy and D_{eq} is the bond-dissociation energy (in wavenumbers). The $(r - r_e)$ term in the exponential is the displacement from the

Table 2.3 AHO frequencies and corrections

Transition	Transition name	Frequency
$\Delta v = +1$	"Fundamental" band	$\overline{\nu}_{1\leftarrow 0} = \omega_e(1 - 2x_e) = G(1) - G(0)$
	(e.g. $1 \leftarrow 0, 2 \leftarrow 1$)	$\overline{\nu}_{2\leftarrow 1} = \omega_e(1 - 4x_e)$
$\Delta v = +2$	First overtone	$\overline{\nu}_{2\leftarrow 0} = 2\omega_e(1 - 3x_e)$
	(e.g. $2 \leftarrow 0, 3 \leftarrow 1$)	
$\Delta v = +3$	Second overtone	$\overline{\nu}_{3\leftarrow 0} = 3\omega_e(1 - 4x_e)$
	(e.g. $3 \leftarrow 0, 4 \leftarrow 1$)	

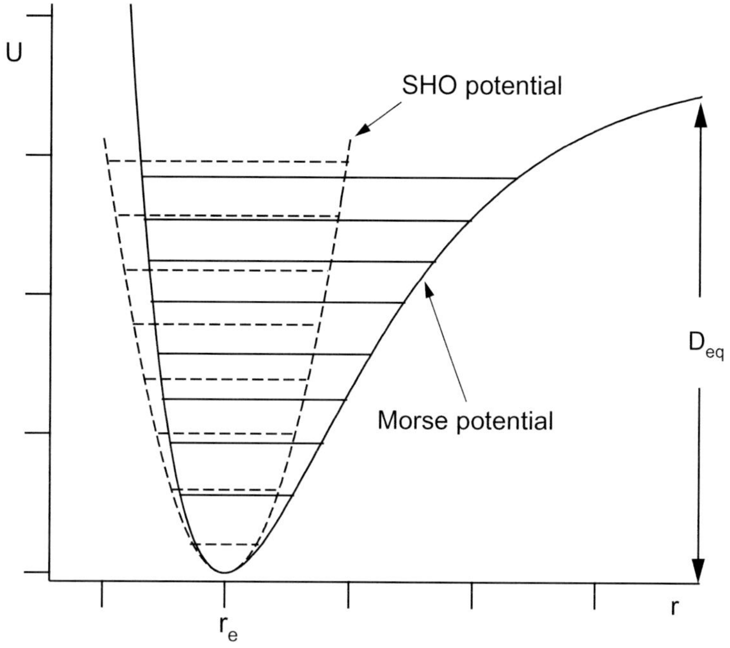

Fig. 2.10 Potential energy and vibrational levels for a diatomic molecule

equilibrium internuclear distance, r_e (in centimeter units). The term β (not to be confused with β_e in Eq. (2.51)) is

$$\beta = 1.2177 \times 10^7 \, \omega_e \sqrt{\mu / D_{\text{eq}}}. \tag{2.54}$$

The term ω_e is in wavenumber units and the reduced mass is in atomic mass units.

2.4.3 Typical Correction Magnitudes

Rotational Correction

As described in Sect. 2.4.1, the improvements to the model for molecular rotation include vibrationally dependent constants B_v and D_v. Examples showing the relative magnitudes of the pertinent parameters are below. In general,

$$\beta_e \ll D_e \ll \alpha_e \ll B_e$$

That is, the centrifugal distortion depends only very weakly on vibrational level and hence both D_e and D_v are small compared to the stretching effect of vibration (α_e). The moment of inertia (and thus the rotational constant, B) is well-reflected by a rigid rotor approximation, but vibrational effects (α_e) can cause small changes (about 1 % as shown in the example below).

1. $(D/B) \ll 1$:
By balancing the force of centrifugal distortion with a restorative force from a harmonic oscillator, it can be shown that[1]

$$D = \frac{4B^3}{\omega_e^2} \ll B \tag{2.55}$$

Hence, D/B is quite small, especially for molecules with "stiff" or high-frequency bonds. For example,

$$(D/B)_{\mathrm{NO}} = 4\left(\frac{B}{\omega_e}\right)^2 \approx 4\left(\frac{1.7}{1900}\right)^2 \approx 10^{-6}$$

2. $(\alpha_e/B_e) \ll 1$
For a potential energy well described by the Morse function,

$$\alpha_e = \frac{6\sqrt{\omega_e x_e B_e^3}}{\omega_e} - \frac{6B_e^2}{\omega_e} \tag{2.56}$$

Frequencies of vibration are quite high compared to rotational constants, so α_e is small compared to B_e. Physically, this means the change in internuclear distance by vibration is small compared to the internuclear distance itself. For example,

$$(\alpha_e/B_e)_{\mathrm{NO}} \approx 0.01$$

3. $(\beta_e/D_e) \ll 1$
Using the Morse function again to describe the molecule's potential, the constant β_e is

$$\begin{aligned} \beta_e/D_e &= \frac{8\omega_e x_e}{\omega_e} - \frac{5\alpha_e}{B_e} - \frac{\alpha_e^2\,\omega_e}{24B_e^3} \ll 1 \\ &\approx \frac{8\omega_e x_e}{\omega_e} \end{aligned} \tag{2.57}$$

The constant β_e is often much smaller than D_e, which itself is small, and may therefore typically be neglected. For example,

$$(\beta_e/D_e)_{\mathrm{NO}} \approx 0.001$$

[1] See Herzberg [4, pp. 103–104], for more details.

Table 2.4 Typical values for vibration (Banwell [5], p. 62, Table 3.1)

Gas	Vibration ω_e [cm^{-1}]	Anharmonicity constant x_e	Force constant k_s [dynes/cm]	Internuclear distance r_e [Å]	Dissociation energy D_{eq} [eV]
CO	2170	0.006	19×10^5	1.13	11.6
NO	1904	0.007	16×10^5	1.15	6.5
H_2[a]	4395	0.027	16×10^5	1.15	6.5
Br_2[a]	320	0.003	2.5×10^5	2.28	1.8

[a] Not IR-active, use Raman spectroscopy!

Vibrational Correction

Typical values for anharmonicity constants as well as some other molecular constants are listed in Table 2.4.

Useful Conversions

$$1\,\text{eV} = 8065.54\,\text{cm}^{-1} = 23.0605\,\text{kcal/mole} = 1.60219\times10^{-19}\,\text{J}$$
$$1\,\text{cal} = 4.1868\,\text{J}$$
$$1\,\text{N} = 10^5\,\text{dynes}$$
$$1\,\text{Å} = 0.1\,\text{nm}$$

Example: NO, Nitric Oxide

$$\begin{aligned} B_e &= 1.7046\,\text{cm}^{-1} \\ \alpha_e &= 0.0178 \\ D_e &\approx 5.8 \times 10^{-6}\,\left({}^2\Pi_{1/2}\right) \\ \beta_e &\approx 0.0014 \times D_e \\ &\approx 8 \times 10^{-9}\,\text{cm}^{-1} \\ \omega_e &= 1904.03\,\left({}^2\Pi_{1/2}\right);\ 1903.68\,\left({}^2\Pi_{3/2}\right) \\ \omega_e\,x_e &= 13.97\,\text{cm}^{-1} \\ r_e &= 1.1508\,\text{Å} \end{aligned}$$

2.5 Rovibrational Spectra: Simple Model

2.5.1 Born–Oppenheimer Approximation

The simplest model for rovibration is a vibrating rigid rotor based on the Born–Oppenheimer approximation, in which vibration and rotation are regarded as independent. These transitions include a simultaneous change in vibrational quantum number, v, and rotational quantum number, J. The total energy for these transitions, $T(v, J)$, is a sum of the energy for a rigid rotor, $F(J)$ (Eq. (2.15)), and SHO, $G(v)$ (Eq. (2.38)).

$$T(v,J) = E_{\rm RR} + E_{\rm SHO}$$
$$= F(J) + G(v)$$
$$= BJ(J+1) + \omega_e(v+1/2) \tag{2.58}$$

The same selection rules that applied for the rigid rotor and SHO, namely that the quantum numbers change only by 1 in a transition, also apply for the combined transition, with only slight reinterpretation [5].

Selection Rules

$$\boldsymbol{\Delta v = +1}$$
$$\boldsymbol{\Delta J = \pm 1}$$

Note that $\Delta J = J' - J'' = \pm 1$ rather than only $+1$. This new selection rule occurs simply because both allowed values of J', namely $J'' \pm 1$, lead to *upper* states with higher energy than the *lower* state of the molecule, i.e.

$$F(J') + G(v') > F(J'') + G(v'')$$

for both $J' = J'' + 1$ and $J' = J'' - 1$. The line positions are based on the differences in total energy for the upper and lower rovibrational states.

$$\overline{\nu} = T' - T'' = T(v', J') - T(v'', J'') \tag{2.59}$$

2.5.2 Spectral Branches

Because the rotational quantum number can either increase or decrease by 1, two branches of line positions emerge (Fig. 2.11). The *R* branch is associated with an increase in rotational quantum number ($J' > J''$) and the *P* branch is associated with a decrease in rotational quantum number ($J' < J''$). There is a gap (the "null gap") between the lowest lines in the *P* and *R* branches, as shown in Fig. 2.12.

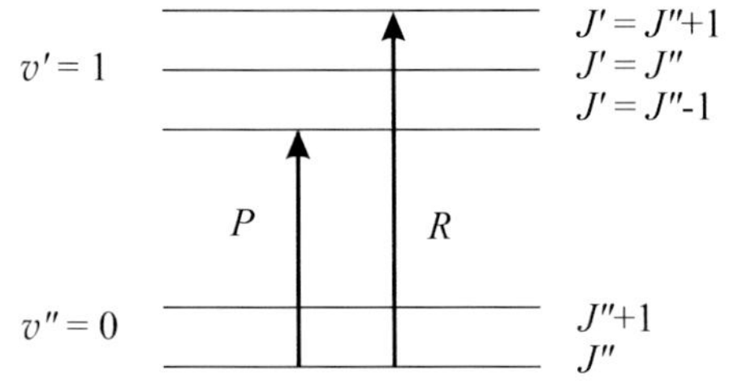

Fig. 2.11 Energy level diagram denoting *P* and *R* absorption transitions from a ground vibrational state for a heteronuclear diatomic molecule

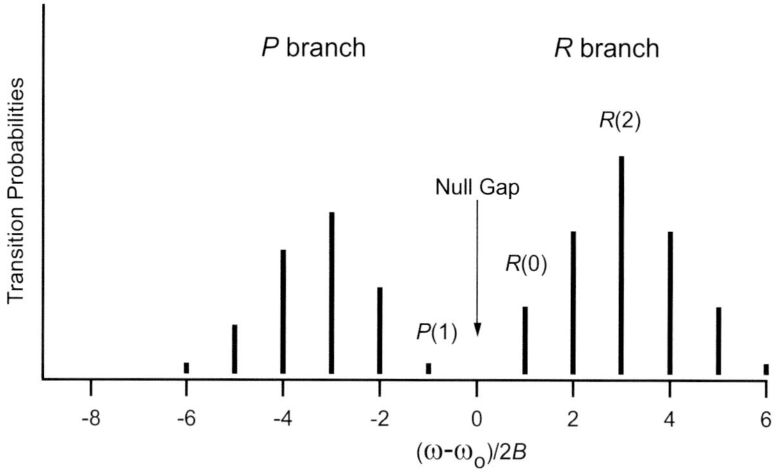

Fig. 2.12 Simulated absorption spectrum of the *P* and *R* branches of a ground state rovibrational transition of a heteronuclear diatomic molecule

$$\begin{aligned} &\boldsymbol{P}\ \textbf{branch:}\ \Delta J = -1 \\ &\boldsymbol{R}\ \textbf{branch:}\ \Delta J = +1 \end{aligned}$$

Note that for this simple model, both branches have constant line spacing equal to $2B$. Figure 2.12 shows unequal strengths between the P and R branches, which arises from Hönl–London factor considerations (more in Chap. 7). This effect, magnified in Fig. 2.12, illustrates the error associated with the principle of "equal probability" introduced in Sect. 2.2.9.

R Branch

The energy in the R branch is denoted $R(v, J)$.

$$R(v'', J'') = \underbrace{[G(v') - G(v'')]}_{\Delta G = \omega_o} + B(J'' + 1)(J'' + 2) - BJ''(J'' + 1)\ [\mathrm{cm}^{-1}] \tag{2.60}$$

In this simple model, the difference in vibrational energy, $\Delta G = G(v') - G(v'')$, known as the rotationless transition frequency, is independent of J. ΔG, often written as either ω_0 or $\overline{\nu}_0$, is given numerically by ω_e in the SHO model. For an AHO, ω_o is a function of ω_e, $\omega_e x_e$, and v''.

$$\Delta G = \begin{cases} \overline{\nu}_0, \omega_0 & \text{rotationless transition wavenumber} \\ \omega_e & \text{SHO model} \\ \omega_e(1 - 2x_e) & \text{AHO model for } v' \leftarrow v'' = 1 \leftarrow 0 \end{cases}$$

Thus,

$$R(v'', J'') = \omega_o + 2B(J'' + 1). \tag{2.61}$$

P Branch

Similarly, for the P branch, the energy, $P(v, J)$, is

$$P(v'', J'') = \omega_o - 2BJ''. \tag{2.62}$$

Note that since the P branch occurs for net changes in rotational quantum number of -1, the $P(0)$ transition is not possible, leaving a gap (the "null gap") between the lowest lines in the P and R branches.

Note: The naming convention is $R(J'')$ or $P(J'')$ for rotational transitions. For example, $R(7)$ indicates a transition involving (in absorption or emission) a lower rotational state of $J'' = 7$ and an upper rotational state of $J' = 8$, while $P(7)$ represents a transition involving a lower rotational state of $J'' = 7$ and an upper rotational state of $J' = 6$.

Branch Separation

The separation between the tallest peaks in the P and R branch absorption intensities is a direct function of temperature due to Boltzmann statistics. Subject to the "equal probability" approximation where the absorption spectrum (in both the P and R branches) maps directly from the Boltzmann distribution over rotational state, the peak-to-peak frequency separation is

$$\Delta\nu = \sqrt{\frac{8BkT}{hc}}. \tag{2.63}$$

As the temperature increases, the most probable transition shifts to higher energy levels due to increasing population of those levels (see Eq. (2.32)). Thus the frequency location of each branch's maximum will move further away from the null gap, leading to larger separation.

A more complete analysis, without the equal probability approximation, leads to a more complex expression for the peak spacing, but the numerical values do not differ greatly except at low temperatures.

2.6 Rovibrational Spectra: Improved Model

2.6.1 Breakdown of Born–Oppenheimer Approximation

By allowing for non-rigid rotation, anharmonic vibration, and interactions between vibration and rotation (i.e., the breakdown of the Born–Oppenheimer approximation), an improved model for rovibrational energy can be established.

$$\begin{aligned} T(v, J) &= G(v) + F(v, J) \\ &= \underbrace{\omega_e(v + 1/2)}_{\text{SHO}} - \underbrace{\omega_e x_e(v + 1/2)^2}_{\text{anharm. corr.}} + \underbrace{B_v J(J + 1)}_{\text{RR(v)}} - \underbrace{D_v J^2(J + 1)^2}_{\text{cent. dist. term}} \end{aligned} \tag{2.64}$$

Recall from Sect. 2.4.1 [Eqs. (2.50) and (2.51)] that the rotational and centrifugal distortion constants, B_v and D_v, respectively, introduce vibrational coupling into the rotational energy:

$$B_v = B_e - \alpha_e(v + 1/2) \tag{2.65}$$

$$D_v = D_e + \beta_e(v + 1/2) \tag{2.66}$$

This coupling, indicated by the subscript v on B and D, signifies that the Born–Oppenheimer approximation is no longer in effect. The rotational constants can be related to previous notation, B' and B'', by noting that $B' = B_v(v')$ and $B'' = B_v(v'')$.

2.6.2 Spectral Branches

Just as before, the possibility for net changes in rotational quantum number of either $+1$ or -1 yields two spectral branches with a null gap separating them. However, by including the correction terms to the simple models, the line spacing of the branches will not be constant at $2B$.

R Branch

The new expression for R-branch energies as a function of the vibrational and rotational quantum numbers is

$$\begin{aligned} R(v'', J'') &= \Delta G + B'(J'' + 1)(J'' + 2) - B''(J'')(J'' + 1) \\ &= \Delta G + 2B' + (3B' - B'')J'' + (B' - B'')(J'')^2, \end{aligned} \tag{2.67}$$

where, as before,

$$\Delta G = G(v') - G(v'') \tag{2.68}$$

For clarity, these expressions are typically written without the $''$ and assumed to be a function of the lower-state quantum numbers only. Thus, Eq. (2.67) can be written as

$$R(v, J) = \Delta G + 2B' + (3B' - B'')J + (B' - B'')J^2, \tag{2.69}$$

where J refers to J''.

P Branch

Similarly, for the P branch,

$$\begin{aligned} P(v'', J'') &= \Delta G + B'(J'' - 1)(J'') - B''(J'')(J'' + 1) \\ &= \Delta G - (B' + B'')J'' - (B'' - B')(J'')^2 \end{aligned} \tag{2.70}$$

Writing with J instead of J'' produces

$$P(v, J) = \Delta G - (B' + B'')J - (B'' - B')J^2 \tag{2.71}$$

2.6.3 Rotational Constant

As shown above, the rotational constant depends on the vibrational state of the molecule.

$$B' = B_e - \alpha_e(v' + 1/2) \tag{2.72}$$

$$B'' = B_e - \alpha_e(v'' + 1/2) \tag{2.73}$$

And, for $v' = v'' + 1$,

$$B' - B'' = -\alpha_e. \tag{2.74}$$

Since $\alpha_e > 0$,

$$B' < B''. \tag{2.75}$$

As a result, the line spacing decreases with J in the R branch and increases with J in the P branch.

2.6.4 Bandhead

The unequal spacing in the P and R branches leads to a *bandhead* in the R branch as the lines "wrap around" on themselves (Fig. 2.13). This bandhead occurs where $dR(J)/dJ = 0$.

$$\frac{dR(J)}{dJ} = \underbrace{(3B' - B'')}_{2B'-\alpha_e} + 2\underbrace{(B' - B'')}_{-\alpha_e} J'' = 0 \tag{2.76}$$

The location of the bandhead is

$$J''_{\text{bandhead}} \approx \frac{2B' - \alpha_e}{2\alpha_e} \approx \frac{B}{\alpha_e} \tag{2.77}$$

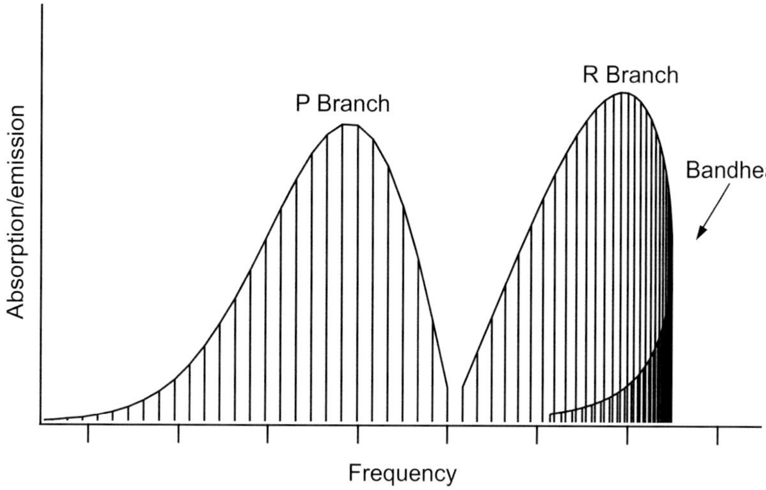

Fig. 2.13 Unequal line spacing due to non-rigid rotation leads to a bandhead in the R branch for diatomic molecules

Example: CO Bandhead
For CO,

$$\frac{B}{\alpha_e} \approx \frac{1.9}{0.018} \approx 106$$

Thus, the bandhead will only be observed in high temperature spectra.

2.6.5 Finding Key Parameters: $B_e, \alpha_e, \omega_e, x_e$

Assuming access to the absorption spectra of a molecule, e.g. a tabular listing of the R- and P-branch line positions for the $v = 1 \leftarrow 0$ and $v = 2 \leftarrow 0$ bands, how would one extract the key parameters?

First Approach

Use measured band origin data for the fundamental and first overtone with $v'' = 0$ to get ω_e and x_e.

$$\Delta G_{1\leftarrow 0} = G(1) - G(0) = \omega_e(1 - 2x_e) \tag{2.78}$$

$$\Delta G_{2\leftarrow 0} = G(2) - G(0) = 2\omega_e(1 - 3x_e) \tag{2.79}$$

Second Approach

Fit rotational transitions to the line spacing equation to get B_e and α_e.

$$\omega = \omega_o + (B' - B'')m^2 + (B' + B'')m \tag{2.80}$$

where

$$m = \begin{cases} -J & \text{for the } P\text{-branch} \\ J + 1 & \text{for the } R\text{-branch} \end{cases} \tag{2.81}$$

Equation (2.80) is known as the Fortrat parabola formula. Finding B' and B'' allows direct determination of B_e and α_e. The Fortrat parabola can also be useful in the analysis of electronic systems (Sect. 2.7.3).

Third Approach

Use the "method of common states" to get B_e and α_e. In general,

$$F(J) = BJ(J + 1) \tag{2.82}$$

Then, for Fig. 2.14, drawn for a "common upper state,"

$$\Delta E = F(J + 1) - F(J - 1) = R(J - 1) - P(J + 1) \tag{2.83}$$

$$\Delta E = B''(J + 1)(J + 2) - B''(J - 1)(J) \tag{2.84}$$

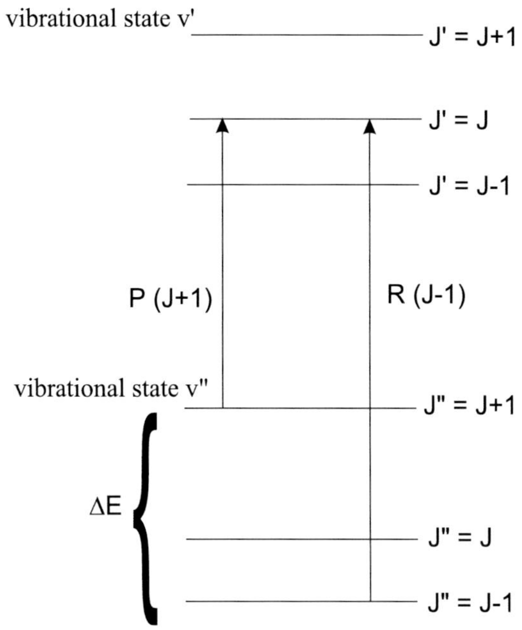

Fig. 2.14 Energy level diagram for the method of common upper states

Therefore,

$$\Delta E = B''(4J + 2) \tag{2.85}$$

The energy difference between R and P branch transitions to a common upper state $J' = J$ leads directly to a value for B''. Simply divide ΔE by $(4J+2)$. A similar approach, with a common lower state, leads directly to a value for B'. Knowledge of B' and B'' can be used to determine the parameters B_e and α in Eq. (2.50). The method of common states will also find use in analysis of electronic spectra where B' and B'' differ in the two electronic states of an absorption or emission spectra.

2.6.6 Effects of Isotopic Substitution

What are the effects of isotopic substitution on absorption or emission spectra? Changes in nuclear mass (neutrons) do not change r_e or bond stiffness since these properties depend primarily on electric binding forces, which are unchanged with mass of nuclei. Since B varies as $1/\mu$,

$$B \propto \frac{1}{I} \propto \frac{1}{\mu},$$

the *spacing* of lines changes as μ changes. Similarly, the fundamental frequency of vibration, ω_e, varies with μ,

$$\omega_e \propto \sqrt{\frac{k_s}{\mu}} \propto \sqrt{\frac{1}{\mu}},$$

therefore the *band origin* also changes as μ changes.

Example: CO Isotope, $^{13}C^{16}O$

What can be learned from the combined IR absorption spectra of $^{12}C^{16}O$ and $^{13}C^{16}O$? After the lines are assigned, the line spacing can be used to infer the B values of both species, yielding:

$$B_{^{12}C^{16}O} = 1.92118 \text{ and } B_{^{13}C^{16}O} = 1.83669\,\text{cm}^{-1} \tag{2.86}$$

1. The change in line spacing from $^{12}C^{16}O$ to $^{13}C^{16}O$ is $\Delta(2B) = -0.17\,\text{cm}^{-1}$
2. The ratio of the B values can be used to calculate the mass of ^{13}C from the known value of $m_{^{12}C} = 12.0$, i.e.

$$\frac{B_{^{12}C^{16}O}}{B_{^{13}C^{16}O}} = \frac{\mu_{^{13}C^{16}O}}{\mu_{^{12}C^{16}O}} \Rightarrow m_{^{13}C} = 13.0006 \tag{2.87}$$

This calculation is within 0.02 % of the actual value, 13.0034!

3. The relative B values can also be used to estimate the shift in the band origin:

$$\frac{\omega_{e,^{13}\mathrm{C}^{16}\mathrm{O}}}{\omega_{e,^{12}\mathrm{C}^{16}\mathrm{O}}} = \sqrt{\frac{B_{^{13}\mathrm{C}^{16}\mathrm{O}}}{B_{^{12}\mathrm{C}^{16}\mathrm{O}}}} = \sqrt{0.956}$$

Using $\omega_{e,^{12}\mathrm{C}^{16}\mathrm{O}} \approx 2200\,\mathrm{cm}^{-1}$, the change in band origin is

$$\Delta\omega_e \approx 50\,\mathrm{cm}^{-1}$$

See Banwell [5] (Fig. 3.7, p. 67) for an example absorption spectrum in which the natural abundance of ^{13}C (about 1.1 %) is evident.

2.6.7 Hot Bands

Hot bands are those that involve excited states, i.e. having a lower state with a vibrational quantum number greater than zero. When are hot bands important? Recall, the Boltzmann fraction for vibrational states is

$$\frac{N_{\mathrm{vib}}}{N} = \frac{g_{\mathrm{vib}} \exp(-v\theta_{\mathrm{vib}}/T)}{Q_{\mathrm{vib}}} \tag{2.88}$$

$$= \exp(-v\theta_{\mathrm{vib}}/T)\,(1 - \exp(-\theta_{\mathrm{vib}}/T)) \tag{2.89}$$

Hence, the necessary condition to allow neglect of hot bands is that $T \ll \theta_{\mathrm{vib}}$. Since the characteristic vibrational temperature, θ_{vib}, often exceeds 10^3 K (see Table 2.2), hot bands can often be neglected in absorption and emission.

Example: CO Hot Bands

$$\theta_{\mathrm{vib,CO}} \approx 3000\,\mathrm{K}$$

$$\frac{N_1}{N} = \begin{cases} e^{-10} \approx 0 & T = 300\,\mathrm{K} \\ e^{-1}(1 - e^{-1}) \approx 0.23 & T = 3000\,\mathrm{K} \end{cases}$$

Therefore, "hot bands" become important only when the temperature is significant relative to the characteristic vibrational temperature.

2.7 Electronic Spectra of Diatomic Molecules

We have so far introduced models to adequately interpret or predict the rovibrational spectra of diatomic molecules. We are now ready to incorporate electronic transitions. Electronic spectra involve transitions between different potential energy wells, each representing a different electronic configuration (and hence energy).

2.7.1 Potential Energy Wells

There is a different potential energy well for each electronic configuration (described by one or more electronic quantum numbers). Potential wells illustrate the variation of electronic forces with internuclear spacing, since

$$F = -\frac{dV}{dr} \tag{2.90}$$

where F is the force, V is the potential energy, and r is a one-dimensional distance (often the internuclear distance for diatomics). As the electronic configurations change, the electronic forces change, and thus the potential wells change in shape, energy minimum (T_e) and equilibrium internuclear distance (r_e) (see Fig. 2.15).

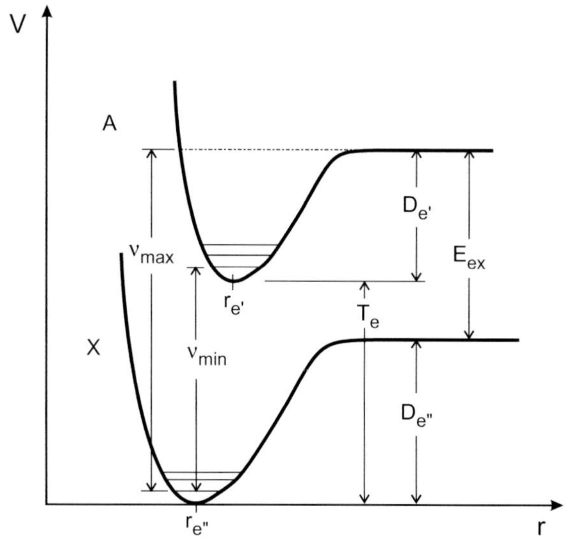

Fig. 2.15 Sample potential wells for the X and A electronic energy states

Example: Potential Energy Wells for N_2

A	is the first excited state.
X	is the ground electronic state.
T_e	is the energy of the A-state with respect to the ground state (measured between the well minimums).
ν_{min}, ν_{max}	are the extremes of photon energies for discrete absorption from $v'' = 0$ (note that the end-points are at $v'', v' = 0$ and at the dissociation limit of the A-state).
E_{ex}	is the difference in electronic energy of atomic fragments.
D_e	is the dissociation energy of the lower ($''$) or upper ($'$) electronic state (not to be confused with the rotational distortion constant, which, unfortunately, sometimes shares the same symbol).

Characteristic Event Times and the Franck–Condon Principle

Absorption and emission associated with molecular transitions from one potential well to another is essentially instantaneous because the time to move or excite electrons is much shorter than the time required to move or excite nuclei during vibrations or rotations.

$$\begin{aligned} \tau_{elec} &\approx 10^{-16}\,\text{s} && \text{time to move/excite electrons} \\ \tau_{vib} &\approx 10^{-13}\,\text{s} && \text{characteristic time for vibration} \\ \tau_{coll} &\approx 10^{-12}\,\text{s} && \text{duration of collision} \\ \tau_{rot} &\approx 10^{-10}\,\text{s} && \text{characteristic time for rotation} \\ \tau_{emiss} &\approx 10^{-6}\text{–}10^{-8}\,\text{s} && \text{“radiative lifetime”} \end{aligned}$$

It is clear that $\tau_{elec} \ll \tau_{others}$.

The Franck–Condon principle reflects the relative characteristic times by approximating that the internuclear distance, r, remains constant during an electronic transition. In other words, during the time it takes for the electronic transition to occur, the molecule's vibration and rotation appear frozen (hence, we draw lines vertically between potential wells to represent an electronic transition at constant r).

Some additional points of note:

1. It is evident that $\tau_{coll} \approx \tau_{vib}$. This can lead to resonant behavior between vibrations and collisions.
2. The "radiative lifetime," τ_{emiss}, is the average time a molecule (or atom) spends in an excited state before undergoing radiative emission.

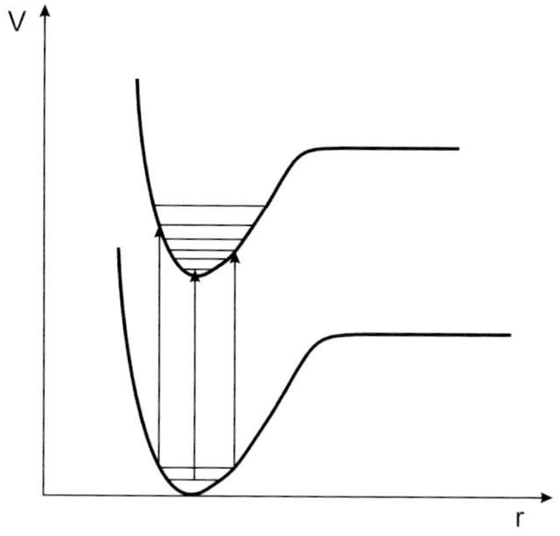

Fig. 2.16 Sample potential wells for discrete electronic spectra

2.7.2 Types of Spectra

Electronic spectra can be discrete or continuous. Sometimes a spectrum can contain both discrete and continuous parts, depending on the potential energy curves of the states involved.

Discrete

When the equilibrium internuclear distance is approximately the same for the upper and lower potential energy wells, $r_e' \approx r_e''$; the result is an electronic spectrum with discrete features (Fig. 2.16).

Recall that:

1. $r \approx$ constant in absorption and emission (Franck–Condon Principle)
2. vibrationally excited molecules ($v \neq 0$) spend more time near the edges of the potential well, so that transitions to and from these locations will be favored
3. lowest v'' levels are the most populated

Continuum

Sometimes one of the states involved in a transition has no equilibrium internuclear distance (the atoms only repel each other), or the transition frequency exceeds the dissociation limit. In these cases, the electronic spectrum is a continuum. Examples of each of these cases are shown in Fig. 2.17.

1. For the left figure, $\nu > \Delta$ leads to a continuous absorption spectrum, and $\nu < \Delta$ results in a discrete spectrum.

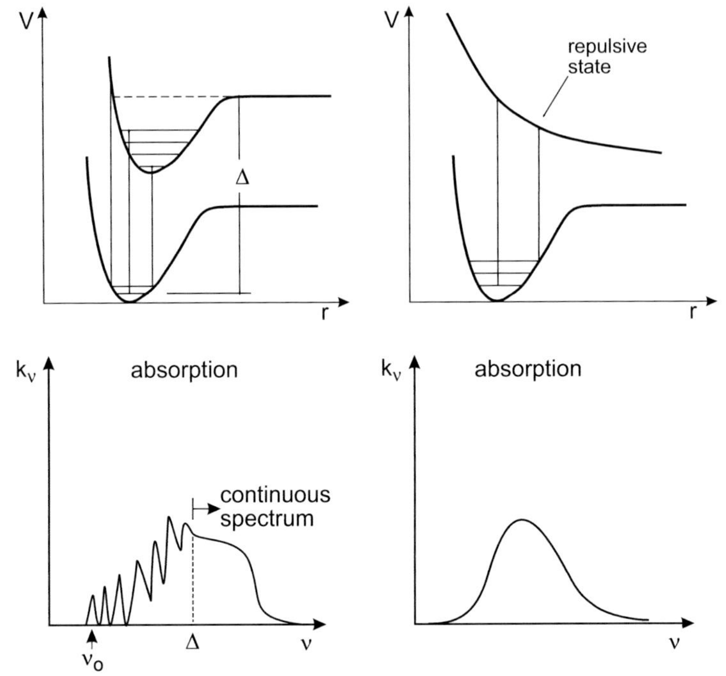

Fig. 2.17 Samples of potential wells (*top*) that result in continuous spectra (*bottom*). The parameter k_ν is an absorption coefficient, i.e. a measure of absorption strength

2. For the right figure, the upper state is always repulsive. That is to say, there is no "well" in the potential curve, and the molecule is equally likely to be excited by any sufficiently energetic photon. The absorption spectrum is thus a continuum.

2.7.3 Rotational Analysis

Here we wish to analyze the rotational transitions within a single band (v', v'') of an electronic system. As with the rovibrational transition analysis, we begin by simply adding the expressions for energy. For the upper state,

$$\begin{aligned} T' &= T_{\text{rot}} + T_{\text{vib}} + T_{\text{elec}} \\ &= F(J') + G(v') + T'_{\text{elec}} \\ &= B'J'(J'+1) + \underbrace{\omega_e{}'(v'+1/2) - \omega_e x_e{}'(v'+1/2)^2 + T'_{\text{elec}}}_{C' \text{(const. for rot. analysis in a single band)}} \end{aligned} \quad (2.91)$$

For the lower state,

$$\begin{aligned} T'' &= T_{\text{rot}} + T_{\text{vib}} + T_{\text{elec}} \\ &= F(J'') + \underbrace{G(v'') + T''_{\text{elec}}}_{C''} \end{aligned} \tag{2.92}$$

Note that if the lower state is in the ground electronic state, $T''_{\text{elec}} \equiv 0$. Defining

$$C = C' - C''$$

and combining Eqs. (2.91) and (2.92) gives

$$T' - T'' = B'J'(J' + 1) - B''J''(J'' + 1) + C \tag{2.93}$$

Similar to the rovibrational analysis, we can simplify Eq. (2.93) to a Fortrat parabola by creating a new variable,

$$m = \begin{cases} -J & \text{for the } P \text{ branch,} \\ J + 1 & \text{for the } R \text{ branch,} \end{cases} \tag{2.94}$$

where $J = J''$. Now, Eq. (2.93) reduces to a parabolic formula.

$$\boxed{T' - T'' = am^2 + bm + C} \tag{2.95}$$

Equation (2.95) is virtually the same as Eq. (2.80), except for the use of the constant C rather than ω_o, and the introduction of constants a and b:

$$\begin{aligned} a &= B' - B'' \\ b &= B' + B'' \end{aligned}$$

The bandhead can be found by taking the derivative of Eq. (2.95) and setting it equal to zero. Letting $T = T' - T''$,

$$\frac{dT}{dm} = 2am + b = 0 \tag{2.96}$$

Therefore,

$$m_{\text{bandhead}} = -\frac{b}{2a} = \frac{B' + B''}{2(B'' - B')} \tag{2.97}$$

Note: 1. If $r'_e > r''_e$, then $B' < B''$, $a < 0$, and the bandhead is in the R branch.
2. If $r'_e < r''_e$, then $B' > B''$, $a > 0$, and the bandhead is in the P branch.

Example: O_2
The $X^3\Sigma_g^-$ ground state has $B'' = 1.44\,\mathrm{cm}^{-1}$, and the $A^3\Sigma_u^+$ upper state has $B' = 1.05\,\mathrm{cm}^{-1}$. The bandhead location is at

$$m_{\mathrm{bh}} = \frac{2.49}{2(0.39)} \approx 3$$

Note that the bandhead can occur at low J owing to the large possible differences in B for different electronic states. This particular electronic system, known as the Herzberg bands, is comprised of weakly "forbidden" transitions (meaning they are not allowed via typical selection rules but occur with low probability due to second-order effects). A much stronger transition system in O_2 is $B^3\Sigma_u^- \leftarrow X^3\Sigma_g^-$, known as the Schumann–Runge system. B_e for $B^3\Sigma_u^-$ is $0.82\,\mathrm{cm}^{-1}$.

Fortrat Parabola

One can graph the Fortrat parabola by plotting line positions, and use it to find rotational constants as well as the bandhead (Fig. 2.18). The Fortrat parabola can be used for rotational analysis by following these steps:

1. separate spectra into bands (v', v'') for detailed analysis
2. tabulate positions of lines in a given band
3. identify null gap and label lines (not always trivial)
4. infer B' and B'' from the Fortrat equation or method of common states

Note: When labelling lines, keep these items in mind:

- If there is no bandhead, then a null gap is obvious.
- If there is a bandhead,then lines overlap.
- If there is a bandhead, it is recommended to start from the wings of the parabola and work backwards, using a *constant second difference*.

The first and second differences are illustrated as follows:

first difference: $T_1(m) = T(m+1) - T(m)$

second difference: $T_2(m) = T_1(m+1) - T_1(m) = 2(B' - B'') = 2a$

Therefore, the second difference is constant in each branch!

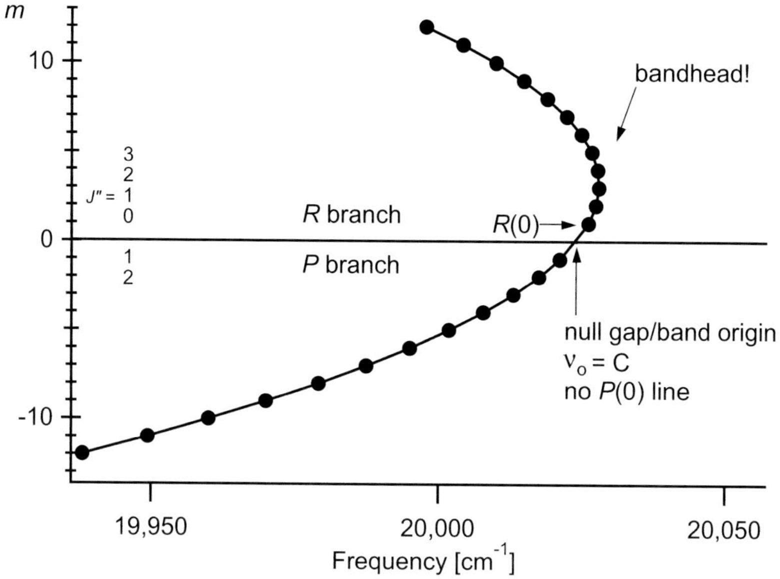

Fig. 2.18 Fortrat parabola for the case with $B' < B''$

Example: Rotational Analysis of Electronic Spectra
The following line positions (in cm^{-1}) were observed in the $(v', v'') = (0, 0)$ band of an electronic transition $(A^3\Pi_{0+u} - X^1\Sigma_g^+)$ in $^{35}Cl_2$; see the spectrum below. Find B_e', B_e'', r_e', r_e'' and the null gap frequency (Figs. 2.19 and 2.20).

1. $\nu_0 = 18{,}147.40\,\mathrm{cm}^{-1}$ (found by inferring the null gap)
2. $2a = -0.173$ (found from average of second differences; note $a < 0$ as the first differences are negative except for small J in R-branch)
3. use common states to get B'' (Fig. 2.21)
   $R(0) = 18147.71$, $P(2) = 18{,}146.25$
   $R(0) - P(2) = 1.46$
   $B'' = 1.46/6 = 0.243\,\mathrm{cm}^{-1}$
   $B' = B'' + a = 0.157\,\mathrm{cm}^{-1}$
4. Solve for r', r'' from B' and B''

Compare the values determined from rotational analysis with those listed in Herzberg [4]:

$$B_e'' = 0.2438, \alpha_e = 0.0017 \Rightarrow B_0'' = 0.2438 - 0.0008 = 0.243$$

$$B_e' = 0.158, \alpha_e = 0.003 \Rightarrow B_0' = 0.158 - 0.0015 \approx 0.157$$

$$T_e = 18310.5,\ r_e'' = 1.988\,\text{Å}, r_e' = 2.47\,\text{Å}$$

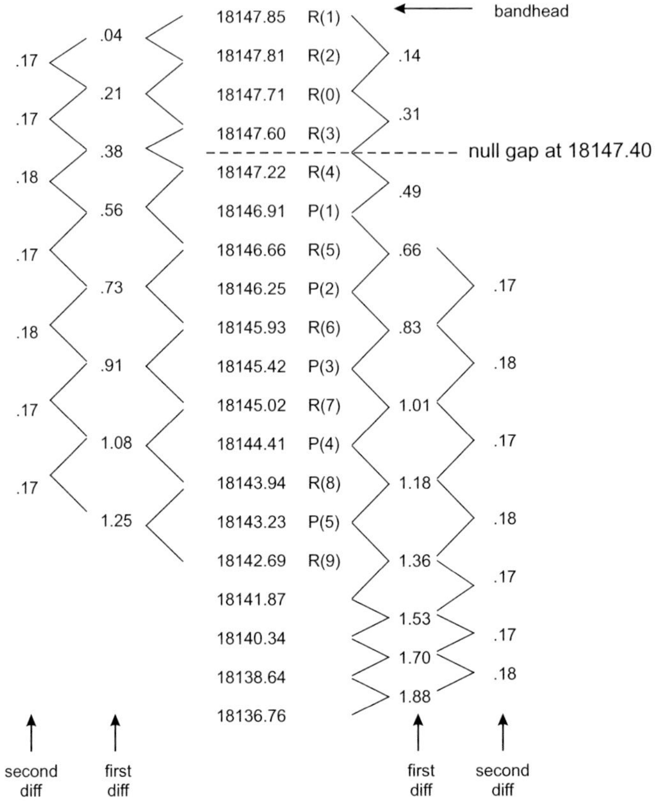

Fig. 2.19 Second difference rotational analysis for $a < 0$

Fig. 2.20 Rotational spectrum in the 0–0 band of $^{35}Cl_2$

2.7.4 Vibrational Analysis

Vibrational analysis can be used to determine ω_e and x_e.

Band Origin Data

Absorption gives information on *upper states*, and emission gives information on *lower states* (Fig. 2.22).

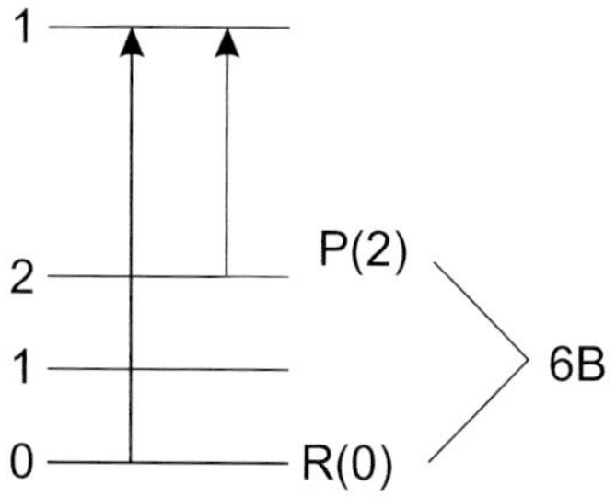

Fig. 2.21 Common upper states

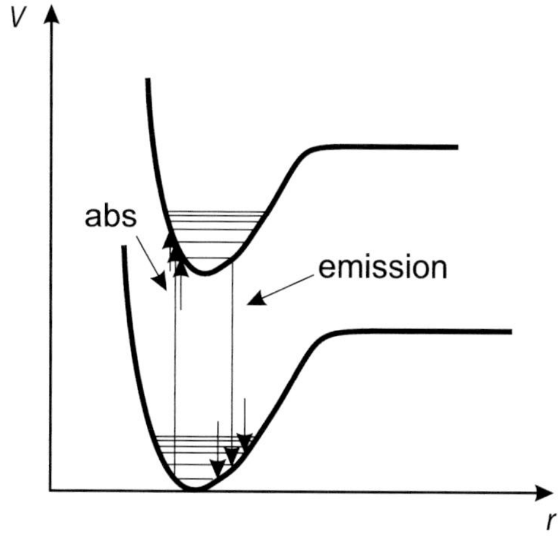

Fig. 2.22 Absorption and emission between two potential wells

Deslandres Table

Tables of band origin values, known as Deslandres Tables, can be used via row analysis to get ω_e'' and $\omega_e x_e''$. With column analysis, information regarding ω_e' and $\omega_e x_e'$ can be retrieved (Fig. 2.23).

Recall:

$$G(v) = \omega_e(v + 1/2) - \omega_e x_e(v + 1/2)^2$$

$$\left.\begin{aligned} G(1) - G(0) &= \omega_e - 2\omega_e x_e \\ G(2) - G(1) &= \omega_e - 4\omega_e x_e \end{aligned}\right\} 2\omega_e x_e$$

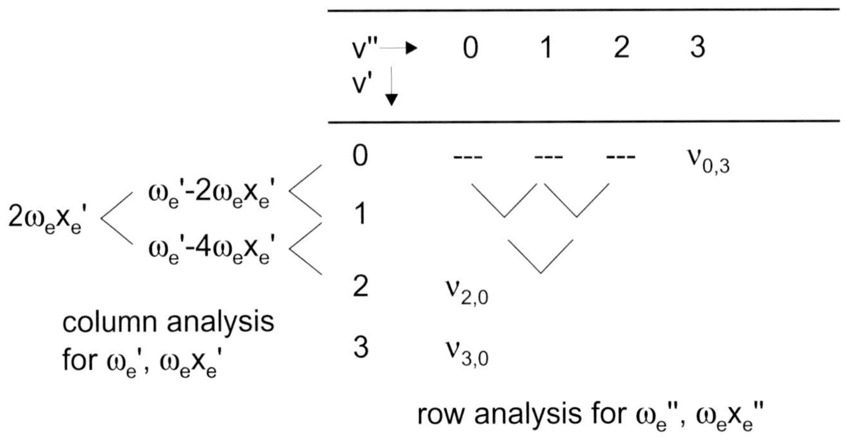

Fig. 2.23 Deslandres table with row and column analysis

Table 2.5 Analysis techniques and their related fundamental quantities

Analysis	Parameters
Rotational analysis	B_e, α_e, D_e, β_e
Vibrational analysis	ω_e, $\omega_e x_e$
Emission analysis	D_e'' and $G(v'')$
Absorption analysis	D_e', T_e and $G(v')$

2.8 Summary

Table 2.5 summarizes the analytical techniques covered thus far and the fundamental quantities that can be determined with them. Rotation is described by a rigid rotor, characterized by the rotational constant, B, however, non rigid corrections due to vibrational (B_e, α_e) and centrifugal (D_e, β_e) distortion are often used to improve the model. Diatomic vibrations are usually described primarily as a harmonic oscillator (ω_e) with a small, anharmonic correction ($\omega_e x_e$) that may become important at high vibrational energies.

Absorption spectra, in general, can provide information on the upper state properties like D_e' (dissociation energy), T_e, and $G(v')$, as shown in Fig. 2.24. *Emission spectra*, conversely, provide information about the lower state, e.g., D_e'' and $G(v'')$, as shown in Fig. 2.25.

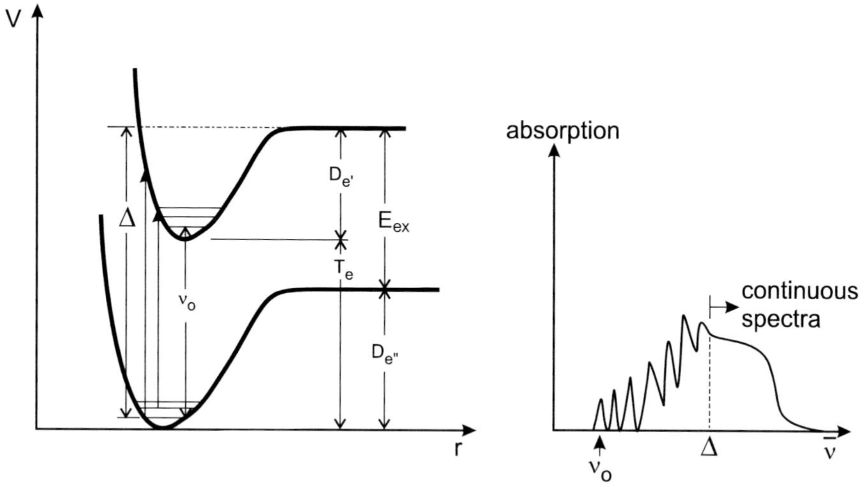

Fig. 2.24 Example potential wells and corresponding absorption spectrum for upper state properties

Typical analyses for absorption include:

1. using band origin data to give $G(v')$ and hence $G(v' = 0)$.
2. using measured $\nu_0 = T_e + G(v' = 0) - G(v'' = 0)$ to find T_e.
3. using measured Δ to give D_e' via $\Delta + G(v'' = 0) = T_e + D_e'$.

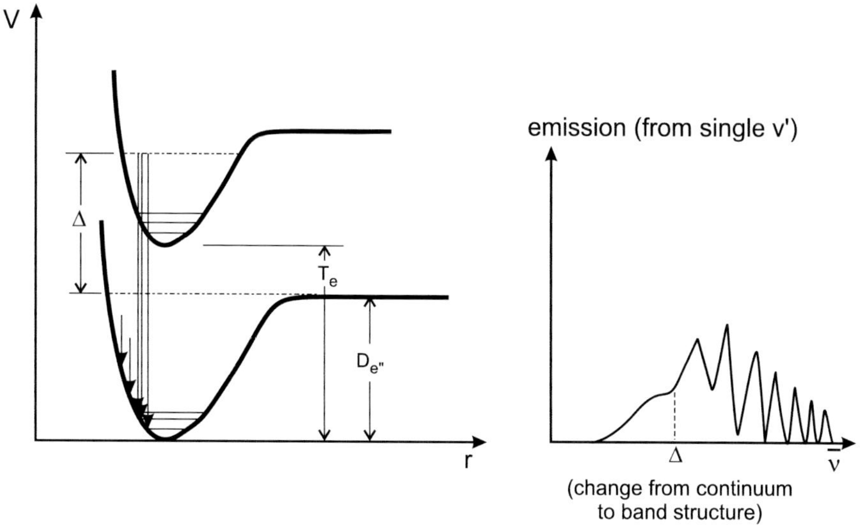

Fig. 2.25 Potential curves and emission spectrum for lower-state properties

Typical analyses for emission include:

1. using band origin data (Deslandres table) from fixed v' to find $G(v'')$.
2. using measured Δ and known T_e and $G(v')$ to find D_e'' via $D_e'' + \Delta = T_e + G(v')$.

2.9 Exercises

1. Which of the following molecules would show (a) a microwave (rotational) spectrum, and (b) an infrared (vibrational) spectrum: Cl_2, HCl, CO_2?
2. The rotational spectrum of $^1H^{127}I$ shows equidistant lines 13.102 cm^{-1} apart. What is the rotational constant, moment of inertia, and bond length for this molecule? What is the wavenumber of the $J = 8 \rightarrow J = 9$ transition? Find which transition gives rise to the most intense spectral line at 300 K. Calculate the angular velocity (in revolutions per second) of an HI molecule when in the $J = 0$ state and when in the $J = 10$ state.
3. Three consecutive lines in the rotational spectrum of $H^{79}Br$ are observed at 84.544, 101.355, and 118.112 cm^{-1}. Assign the lines to their appropriate $J'' \rightarrow J'$ transitions, then deduce values for B and D, and hence evaluate the bond length and approximate vibrational frequency of the molecule.
4. The carbon monoxide molecule, $^{12}C^{16}O$, has a rotational constant, B_v, of 1.9226 cm^{-1}. Boltzmann's equation gives the ratio of the population in rotational energy level J to the total number of molecules as shown below.

$$\frac{N_J}{N} = \frac{g_J}{Q_{\text{rot}}} \exp\left(\frac{-E}{kT}\right)$$

The rotational degeneracy (i.e., the number of states with the same energy level), g_J, is given by $2J + 1$, and the partition function Q_{rot} is given by T/θ_{rot}, where $\theta_{\text{rot}} = B_v\left(\frac{hc}{k}\right)$.

   (a) Find the rotational level that has the maximum population if $T = 1000$ K.
   (b) Calculate the temperature which maximizes the population fraction N_J/N for the J value found in part (a).
   (c) Plot N_J/N as a function of J for the two temperatures in (a) and (b).

5. The following are the line positions in wavenumber units of the fundamental and first overtone bands of BBr, with $v'' = 0$.

674.31	1344.6
675.30	1345.5
676.28	1346.5
678.23	1348.4
679.19	1349.4
680.15	1350.4

   (a) Assign proper labels to all of the lines and calculate B_e, α, ω_e, $\omega_e x_e$.
   (b) Estimate the centrifugal distortion coefficient D_e and use it to determine the centrifugal correction to the position of the $P(3)$ line of the fundamental band. Assume $D_0 = D_1 = D_e$.
   (c) Calculate the position of the $P(1)$ and $R(0)$ lines for the second overtone band of BBr.

6. The band origin of a transition in C_2 is observed at 19,378 cm^{-1}, while the rotational fine structure indicates that the rotational constants in excited and ground states are, respectively, $B' = 1.7527\,\text{cm}^{-1}$ and $B'' = 1.6326\,\text{cm}^{-1}$; the centrifugal distortion parameters D' and D'' are negligible. Determine the position of the bandhead, i.e. the branch, the value of J'', and the frequency of the transition. Which state has the larger equilibrium internuclear distance, r_e?
7. The following lines (wavenumber units) were observed in the $4'$–$0''$ band of the Lyman series of H_2 $[B^1\Sigma_u^+ \leftarrow X^1\Sigma_g^+]$:

95,253.64	95,193.60	95,105.72	95,044.22
94,897.76	94,805.51	94,600.47	94,477.47
94,213.94	94,060.10	93,737.88	93,553.38
93,172.58	92,957.34	92,517.96	92,271.96
91,773.99			

Determine B_4', B_0'', and the null gap.

Helpful Hints:

(a) $^1\Sigma -^1\Sigma$ bands have only two branches: P and R.
(b) Since the H atom nuclear spin is 1/2, Fermi statistics apply and all J states are populated.
(c) It is often helpful in sorting out a spectrum to plot the line positions along the frequency axis.
(d) If a bandhead is apparent, you may wish to begin at the opposite end of the spectrum and try to find a pattern with constant second differences.

References

1. C.H. Townes, A.L. Schawlow, *Microwave Spectroscopy* (Dover Publications, New York, NY, 1975)
2. W.G. Vincenti, C.H. Kruger, *Physical Gas Dynamics* (Krieger Publishing Company, Malabar, FL, 1965)
3. M. Diem, *Introduction to Modern Vibrational Spectroscopy* (Wiley, New York, NY, 1993)
4. G. Herzberg, *Molecular Spectra and Molecular Structure. Volume I. Spectra of Diatomic Molecules*, 2nd edn. (Krieger Publishing Company, Malabar, FL, 1950)
5. C.N. Banwell, E.M. McCash, *Fundamentals of Molecular Spectroscopy*, 4th edn. (McGraw-Hill International (UK) Limited, London, 1994)

Bond Dissociation Energies

3

The bond dissociation energy, D_e, is a critical parameter in thermodynamics, spectroscopy, and kinetics. For example, in thermodynamics, bond energies directly affect heats of formation and reaction. Dissociation energies thus also play key roles determining rates of reaction as a function of temperature. This chapter will give several examples of how spectroscopic information can reveal D_e (of ground and excited electronic states).

3.1 Birge–Sponer Method

A simple (but highly approximate) model, known as the Birge–Sponer method, can be used to directly convert spectroscopic parameters to dissociation energies [1]. It is based on a model of constant anharmonicity. If the anharmonicity is constant, the vibrational level spacing decreases to zero in the limit of dissociation.

$$G(v) = \omega_e(v + 1/2) - \omega_e x_e(v + 1/2)^2 \tag{3.1}$$

$$G(v + 1) = \omega_e(v + 3/2) - \omega_e x_e(v + 3/2)^2 \tag{3.2}$$

$$\Delta G(v) = G(v + 1) - G(v) = -2\omega_e x_e v + (\omega_e - 2\omega_e x_e) \tag{3.3}$$

The expression above [Eq. (3.3)] is linear with the form

$$\Delta G(v) = av + b \tag{3.4}$$

where the slope, a, is $-2\omega_e x_e$ and the intercept, b, is $\omega_e - 2\omega_e x_e$. ΔG, the separation between vibrational transitions, goes to zero when $v = v_D$, as shown in Fig. 3.1, and the corresponding dissociation energy is $G(v_D)$. (Note that $G(v_D)$ is also proportional to the integrated area under the curve of Fig. 3.1.)

R.K. Hanson et al., *Spectroscopy and Optical Diagnostics for Gases*,
DOI 10.1007/978-3-319-23252-2_3

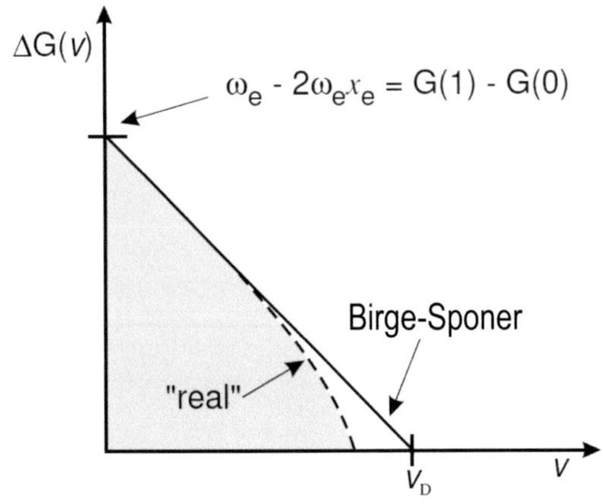

Fig. 3.1 Illustration of Birge–Sponer method for finding the dissociative vibrational level, v_D, and the dissociation energy, D_e

The "real" values of $\Delta G(v)$ are not perfectly linear with quantum number v because the anharmonicity of a real molecule tends to increase near the dissociation limit. The estimation by the Birge–Sponer method thus tends to overpredict the dissociation energy, D_e.

At the dissociation limit, $v = v_D$ and $\Delta G(v_D) = 0$.

$$\Delta G(v_D) = 0 = -2\omega_e x_e v_D + (\omega_e - 2\omega_e x_e) \tag{3.5}$$

Solving for v_D yields

$$\boxed{v_D = \frac{\omega_e}{2\omega_e x_e} - 1} \tag{3.6}$$

Substituting Eq. (3.6) into

$$D_e = \omega_e(v_D + 1/2) - \omega_e x_e (v_D + 1/2)^2 \tag{3.7}$$

gives

$$\boxed{D_e = \frac{\omega_e{}^2}{4\omega_e x_e} - \frac{\omega_e x_e}{4} \approx \frac{\omega_e{}^2}{4\omega_e x_e} = \frac{\omega_e}{4x_e}} \tag{3.8}$$

Example: HCl

For HCl, $\omega_e = 2990\,\text{cm}^{-1}$ and $x_e = 0.0174$. Thus,

$$v_D = 27.7 \rightarrow 27 \text{ next lowest integer}$$

and

$$D_e = 513\,\text{kJ/mole.}$$

A more accurate value, based on known thermochemistry, is $D_e = 427\,\text{kJ/mole}$. Thus, Birge–Sponer gives a good first estimate, but overpredicts D_e by about 20 %.

3.2 Thermochemical Approach

The thermochemical approach is another method for finding D_e but can also depend indirectly on spectroscopic data. This method is based on evaluating the extent of a reaction as a function of temperature. For example, if the reaction were $I_2 \rightarrow 2I$, then the equilibrium constant is given by

$$K_p = \frac{P_{\mathrm{I}}^2}{P_{\mathrm{I}_2}}$$

and its change with temperature may be given by Van't Hoff's equation:

$$\frac{d(\ln K_p)}{dT} = \frac{\Delta H}{RT^2}$$

where

$$\Delta H = \sum v_i H_i = H_{\text{prod}} - H_{\text{react}} = D_e + 2\int_{\mathrm{I}} \hat{c}_p dT - \int_{\mathrm{I}_2} \hat{c}_p dT$$

Thus, measurements of partial pressures can be used to infer $K_p(T)$ and hence both ΔH and D_e''. Partial pressures or species concentrations are often measured spectroscopically (e.g., by laser absorption) because these techniques provide an experimentalist with the ability to accurately measure a single species within a mixture.

3.3 Predissociation

Another way of establishing key energies is with the curve-crossing method, so named because it refers to a predissociative excited electronic state whose potential curve can cross the potential curve of the ground electronic state [1]. Two examples are shown below: HNO (nitroxyl) and N_2O (nitrous oxide).

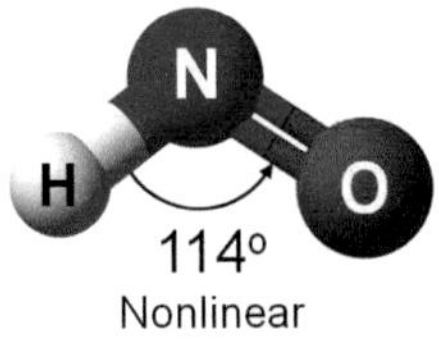

Fig. 3.2 HNO structure

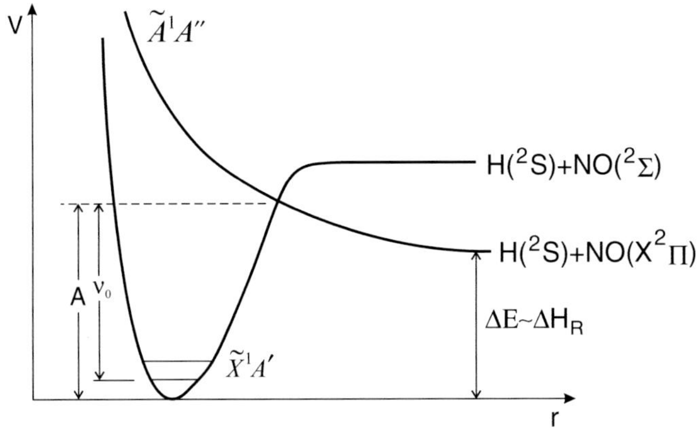

Fig. 3.3 Curve-crossing method for HNO dissociation

3.3.1 HNO

Even though HNO is nonlinear (Fig. 3.2), we can plot its energy level diagram versus H–NO bond distance (i.e., for dissociation of HNO to H+NO) (Fig. 3.3).

$$\Delta E \approx \Delta H_R,$$

but

$$\Delta H_R = \Delta H_f^{\mathrm{H}} + \Delta H_f^{\mathrm{NO}} - \Delta H_f^{\mathrm{HNO}}$$

Therefore, we can solve for $\Delta H_f^{\mathrm{HNO}}$ from an estimate of ΔE, and knowledge of ΔH_f^{H} and ΔH_f^{NO}.

Recall: $1\,\frac{\mathrm{kcal}}{\mathrm{mol}} = 349.7\,\mathrm{cm}^{-1}$
$1\,\mathrm{cal} = 4.187\,\mathrm{J}$

Note: 1. Dissociation (without curve-crossing) of the ground state HNO leads to NO($^2\Sigma$), rather than the lower energy state NO($^2\Pi$).

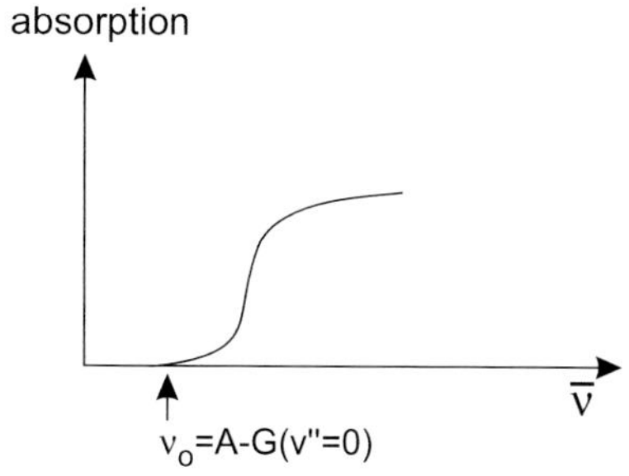

Fig. 3.4 Allowed absorption spectrum for HNO

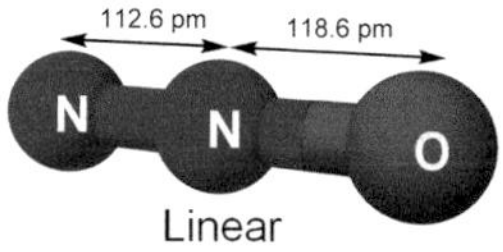

Fig. 3.5 N_2O structure

2. Predissociation occurs at 590 nm (17,000 cm^{-1}) in absorption spectra, corresponding to $A \approx 49$ kcal/mol. This is then an upper bound on ΔH_R.
3. $\Delta H_R(0\ \text{K}) \approx \Delta H_f^{\text{H}} + \Delta H_f^{\text{NO}} - \Delta H_f^{\text{HNO}}$, and hence we can use the upper bound value of ΔH_R, and known values for ΔH_f^{H} and ΔH_f^{NO} to establish a value for the heat of formation of HNO, ΔH_f^{HNO}.
4. Because both electronic states of HNO have the same multiplicity (singlet states with spin = 0), there is an allowed absorption spectrum (Fig. 3.4).
5. For polyatomic molecules, the electronic term symbols include a tilde ($\sim$) over the initial symbol; Roman symbols are used to denote the electronic structure (e.g., $^1A''$ and $^1A'$) unless the molecule is linear, in which case Greek symbols are used (e.g., $^1\Sigma$ or $^3\Pi$).

3.3.2 N_2O

Nitrous oxide is important in combustion chemistry and is linked to NO production and the greenhouse effect. N_2O is also a source of atomic oxygen in shock tube kinetics experiments (Fig. 3.5).

There are three relevant energies on the diagram in Fig. 3.6: the depth of the bound potential well, D_e''; the energy of the curve intersection, E_{act}; and the difference in energy between the repulsive state products and the bottom of the ground state potential well.

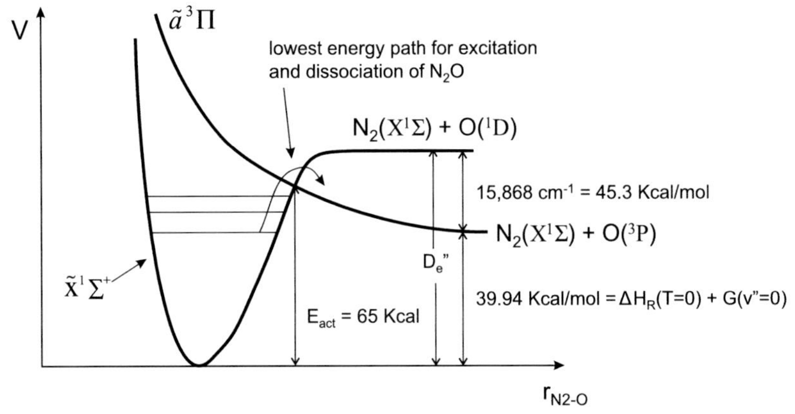

Fig. 3.6 Potential energy wells for N_2O

Note:
1. There is a spin change between the ground electronic state of N_2O and the excited repulsive state shown. As a result, there is no strong absorption process between these electronic states.
2. The dissociation products may result in either O(^1D) or O(^3P), but the latter is lower in energy, and hence more likely.
3. A measurement of the dissociation rate for

$$N_2O + M \rightarrow N_2 + O + M,$$

e.g. in a shock tube gives E_{act}, i.e. $k \propto \exp(-65\ [\text{kcal/mole}]/RT)$.
4. The observed activation energy of 65 kcal/mol provides a lower bound on the dissociation energy, D_e''.

3.4 Exercises

1. A banded structure is observed in the absorption spectrum of ground-state oxygen, which changes to a continuum at a wavelength corresponding to 7.047 eV. The upper electronic state of molecular oxygen dissociates into one ground state (^3P) atom and one excited (^1D) atom; the excitation energy of the (^1D) atom relative to the (^3P) atom is 1.967 eV. Determine D_o'' for O_2 in kcal/mole.
2. The zero-point energy of the ground state of O_2 is 793 cm^{-1}, and the difference in energy between the potential-energy minima of the two electronic states, T_e, of Problem 1 is 49,800 cm^{-1}. Determine D_e for the upper and ground states of O_2 in cm^{-1}.
3. Partial electronic band origin data for an absorption spectrum from the ground electronic state is listed below.

v'	$v'' = 0$
1	$31{,}800\ \mathrm{cm}^{-1}$
2	$32{,}400\ \mathrm{cm}^{-1}$
3	$32{,}800\ \mathrm{cm}^{-1}$

(a) Estimate the bond dissociation energy of the upper electronic state
(b) If $G(v'' = 0)$ is $575\ \mathrm{cm}^{-1}$, what is T_e in cm^{-1}?

4. Given the following band origin data (in cm^{-1}) for an electronic system:
   (a) Determine ω_e', $\omega_e x_e'$, ω_e'', $\omega_e x_e''$
   (b) Calculate v_D' and D_e' using the Birge–Sponer method

v'	$v'' = 0$	$v'' = 1$	$v'' = 2$	$v'' = 3$
0	25,000			
1		24,700		
2	26,300			
3	26,800	25,800	24,800	23,800

Reference

1. C.N. Banwell, E.M. McCash, *Fundamentals of Molecular Spectroscopy*, 4th edn. (McGraw-Hill International (UK) Limited, London, 1994)

Polyatomic Molecular Spectra

4

The mechanism for electromagnetic radiation to interact with polyatomic molecules is similar to the process discussed previously for diatomics. Molecular vibrations and rotations cause changes in electric dipole moments that occur at resonant frequencies. At these frequencies, molecules can interact with radiation (via emission, absorption, or scattering). The difference is that polyatomic molecules have more rotational and vibrational modes, and each of these modes gives rise to additional possible resonances.

This chapter will introduce the fundamental concepts necessary to understand the spectra of various groups of polyatomic molecules. In so doing, we will also present how groups of molecules are classified by their geometry, e.g. linear molecules, symmetric tops, spherical tops, and asymmetric rotors. Similar to our presentation of the material on diatomics, we will first present the aspects of rotational lines and vibrational bands separately before combining them into rovibrational spectra.

4.1 Rotational Spectra of Polyatomic Molecules

A body or molecule is characterized by three principal axes of molecular rotation, about which the three principal moments of inertia, I_A, I_B, and I_C, are defined. These axes pass through the center of mass and are orthogonal to each other (see Fig. 4.1). Molecules are classified in terms of the relative values of I_A, I_B, and I_C. There is more than one convention for assigning the *A*-axis, but we will say that the *A*-axis is the "unique" or "figure" axis, along which lies the molecule's defining symmetry.

R.K. Hanson et al., *Spectroscopy and Optical Diagnostics for Gases*,
DOI 10.1007/978-3-319-23252-2_4

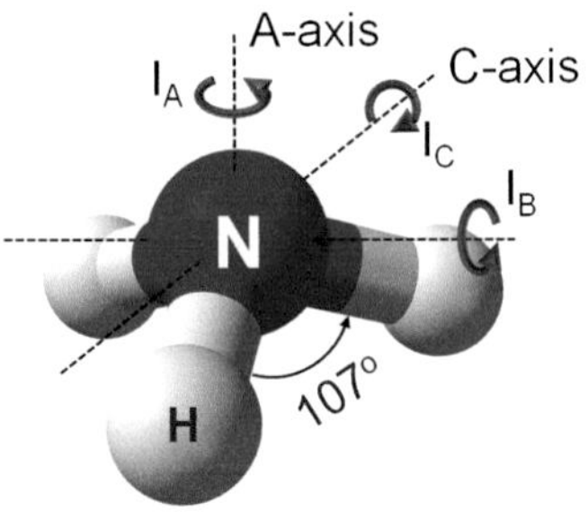

Fig. 4.1 Orthogonal axes and moments of inertia for ammonia, a polyatomic molecule

4.1.1 Linear Molecules

Linear molecules are those for which all the atoms are "on a line," including diatomics. The principal moments of inertia are $I_B = I_C$ and $I_A \approx 0$, where the A-axis passes through each atom. I_A is actually finite, but quantized momentum means that the molecule will remain in the lowest state of rotation about the A-axis, so that effectively, $I_A = 0$. Examples of linear molecules include OCS, HCN, and HC_2Cl.

Linear polyatomic molecules must be asymmetric to have a permanent electric dipole moment and resulting rotational spectra. Linear molecules that do not have a permanent dipole moment (i.e., molecules that are both linear and symmetric), such as CO_2 and C_2H_2, are not microwave active. Isotopic substitution in symmetric molecules does not alter bond lengths or charge distributions, and thus does not yield dipole moments.

Linear polyatomics can be treated the same as diatomics, with one value of I. Therefore, the linear polyatomic rotational constant, rotational energy levels, and transition frequencies are the same as for diatomics.

$$B = \frac{h}{8\pi^2 I_B c} \tag{4.1}$$

$$F(J) = BJ(J+1) - DJ^2(J+1)^2 \tag{4.2}$$

$$\overline{\nu}(J) = 2B(J''+1) - 4D(J''+1)^3 \tag{4.3}$$

Compared to diatomics, polyatomic molecules typically have a larger moment of inertia, I, and thus have smaller rotational constants, B, and smaller line spacing.

For linear molecules with N atoms, there are $N-1$ bond lengths that can be found with spectroscopy. Absorption/emission spectra yield the rotational constant, B, and its associated moment of inertia, I_B. Using $N-1$ isotopes yields measurements of $N-1$ different values for I_B; the resulting $N-1$ equations can be solved for the $N-1$ bond lengths. Carbon oxysulfide provides one such example (Fig. 4.2).

Example: OCS, Carbon Oxysulfide

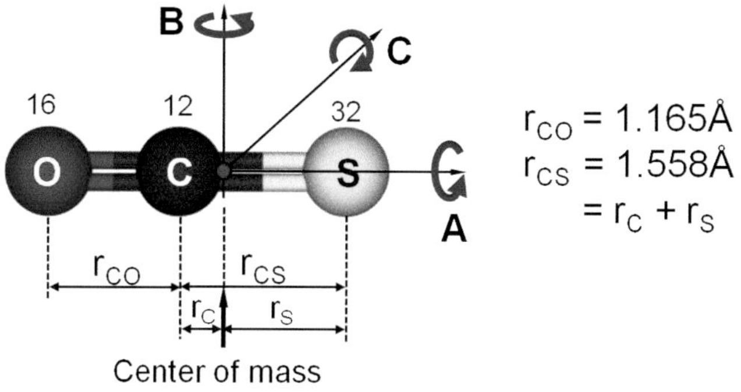

Fig. 4.2 Model of the linear polyatomic molecule OCS

There are two bond lengths, r_{CO}, and r_{CS}, which can be determined from the measured moment of inertia I from two isotopes, e.g.

$$I_{^{16}O^{12}C^{32}S} = \mathcal{F}(\text{masses}, r_{CO}, r_{CS})$$
$$I_{^{18}O^{12}C^{32}S} = \mathcal{F}(\text{masses}, r_{CO}, r_{CS})$$

4.1.2 Symmetric Top

Molecules with symmetric top structure are those that have two equivalent moments or inertia, both of which are different from the third moment. Molecules such as boron trichloride (BCl_3) and ammonia (NH_3) are symmetric tops. The A-axis corresponds to the figure axis of a symmetric top, and thus is also the main axis of symmetry. For symmetric tops, $I_A \neq I_B = I_C$, and $I_A \neq 0$ (Fig. 4.3). Similar to the description for linear molecules, the rotational constants are related to the moments of inertia by

$$A = \frac{h}{8\pi^2 I_A c}, \; B = \frac{h}{8\pi^2 I_B c}, \; C = \frac{h}{8\pi^2 I_C c} \quad [\text{cm}^{-1}] \tag{4.4}$$

Typically, the equivalent moments of inertia for symmetric top molecules are described simply as I_B, and the main axis moment is I_A. The relative magnitudes of

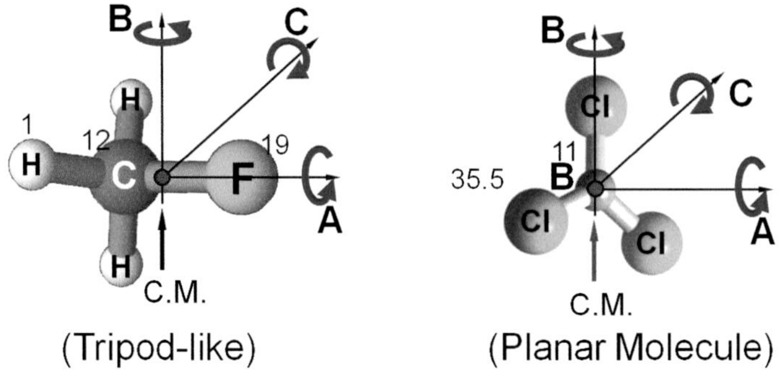

Fig. 4.3 Symmetric top structure. The A-axis passes through C–F for CH_3F; for BCl_3, it is perpendicular to the plane formed by the atoms and passes through B

the inertial moments or rotational constants can be used to further distinguish the tops as *Prolate* or *Oblate*:

Prolate: $I_A < I_B = I_C$
$A > B = C$ Example: CH_3F

Oblate: $I_A > I_B = I_C$
$A < B = C$ Example: BCl_3

Since symmetric tops have two main directions of rotation, they have two quantum numbers to describe rotational energy, J and K, where J represents the total angular momentum $(0, 1, 2, \ldots)$, and K represents the angular momentum about the A-axis. There are $2J + 1$ possible values of K for each value of J, with K restricted to $J, J - 1, \ldots, 1, 0, -1, \ldots, -J$. All non-zero values of K are doubly degenerate since states, that differ only by the sign of K, have different directions of motion but the same energy.

J Total angular momentum: $0, 1, 2, \ldots$
K Angular momentum about the A axis: $J, J - 1, \ldots, 1, 0, -1, \ldots, -J$
There are $2J + 1$ possibilities of K for each J. Positive and negative values of K are allowed, without a change in energy.

As before, the quantized total angular momentum is (see Eq. (2.5))

$$I_A{}^2\omega_A^2 + I_B{}^2\omega_B^2 + I_C{}^2\omega_C^2 = J(J+1)\hbar^2 \tag{4.5}$$

Here ω_i is the rotational angular velocity. In addition, the angular momentum about the A-axis is quantized.

$$I_A{}^2\omega_A^2 = K^2\hbar^2 \tag{4.6}$$

Energy Levels

The energy levels, assuming rigid rotation, from Eqs. (4.5) and (4.6) are given by

$$E_{J,K} = \frac{1}{2}\sum_i I_i\omega_i^2 \tag{4.7}$$

which leads, after some algebra, to

$$F(J,K) = BJ(J+1) + (A-B)K^2 \tag{4.8}$$

Note that the sign of K has no effect on the rotational energy, i.e., the direction of rotation does not affect the rotational energy.

Selection Rules

The selection rules that result from quantum mechanics are:

$\Delta J = +1$ As before for pure rotation
$\Delta K = 0$

The interpretation is that since there is no dipole moment for rotation about the A-axis, no changes in K will occur with absorption or emission.

Line Positions

The transition frequencies for the polyatomic rigid rotor are

$$\overline{\nu}_{J,K} = F(J+1,K) - F(J,K) = 2B(J+1)\ [\text{cm}^{-1}] \tag{4.9}$$

Again, since the line positions are differences in energy levels, they are independent of K for a rigid rotor. K-dependency is introduced for non-rigid rotation.

Non-rigid Rotation

Energy level descriptions for non-rigid rotation include higher-order correction terms with centrifugal distortion constants that are J-dependent (D_J), K-dependent (D_K), and J–K dependent (D_{JK}).

$$\begin{aligned} F(J,K) = {} & BJ(J+1) + (A-B)K^2 - D_J J^2(J+1)^2 \\ & - D_{JK}J(J+1)K^2 - D_K K^4 \end{aligned} \tag{4.10}$$

$$\overline{\nu}_{J,K} = 2(J+1)\left[B - 2D_J(J+1)^2 - D_{JK}K^2\right] \tag{4.11}$$

Note that there are $2J+1$ components (K values) for each total angular momentum quantum number J but only $J+1$ frequencies, since $-K$ and K are degenerate. The last two terms in Eq. (4.11) are small except at very high J and K values. The reader should gain some idea of the magnitude of these corrections from the following example.

Example: CH_3F, **Methyl Fluoride**

$$B = 0.851\ \text{cm}^{-1}$$
$$D_J = 2 \times 10^{-6}\ \text{cm}^{-1}$$
$$D_{JK} = 1.47 \times 10^{-5}\ \text{cm}^{-1}$$

4.1.3 Spherical Top

Spherical tops are characterized by three equal moments of inertia.

$$I_A = I_B = I_C \tag{4.12}$$

Molecules such as CH_4 are spherically symmetric (Fig. 4.4). There is no permanent electric dipole moment along any of the principal axes; therefore, spherical tops have no rotational spectra.

4.1.4 Asymmetric Rotor

Asymmetric rotors (also known as asymmetric tops) are characterized by three moments of inertia, none of which are equal to each other.

$$I_A \neq I_B \neq I_C \tag{4.13}$$

This category is the most complex and will not be addressed here. Examples of asymmetric rotors include H_2O and NO_2 (Fig. 4.5).

4.1.5 Rotational Partition Function

The classical expressions for polyatomic rotational partition functions are different for the various molecular structures. Linear polyatomics can be treated as diatomics, and thus, from before,

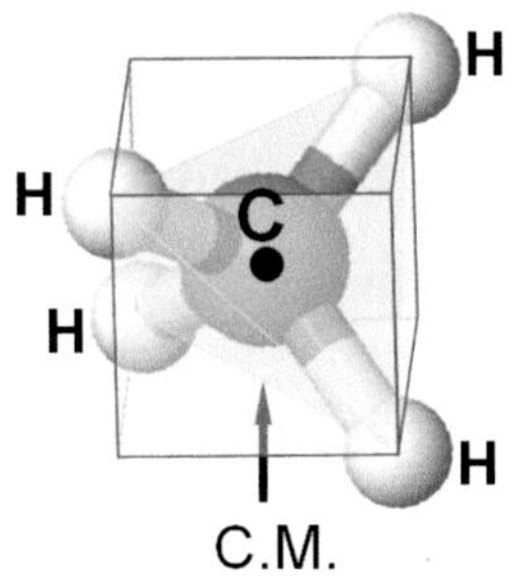

Fig. 4.4 Molecular structure of CH_4

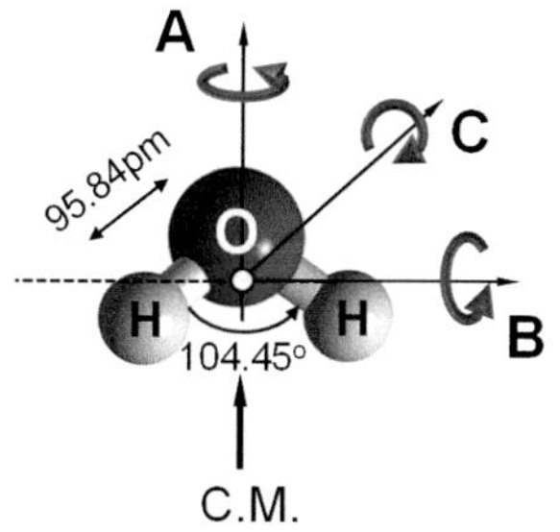

Fig. 4.5 Molecular structure of H_2O

Table 4.1 Symmetry factors for a few polyatomic molecules

Molecule	Symmetry factor, σ	Molecule type
CO_2	2	Linear
NH_3	3	Symmetric top
CH_4	12	Spherical top
H_2O	2	Asymmetric rotor

$$Q_{\mathrm{rot}} = \frac{kT}{\sigma hcB}, \tag{4.14}$$

where σ is the molecule-dependent symmetry factor (defined in Sect. 2.2.7) (Table 4.1). For symmetric top molecules, where $B = C$,

$$Q_{\mathrm{rot}} = \frac{1}{\sigma}\sqrt{\frac{\pi}{AB^2}\left(\frac{kT}{hc}\right)^3}. \tag{4.15}$$

For spherical top molecules, the rotational partition function simplifies even further to

$$Q_{\mathrm{rot}} = \frac{1}{\sigma}\sqrt{\frac{\pi}{B^3}\left(\frac{kT}{hc}\right)^3}. \tag{4.16}$$

Similarly, for an asymmetric rotor:

$$Q_{\mathrm{rot}} = \frac{1}{\sigma}\sqrt{\frac{\pi}{ABC}\left(\frac{kT}{hc}\right)^3}. \tag{4.17}$$

4.2 Vibrational Bands of Polyatomic Molecules

Polyatomic molecules have multiple and different vibrational modes depending on their structure and number of atoms. The existence and nature of these modes affect the vibrational bands in the spectra of these molecules.

4.2.1 Number of Vibrational Modes

For polyatomic molecules with N atoms, a total of $3N$ dynamical coordinates are needed to specify the instantaneous location and orientation of the molecule (i.e., the nuclei). This total corresponds to the sum of coordinates needed to specify the molecular center of mass, the angular rotation of the molecule, and vibrational motion (bond lengths), and hence provides an easy accounting method for identifying the number of vibrational modes, as follows:

Center of Mass: requires 3 coordinates (has 3 translational modes)

Rotation: 2 angular coordinates (hence rotational modes) for linear molecules
3 angular coordinates (rotational modes) for nonlinear molecules

Vibrations: the remaining number of vibrational coordinates/modes is:
$3N - 5$ for linear molecules
$3N - 6$ for nonlinear molecules

4.2.2 Parallel and Perpendicular Modes

H_2O and CO_2 are good examples of triatomic molecules with different vibrational modes. The numbering (i.e., identification) convention for the vibrational modes (and their resultant bands in a spectrum) is based first on symmetry, and second on decreasing energy [1]. That is, ν_1 is the highest-frequency symmetric vibrational mode, ν_2 the next highest symmetric mode, and so on, followed by the asymmetric modes in order of declining frequency. (See [1, p. 272] for an exception to this rule.) There are two types of vibrational modes:

1. Parallel ($\|$), where the vibrations are those that occur parallel to the main axis of symmetry
2. Perpendicular ($\perp$), where the vibrations are those that occur perpendicular to the main axis of symmetry

In order for vibrational motion of a molecule to result in an absorption/emission spectrum, some change must occur in the electric dipole of the molecule during

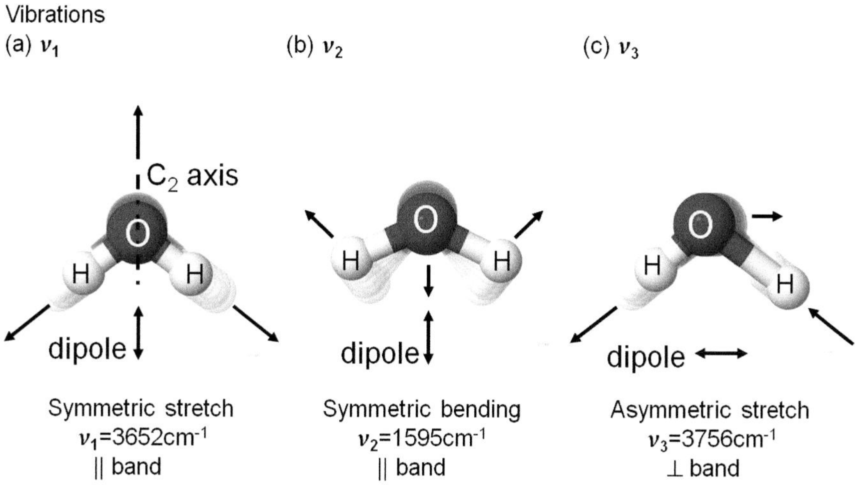

Fig. 4.6 Structure, symmetry, and vibrational modes for H_2O

this motion (i.e., a difference must exist in the dipole moment between vibrational quantum states). If this change in dipole occurs along the axis of symmetry of the molecule, the absorption/emission spectrum is termed a "parallel (||) band," while if the change in dipole moment occurs perpendicular to the axis of symmetry, a "perpendicular (⊥) band" will occur. It follows that for a symmetric molecule, such as CO_2, the symmetric stretch vibration will not produce a change in dipole moment, and hence there will be no active parallel band for this mode of vibration (see the discussion below).

4.2.2.1 Water, H_2O

H_2O is a nonlinear triatomic molecule, so it has three vibrational modes. The C_2 notation in Fig. 4.6 means that there is a twofold symmetry of rotation about this axis (which is the unique *A*-axis of H_2O). The symmetric stretch vibration, known as ν_1 (as it is the highest-frequency symmetric motion), results in a || band, as does the symmetric bending vibration, ν_2. The remaining vibrational mode, ν_3, involves asymmetric stretching, and produces dipole variations perpendicular to the axis of symmetry. Hence ν_3 is a ⊥ band.

Carbon Dioxide, CO_2

For a linear molecule, such as CO_2, there are two stretching modes: symmetric and asymmetric. The stretching and compression of these bonds are illustrated in the following figure with exaggerated amplitude (Fig. 4.7).

For the "symmetric stretch" vibrational mode, the two **C–O** bonds are stretched or compressed simultaneously, preventing formation of a dipole moment. Since the dipole moment remains zero for this mode, no direct light interaction (i.e., absorption or emission) is possible. Thus, the "symmetric stretch" vibration for

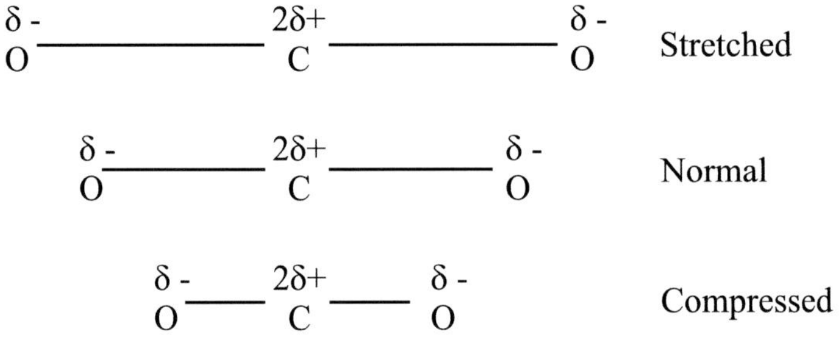

Fig. 4.7 Symmetric stretching of carbon dioxide (Fig. 1.6, Banwell)

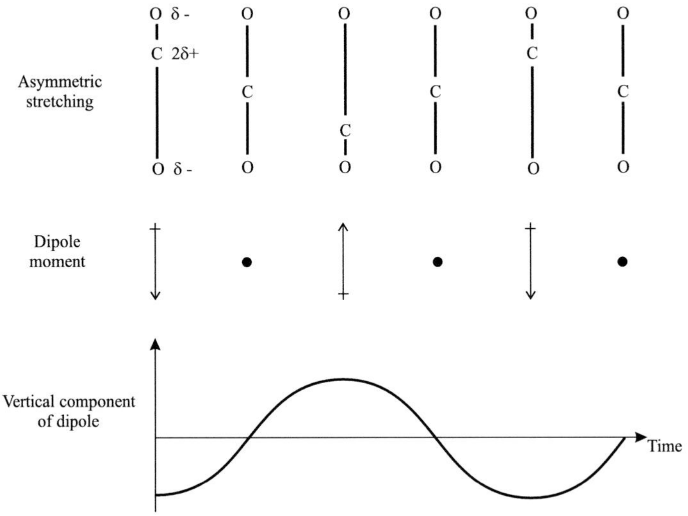

Fig. 4.8 Asymmetric stretching vibrational mode for carbon dioxide

CO_2 is "infra red inactive." For anti-symmetric stretching (see Fig. 4.8), one bond is stretched while the other is compressed, giving rise to a changing dipole moment. This vibrational mode is "infra-red active," and there will be an absorption band at the characteristic frequency of vibration. Since CO_2 is a linear triatomic molecule, it has 4 vibrational modes, two of which are degenerate. Only a single absorption band will appear at the degenerate frequency. Hence there are three fundamental frequencies (ν_1, ν_2, and ν_3) as shown in Fig. 4.9.

Interestingly, for symmetric molecules we will find (Chap. 6) that vibrational modes are either IR-active or Raman-active (see Table 4.2) for symmetric molecules.

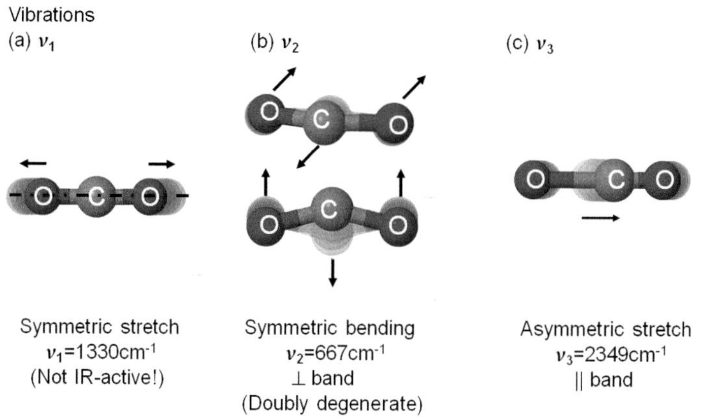

Fig. 4.9 Structure, symmetry, and vibrational modes for CO_2

Table 4.2 Active modes of CO_2 (*S* strong, *VS* very strong); *N.A.* not active

Mode	IR	Raman
ν_1	N.A.	Active
ν_2	S	N.A.
ν_3	VS	N.A.

Table 4.3 Fundamental vibrations, frequencies, types, and descriptions for NH_3

Vibration	Frequency [cm^{-1}]	Type	Description
ν_1	3337	$\parallel$	Symmetric stretch
ν_2	950	$\parallel$	Symmetric bend
ν_3	3444	$\perp$	Asymmetric stretch (degenerate)
ν_4	1627	$\perp$	Asymmetric bend (degenerate)

Ammonia, NH_3

A listing of the fundamental vibrational modes of ammonia is given in Table 4.3. Note that ammonia has six vibrational modes, but two (ν_1 and ν_2) are doubly degenerate.

4.2.3 Types of Bands

There are various types of bands that appear in a spectrum; they are all based on the vibrational modes discussed above. The terminology for these different types of bands is as follows:

Fundamental Bands ν_i, the ith vibrational mode

$\Delta v = v' - v'' = 1$ for the ith mode

First Overtone $2\nu_i$

$\Delta v = v' - v'' = 2$ for the ith mode

Second Overtone $3\nu_i$

$\Delta v = v' - v'' = 3$ for the ith mode

Combination Bands Changes in multiple quantum numbers, e.g.

$\nu_1 + \nu_2$ $\Delta v_1 = \Delta v_2 = 1$; i.e. v_1 and v_2 both increase by 1 for absorption or decrease by one for emission.

$2\nu_1 + \nu_2$ $\Delta v_1 = +2$ and $\Delta v_2 = +1$

Difference Bands Quantum number changes with mixed sign

$\nu_1 - \nu_2$ This means that $v_{1\text{final}} - v_{1\text{initial}} = \pm 1$ and $v_{2\text{final}} - v_{2\text{initial}} = \mp 1$, i.e., a unit increase in v_1 is accompanied by a unit decrease in v_2, and vice versa.

4.2.4 Relative Strengths

The fundamental bands are generally much stronger than the combination, difference, and overtone bands. For fairly harmonic molecules such as CO, the difference in relative strengths between the fundamental and overtone bands is approximately two orders of magnitude. However, for very anharmonic molecules such as NH_3, the difference between the fundamental and overtone or combination bands is often just one order of magnitude or less. Nearly harmonic molecules have much weaker overtones and combination bands because they closely approximate the SHO model that was presented earlier, namely that overtone bands are forbidden. In reality, these bands are not forbidden, but rather have low transition probabilities (that is, they are *nearly* forbidden). For highly anharmonic molecules, the bands are much more likely to occur (that is, they are less forbidden because the SHO solutions are bad approximations), and thus have strengths that are closer to the fundamental bands.

Relative Strengths Exception

Accidental degeneracies (i.e., near resonances) can strengthen weak processes: e.g.,

$$2\nu_{2,CO_2} \text{ at } 1334\,\text{cm}^{-1} \approx \nu_{1,CO_2}$$

Since the first overtone of the ν_2 band is nearly resonant with the fundamental of the ν_1 band, the two vibrational modes are strongly coupled by radiative and collisional exchanges. This case is called Fermi resonance.

4.2.5 Vibrational Partition Function

For polyatomics whose vibrational potential energy can be approximated by the SHO model [see Eqs. (2.41) and (2.42) and note the choice of zero-point energy], the vibrational partition function, Q_{vib}, can be written as the product of the harmonic partition function for each vibrational mode

$$Q_{\text{vib}} = \prod_{i}^{\text{modes}} \left[1 - \exp\left(\frac{-hc\omega_{e,i}}{kT}\right)\right]^{-g_i}, \tag{4.18}$$

where $\omega_{e,i}$ is the harmonic frequency of the ith vibrational mode, and g_i is the degeneracy of that mode. For example, ammonia's molecular formula is NH_3, and thus it has $3N - 6 = 6$ possible vibrations. However, two vibrations are degenerate, so the molecule has four different vibrational frequencies (see Table 4.3). Thus, ammonia's vibrational partition function is

$$\begin{aligned} Q_{\text{vib}}(T) = &\left[1 - \exp\left(\frac{-hc\omega_{e,1}}{kT}\right)\right]^{-1}\left[1 - \exp\left(\frac{-hc\omega_{e,2}}{kT}\right)\right]^{-1} \\ &\times \left[1 - \exp\left(\frac{-hc\omega_{e,3}}{kT}\right)\right]^{-2}\left[1 - \exp\left(\frac{-hc\omega_{e,4}}{kT}\right)\right]^{-2} \end{aligned} \tag{4.19}$$

4.3 Rovibrational Spectra of Polyatomic Molecules

4.3.1 Linear Polyatomic Molecules

There are two types of vibrational bands, parallel and perpendicular, defined in terms of the orientation of the electric dipole moment associated with a specific vibrational mode, as discussed in Sect. 4.2.2. Here we limit consideration to fundamental transitions within that mode (i.e., we exclude overtone and combination bands).

Case I: Parallel Bands

There are two types of parallel band vibrations: *symmetric* and *asymmetric* stretch. For fundamental transitions, we consider only the vibrational energy stored in mode i and rotational energy.

Energy	$T(v_i, J) = G(v_i) + F(J)$
Selection Rules	$\Delta v_i = 1$
	$\Delta J = \pm 1$ (R and P branches)
	$\Delta v_j = 0,\ j \neq i$
Absorption Spectrum	P and R branches only (see Fig. 2.12)

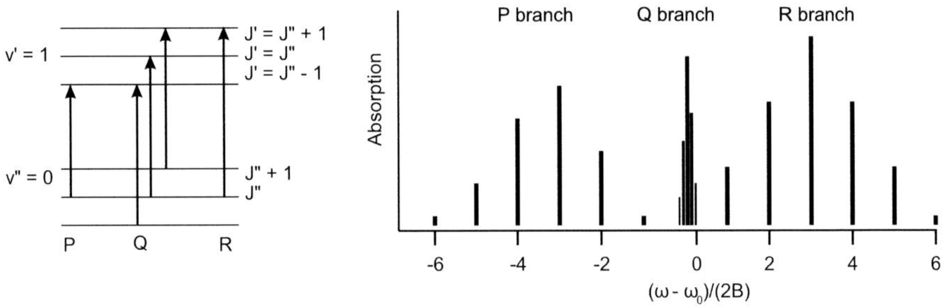

Fig. 4.10 Energy levels and absorption spectrum for the *P*, *Q*, and *R* branches of a linear polyatomic perpendicular band

Table 4.4 Band strength comparison: HCN

Mode	ω_e [cm^{-1}]	ν_i	IR[a]	Raman [a]
Symmetric stretch	3310	ν_1	S	W
Bending (2)[b]	715	ν_2	VS	W
Asymmetric stretch	2097	ν_3	W	S

[a] *W* weak, *S* strong, *VS* very strong
[b] (2) denotes two degenerate (indistinguishable) bending modes

Note:

The IR and Raman band strengths are often complementary, i.e. if IR is strong, Raman is weak and vice versa.

Examples CO_2 (ν_3), HCN (ν_1, ν_3)
i.e., the ν_3 mode of CO_2 (asymmetric stretch), and the ν_1 (symmetric stretch) and ν_3 (asymmetric stretch) modes of HCN are all *parallel* bands.

Note: The ν_1 mode of CO_2 is not IR-active, since symmetric stretching of a symmetric molecule causes no oscillating electric dipole moment. Hence, there is no ν_1 parallel band.

Case II: Perpendicular Bands

Selection Rules

$$\Delta v_i = 1$$
$$\Delta J = \pm 1,\ 0\ (R, P \text{ and } Q \text{ branches})$$
$$\Delta v_j = 0,\ j \neq i$$

1. If $B' = B''$, all Q branch lines occur at the same frequency.
2. If $B' \neq B''$, then $Q(J'') = \omega_o + (B' - B'')J''(J'' + 1)$

   Since $B' - B'' = -\alpha$, the Q branch "degrades" to lower frequencies (i.e., to the "red" in wavelength) (Fig. 4.10).

Table 4.4 summarizes the bands and relative strengths for the moles of HCN.

4.3.2 Symmetric Top Molecules

A symmetric top molecule has two, equal, principal moments of inertia, i.e. $I_B = I_C \neq I_A$. Example molecules include CH_3F and BCl_3. The total energy for symmetric top molecules is simply a summation of the vibrational and rotational energies.

$$\begin{aligned} T(v_i, J, K) &= G(v_i) + F(J, K) \\ &= (v_i + 1/2)\omega_e{}^i - \omega_e{}^i x_e{}^i (v_i + 1/2)^2 \\ &\quad + BJ(J+1) + (A - B)K^2 \end{aligned} \tag{4.20}$$

Recall that K is the quantum number for angular momentum around axis A. Again we limit consideration to fundamental transitions within a single vibrational mode i.

Case I: Parallel Bands

Selection Rules	$\Delta v_i = 1$
a. if $K \neq 0$	$\Delta K = 0,\ \Delta J = \pm 1,\ 0$ (P, Q, R branches)
b. if $K = 0$	$\Delta K = 0,\ \Delta J = \pm 1$ (no Q-branch), except $\Delta J = +1$ for $J = 0$

Since K is the same in the upper and lower states, we have P, Q, and R branches for each value of K. That is to say, there are $2J + 1$ values of K ($K = J, J - 1, \ldots, 0, \ldots, -J$), each of which produces a P, Q, and R branch. The resultant absorption spectrum can become somewhat complex, although the general features are recognizable, as shown in Fig. 4.11. The intensity of the Q branch is a function of (I_A/I_B). As (I_A/I_B) approaches zero, the symmetric top approaches the structure of a linear molecule, and the strength of the Q branch approaches zero.

Referring to Fig. 4.11, the following observations can be made:

1. For $K = 0$, the spectrum reduces to that of a linear molecule, i.e. there is no Q-branch.
2. For each value of K, the minimum value of J in the P-branch is $K + 1$, since $J' = J'' - 1$, and K (unchanged in the transition) cannot exceed J.

Case II: Perpendicular Bands

Selection Rules	$\Delta v_i = 1$
	$\Delta J = \pm 1,\ 0$
	$\Delta K = \pm 1$
***R* Branch**	$\Delta J = +1, \Delta K = \pm 1$
	$\overline{\nu}_R = \omega_o + 2B(J+1) + (A - B)(1 \pm 2K)$
***P* Branch**	$\Delta J = -1, \Delta K = \pm 1$
	$\overline{\nu}_P = \omega_o - 2BJ + (A - B)(1 \pm 2K)$

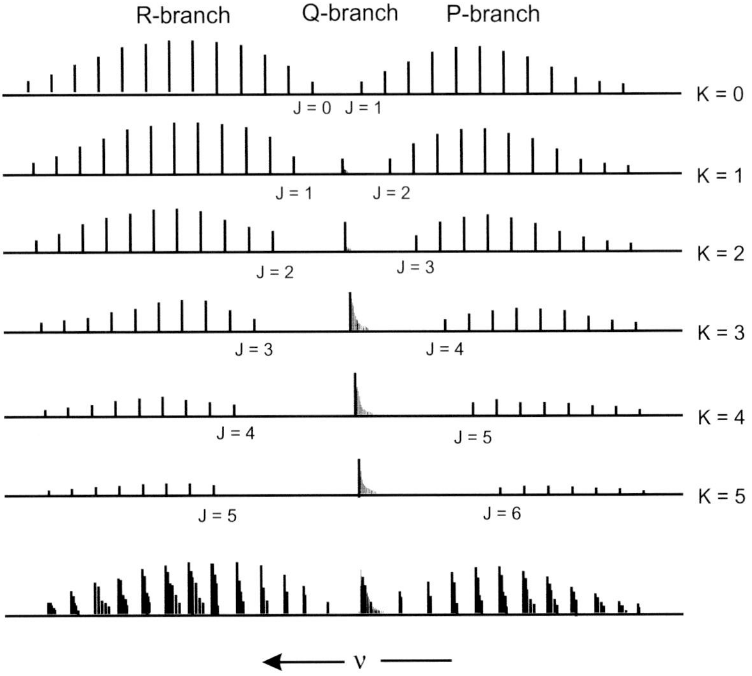

Fig. 4.11 The resolved components of a parallel band showing the contributions from each of the K levels of the $v = 0$ state. The small but discernible splitting evident in the superposed P- and R-branch spectra (*bottom row* of this figure) is due to a difference in the magnitude of $(A - B)$ in the upper and lower vibrational levels; see Eq. (4.20). The discernible splitting in the superposed Q-branch is due mostly to the difference in B in the upper and lower vibrational levels ($B' < B''$, so spectra degrade to lower frequencies)

$$\textbf{\textit{Q} Branch} \qquad \Delta J = 0, \Delta K = \pm 1$$
$$\overline{\nu}_Q = \omega_o + (A - B)(1 \pm 2K)$$

Thus, we have two sets of R, P, and Q branches for each lower-state value of K. The result is generally a very complex spectrum, which is simplest at low temperatures (fewer J levels are populated) and low pressures (individual lines are narrow—more on this in Chap. 8).

An example of the J, K energy levels for a symmetric top molecule and the allowed transitions for a perpendicular band are shown in Fig. 4.12. The complete band can be understood in terms of a summation of sub-bands. These sub-bands consist of all the $\Delta J = 0, \pm 1$ that occur for a given change in K [1]. A resulting spectrum, decomposed to illustrate the component sub-bands, is given in Fig. 4.13. These two figures are from Barrow [2, pp. 151–152].

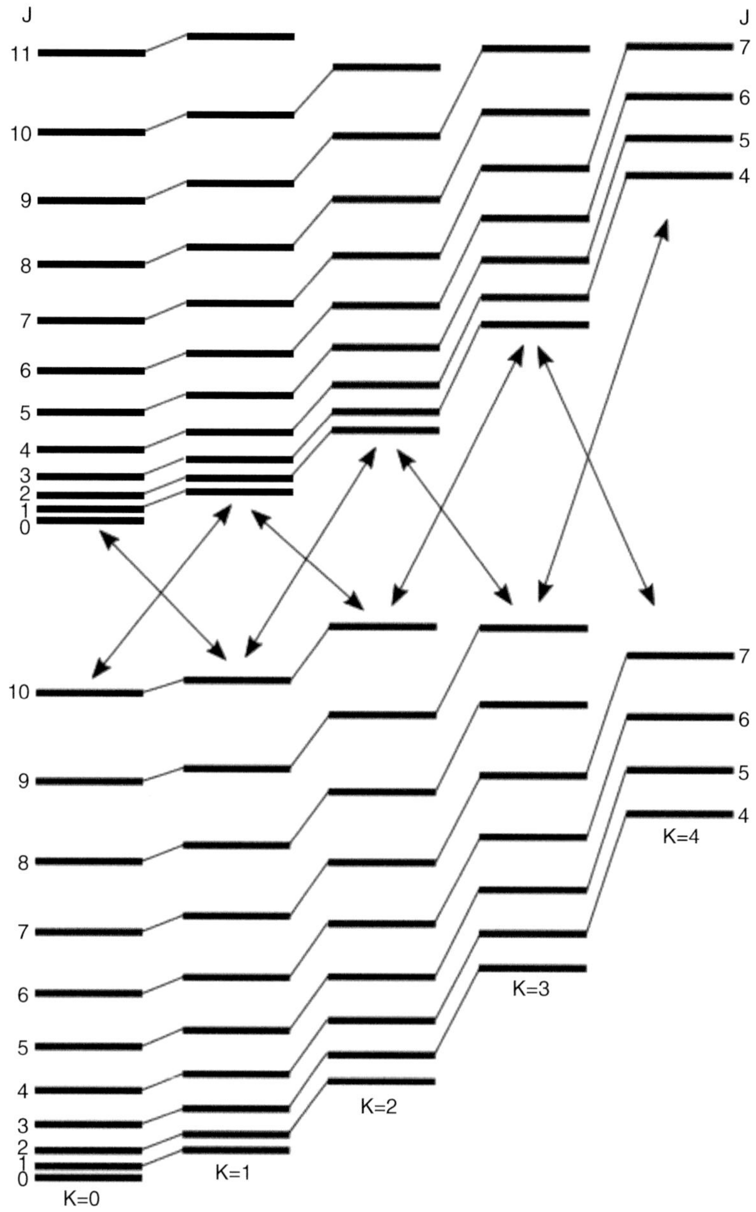

Fig. 4.12 The energy levels of a symmetric top molecule showing the transitions that are allowed for a perpendicular band. Figure from Barrow [2, p. 151]

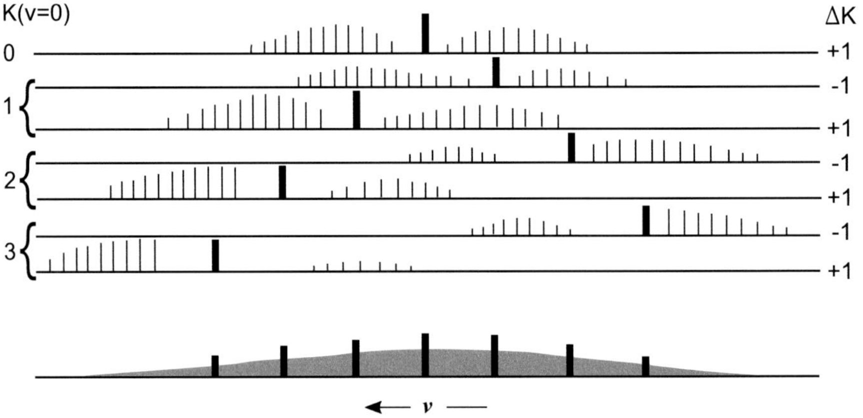

Fig. 4.13 The components of a perpendicular band of a symmetric top molecule. Note that the lines with $\Delta J = \Delta K$ have greater intensity than those with $\Delta J = -\Delta K$, i.e., R-branch lines with $\Delta J = \Delta K = +1$ are stronger than the P-branch lines of $\Delta J = -1$, when $\Delta K = +1$. See Herzberg [1, pp. 424–426], for the selection rules that characterize this effect. Figure from Barrow [2, p. 152]

4.4 Exercises

1. List the vibrational modes of the following molecules, and indicate which of the modes are IR active, and/or Raman active? (Note: the molecular structure can be found in [1])
   (a) HBr
   (b) OCS (linear)
   (c) SO_2 (bent)
   (d) C_2H_4 (only determine the IR- and Raman-activity of the modes: C–C stretch, C–H symmetric stretch)
2. Calculate the vibrational partition function of CO and CO_2 from 300 to 3000 K. Explain the trends and differences you observe.
3. What is the ratio of NH_3's rotational energy for the $J = 1$, $K = 0$, and $J = 4$, $K = 0$ levels?
4. Absorption spectra of three species (CO_2, H_2O, and CO) important to combustion and atmospheric science are shown below. By visual inspection of the spectra and knowledge of bond structure, address the following:
   (a) How many vibrational modes exist for each species?
   (b) What type of absorption spectra are shown (rotational, rovibrational, rovibronic)? Why?
   (c) Label the spectra by species name.
   (d) Which types of bands exist within each spectra (hint: this should help you with part iii)?
   (e) Identify and label the fundamental bands of CO and CO_2.

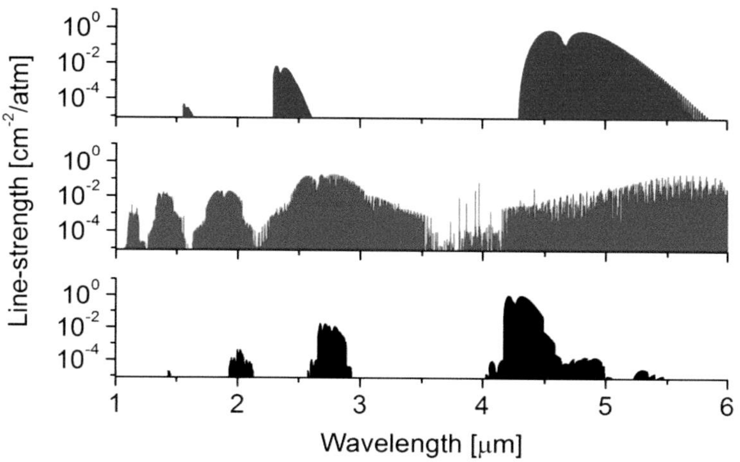

5. A student measured the absorption spectrum (2280–2400 cm^{-1}) of a mixture of 0.2 % CO_2 in argon, at 1 atm and 296 K in a gas cell of length 10 cm using an equipment with very poor resolution. The absorption band was known to be centered at 2349 cm^{-1}. Having a strong background in mathematics, this student is able to describe the absorption profile as:

$$\alpha(\nu) = \frac{0.93}{1 + 0.005(\nu - 2333)^2} + \frac{1.5}{1 + 0.0152(\nu - 2362)^2}$$

where ν is the wavenumber.

(a) Draw the graph of $\alpha(\nu)$. Label P, Q, and R branches, if they exist. What are the allowed and forbidden transitions of this band?

(b) Now we will perform a simple calculation of CO_2 spectrum based on what we learned in class. Assume a simple harmonic oscillator model. All the transitions are known to be in the CO_2 fundamental band with lower-state $v'' = 0$.

Step 1: Given the moment of inertia 7.175×10^{-46} kg m^2, calculate the rotational constant B of CO_2 and give expressions for the transitions in each branch of the band. What is the highest energy transition of P branch? What is the lowest energy transition of R branch?

Step 2: The probability of each transition is nearly proportional to the lower-state population, which we assume to be given by a Boltzman distribution with a degeneracy of $(2J + 1)$. Here we approximate the probability as:

$$\text{transition probability} \propto (2J + 1) \exp\left(\frac{-S(0, J)hc}{kT}\right)$$

where $S(0, J)$ is the energy of the $(0, J)$ state. What is the rotational level that has the maximum transition population, J_{max}? With J_{max}, you can normalize the predicted structure by examining the transition probability ratio. Give the expression for the normalized probability.

Step 3: Plot the above ratio as a function of wavenumber and compare your calculation with the ratio $\alpha(\nu)/\alpha_{max}$, where α_{max} is the maximum value of $\alpha(\nu)$.

References

1. G. Herzberg, *Molecular Spectra and Molecular Structure, Volume II. Infrared and Raman Spectra of Polyatomic Molecules*, 2nd edn. (Krieger Publishing Company, Malabar, FL, 1945)
2. G.M. Barrow, *Introduction to Molecular Spectroscopy*, 1st edn. (McGraw-Hill International (UK) Limited, New York, 1962)

Effects of Nuclear Spin: Rotational Partition Function and Degeneracies

5

In previous chapters on diatomic and polyatomic spectra, we discussed quantized angular momentum as it relates to a rotating molecule and its effect on Boltzmann statistics. Rotational angular momentum is, in fact, only one kind of angular momentum that must be considered when evaluating real molecules and their spectra. For example, electrons also have quantized angular momentum as they orbit nuclei (called *orbital* angular momentum, which we will discuss further in Chaps. 9 and 10). Additionally, individual subatomic particles like nuclei and electrons have their own intrinsic, quantized, angular momentum, analogous to the angular momentum of a spinning top or planet; hence, the common name for this property is *spin*. Because of the added complexity that nuclear spin can introduce, we have not discussed its effects up to this point; however, nuclear spin can, in symmetric molecules, change the degeneracy of some states and thus affect molecular statistics. This chapter will introduce the effects of nuclear spin, when they need to be considered, and how one can incorporate them into the statistics of linear and nonlinear molecules.

5.1 Introduction

Recall that the population fraction, F_i, of a given molecular energy level E_i is

$$F_i = \frac{g_i \exp\left(-\frac{\epsilon_i}{kT}\right)}{Q}, \tag{5.1}$$

where the partition function, Q, is given by the summation over all energy levels

$$Q = \sum_i g_i \exp\left(-\frac{\epsilon_i}{kT}\right) \tag{5.2}$$

When evaluating partition functions and energy level degeneracies, g_i, for asymmetric molecules, e.g. heteronuclear diatomics like NO and polyatomics like

R.K. Hanson et al., *Spectroscopy and Optical Diagnostics for Gases*,
DOI 10.1007/978-3-319-23252-2_5

HCN, we usually consider only three internal energy modes: electronic, vibrational, and rotational. Thus, the partition function is

$$Q = Q_{\text{rot}} Q_{\text{vib}} Q_{\text{elec}} \tag{5.3}$$

We have thus far neglected nuclear spin (even though it exists) for such cases because there is no coupling between nuclear spin and rotation and because the nuclear terms cancel in expressions for population fraction in specific rotational states.

However, in molecules with certain *symmetric* placement of equivalent nuclei, e.g. O_2 and NH_3, we must consider the total partition function, including the nuclear terms:

$$Q = Q_{\text{rot}} Q_{\text{vib}} Q_{\text{elec}} Q_{\text{nuc}} \tag{5.4}$$

Here again, we have invoked the Born–Oppenheimer approximation: the energy associated with each mode is independent or separable. As outlined below, we must include the nuclear terms for symmetric molecules because of their coupling to the rotational levels. In these cases, we use an effective rotational partition function defined by

$$Q'_{\text{rot}} = Q_{\text{rot}} Q_{\text{nuc}}, \tag{5.5}$$

and likewise, an effective rotational degeneracy

$$g'_{\text{rot}} = g_{\text{rot}} g_{\text{nuc}}. \tag{5.6}$$

5.2 Nuclear Spin and Symmetry

Since nuclear energies are quite large for most if not all practical conditions ($E_{\text{nuc}} \gg kT$ for excited nuclear states), we are only interested in the lowest (ground) nuclear energy level. It follows that

$$Q_{\text{nuc}} = g_{\text{nuc}} \tag{5.7}$$

since $\exp(-E/kT) \approx 0$ for all but the ground level. In general, the ground nuclear energy level has a degeneracy associated with the spin quantum number, I (not to be confused with the moment of inertia I_A, I_B, etc.) of the nuclei (Table 5.1). For a single nucleus, the number of degenerate spin states is given by

$$g_{\text{nuc}} = 2I + 1 \tag{5.8}$$

For example, a ^{14}N nucleus has $I = 1$ and a corresponding nuclear spin degeneracy $g_{\text{nuc}} = 3$. For a molecule made up of L atoms, the total nuclear partition function is formed by the product of the terms for the individual nuclei

Table 5.1 Spin of several nuclei

Nucleus	Spin, I
^{1}H	1/2
^{2}H	1
^{11}B	5/2
^{12}C	0
^{13}C	1/2
^{14}N	1
^{16}O	0
^{35}Cl	3/2

$$Q_{\text{nuc}} = \prod_{n=1}^{L} (2I_n + 1) \tag{5.9}$$

where I_n is the spin of the nth nucleus.

Nuclear spin states and rotational states couple through the symmetry properties of their respective wave functions. A wave function that describes any state composed of two or more identical particles is either *symmetric* or *antisymmetric* with respect to those particles. For rotational states, the symmetry property is determined by the resulting wave function if a pair of identical nuclei were interchanged. For symmetric states, the sign of the wave function remains unchanged [+] upon interchange of the nuclei; for antisymmetric states, the sign changes [−]. For example, if a molecule had two nuclei, x and y, its rotational wave function is *symmetric* if $\Psi_{\text{rot}}(x, y) = \Psi_{\text{rot}}(y, x)$; it would be *antisymmetric* if $\Psi_{\text{rot}}(x, y) = -\Psi_{\text{rot}}(y, x)$.

In rotationally symmetric molecules like O_2 and NH_3, nuclear spin states (with their own symmetry properties) can only pair with rotational states having a compatible symmetry, either symmetric or antisymmetric. Which one is compatible depends on the nature of nuclei (i.e., whether the system is a *boson* or *fermion*).

The rules governing symmetry compatibility depend on which statistics apply to the nuclear system. For nuclear systems that follow Fermi–Dirac statistics (fermions), the total molecular wave function must be antisymmetric. This symmetry is made up of a combination of a symmetric and antisymmetric states (an overall sign change comes from multiplying a function having no sign change with a function having a sign change, i.e. $[+] \times [-] = [-]$). For example, antisymmetric nuclear spin states combine (only) with symmetric rotational levels to produce an overall antisymmetric state. For Bose–Einstein nuclear systems (bosons), the overall wave function must be symmetric. Thus symmetric states combine only with symmetric states, and antisymmetric with antisymmetric ($[+] \times [+] = [-] \times [-] = [+]$).

For a single nucleus, Fermi–Dirac statistics are associated with nuclei having half-integral spins ($I = 1/2, 3/2, \ldots$) and Bose–Einstein statistics hold for nuclei with integral spins (0, 1, 2, ...). The important question now becomes, what is the symmetry character of the rotational levels? The most powerful tool for attacking symmetry questions like these is called group theory. Since group theory requires more explanation than can be given here, results for some of the most important (and common) molecular configurations are presented below.

5.3 Case I: Linear Molecules

5.3.1 Asymmetric (e.g., CO and N_2O)

For linear molecules without symmetry, there is no coupling between nuclear spin and molecular rotation. Thus, the effective rotational expressions are simply the standard expressions multiplied by a constant term,

$$Q'_{\text{rot}} = \frac{T}{\sigma\theta_r}\prod_{n=1}^{L}(2I_n + 1) \tag{5.10}$$

and

$$g'_{\text{rot}} = (2J + 1)\prod_{n=1}^{L}(2I_n + 1) \tag{5.11}$$

For asymmetric linear molecules, the symmetry factor $\sigma = 1$; this factor represents the number of ways that the molecule may be rotated into an identical configuration. As stated initially, the nuclear spin terms can be omitted from the partition functions and degeneracies for asymmetric molecules since they cancel in expressions for the population fraction.

Consider CO

$$Q'_{\text{rot}} = \frac{T}{\theta_r}\prod_{n=1}^{2}(2I_n + 1) = \frac{T}{\theta_r}(2I_{\text{C}} + 1)(2I_{\text{O}} + 1) = \frac{T}{\theta_r}$$

$$g'_{\text{rot}} = (2J + 1)(2I_{\text{C}} + 1)(2I_{\text{O}} + 1) = 2J + 1$$

Thus the fractional population in J is, as we found for an asymmetric rigid rotor without consideration of nuclear spin,

$$\frac{N_J}{N} = \frac{(2J + 1)\exp(-F(J)hc/kT)}{T/\theta_r}$$

Consider N_2O

$$Q'_{\text{rot}} = \frac{T}{\theta_r}\prod_{n=1}^{3}(2I_n + 1) = \frac{T}{\theta_r}(2I_{\text{N}} + 1)^2(2I_{\text{O}} + 1) = 9\frac{T}{\theta_r}$$

$$g'_{\text{rot}} = (2J + 1)(2I_{\text{N}} + 1)^2(2I_{\text{O}} + 1) = 9(2J + 1)$$

Both Q_{rot} and g_{rot} are larger by a factor of 9 owing to nuclear spin, but these terms cancel in forming the Boltzmann fraction,

$$\frac{N_J}{N} = \frac{(2J+1)\exp(-F(J)hc/kT)}{T/\theta_r}.$$

Recall that the term $2J+1$ is the degeneracy for energy level $F(J)$ associated with the number of possible directions (orientations) of the angular momentum vector of the molecular rotation. In the presence of an applied electric or magnetic field, these $2J+1$ states may be slightly separated in energy. In the absence of such fields, the states all have the same energy, but the degeneracy remains.

5.3.2 Symmetric (e.g., O_2, CO_2, and C_2H_2)

Now the rotational partition function contains a non-unity value for the symmetry factor, σ. For linear and symmetric molecules, $\sigma = 2$ because these molecules are indistinguishable upon rotation about the B axis.

$$Q'_{\text{rot}} = \frac{1}{2}\frac{T}{\theta_r}\prod_{n=1}^{L}(2I_n+1) \tag{5.12}$$

The degeneracy is determined as follows. The overall symmetry required by the nuclear statistics must first be determined. Then we multiply the degeneracy $(2J+1)$ of a rotational level, with a given symmetry, by the degeneracy of the appropriate nuclear spin state.

In a linear molecule, the nuclear statistics of the overall nuclear system are controlled by the number of fermions (half-integral spin nuclei) on one side of the center of the molecule (if a central nucleus exists, it is ignored). For an odd number of fermions, the system behaves according to Fermi–Dirac statistics and the overall symmetry is antisymmetric. Conversely, for an even number of fermions (including zero or none), Bose–Einstein statistics and an overall symmetric wave function are required.

The degeneracy of the symmetric nuclear spin states is given by

$$g_{\text{nuc,symm}} = \frac{(2I_C+1)}{2}\left[\prod_{m=1}^{M}(2I_m+1)^2 + \prod_{m=1}^{M}(2I_m+1)\right] \tag{5.13}$$

and the degeneracy of the antisymmetric states is given by

$$g_{\text{nuc,asymm}} = \frac{(2I_C+1)}{2}\left[\prod_{m=1}^{M}(2I_m+1)^2 - \prod_{m=1}^{M}(2I_m+1)\right] \tag{5.14}$$

where I_m is the nuclear spin of the mth nucleus on one side of the center of the molecule, $M = (L - 1)/2$, and I_C is the nuclear spin of the central nucleus if one is present (not to be confused with either the nuclear spin of a carbon atom, I_{C}, or the moment of inertia about the C-axis of a nonlinear molecule, I_C). If no central nucleus exists (e.g., in a homonuclear diatomic), the term $(2I_C + 1)$ is replaced by unity and $M = L/2$ [see Eq. (5.16)]. For a symmetric molecule made up of nuclei with spin $I = 0$, it can be seen that only symmetric spin states exist.

The symmetry of a rotational state depends on its rotational quantum number as well as the structure of the electronic manifold within which it exists (Table 5.3). For rotational levels in electronic manifolds designated Σ_g^+ and Σ_u^-, levels with even N are symmetric, and those with odd N are antisymmetric. (Note, here the quantum number N describes molecular rotations, while J includes contributions from electron spin. For molecular states with no electron spin, $J = N$.) For rotational levels in Σ_u^+ and Σ_g^- electronic manifolds, the reverse is true; even N are antisymmetric and odd N are symmetric. These four electronic configurations represent the ground electronic structures of most common linear symmetric molecules. For other electronic configurations (e.g., Π and Δ), it turns out that each rotational level consists of two nearly degenerate states, one symmetric and the other antisymmetric, and like asymmetric molecules, the nuclear spin effects can usually be ignored.

Combining the above, we have

$$g'_{\mathrm{rot}} = (2J + 1)\frac{(2I_C + 1)}{2}\left[\prod_{m=1}^{M}(2I_m + 1)^2 \pm \prod_{m=1}^{M}(2I_m + 1)\right] \tag{5.15}$$

where the choice between adding or subtracting $(+/-)$ the two products is determined from Table 5.2. For homonuclear diatomics, as described above, the rotational degeneracy including nuclear spin effects reduces to

$$g'_{\mathrm{rot}} = (2J + 1)\frac{1}{2}[(2I + 1)^2 \pm (2I + 1)]. \tag{5.16}$$

Consider H_2 (molecular hydrogen comprised of two ^{1}H atoms)

$$Q'_{\mathrm{rot}} = \frac{1}{2}\frac{T}{\theta_r}(2I_{\mathrm{H}} + 1)^2 = 2\frac{T}{\theta_r}$$

Since there is one fermion on either side of the axis of symmetry, Fermi statistics apply. And, since the ground state of H_2 is $X^1\Sigma_g^+$,

Table 5.2 Key to addition or subtraction of nuclear degeneracies in Eqs. (5.15) and (5.16) for different rotational statistics and electronic manifold configurations

Statistics (symmetry)	Electronic configuration	N	Addition or subtraction
Fermi (*antisymmetric*)	Σ_g^+, Σ_u^-	Odd	$+$
		Even	$-$
	Σ_u^+, Σ_g^-	Odd	$-$
		Even	$+$
Bose (*symmetric*)	Σ_g^+, Σ_u^-	Odd	$-$
		Even	$+$
	Σ_u^+, Σ_g^-	Odd	$+$
		Even	$-$

Table 5.3 Sample species and ground state configurations

Ground state configurations	Species
$X^1\Sigma_g^+$	$^1H_2, ^2H_2, ^{14}N_2, ^{15}Cl_2$
$X^3\Sigma_g^-$	$^{16}O_2$
$X^3\prod_u$	$^{12}C_2$
$\tilde{X}^1\Sigma_g^+$	CO_2, C_2H_2, C_2N_2

$$g'_{\text{rot}} = (2J+1)\frac{1}{2}\left[(2I_{\text{H}}+1)^2 \pm (2I_{\text{H}}+1)\right]\begin{Bmatrix} -,\ J \text{ even} \\ +,\ J \text{ odd} \end{Bmatrix}$$

$$= (2J+1)\frac{1}{2}\begin{Bmatrix} 2,\ J \text{ even} \\ 6,\ J \text{ odd} \end{Bmatrix}$$

$$= (2J+1)\begin{Bmatrix} 1,\ J \text{ even} \\ 3,\ J \text{ odd} \end{Bmatrix}$$

so that the effective rotational degeneracy alternates with even and odd J.

Consider O_2

$$Q'_{\text{rot}} = \frac{1}{2}\frac{T}{\theta_r}(2I_{\text{O}}+1)^2 = \frac{T}{2\theta_r} \quad (I_{\text{O}} = 0)$$

Since there are no fermions ($I = 0$), Bose–Einstein statistics apply. And, since we have a $^3\Sigma_g^-$ state,

$$g'_{\text{rot}} = (2J+1)\left(\frac{1}{2}\right)\begin{Bmatrix} 1-1, & J \text{ even} \\ 1+1, & J \text{ odd} \end{Bmatrix}$$

$$= \begin{Bmatrix} 0, & J \text{ even} \\ 2J+1, & J \text{ odd} \end{Bmatrix}$$

Therefore, only the *odd J* states are populated! *Even J* states do not exist for O_2 in its ground electronic state.

Consider CO_2

$$Q'_{\text{rot}} = \frac{1}{2}\frac{T}{\theta_r}(2I_{\text{C}}+1)(2I_{\text{O}}+1)^2 = \frac{T}{2\theta_r}$$

The number of fermions on one side of the axis is zero, therefore the molecule obeys Bose statistics.

$$g'_{\text{rot}} = (2J+1)\frac{1}{2}(2I_{\text{C}}+1)\left[(2I_{\text{O}}+1)^2 \pm (2I_{\text{O}}+1)\right]\begin{Bmatrix} +, & J \text{ even} \\ -, & J \text{ odd} \end{Bmatrix}$$

$$= (2J+1)\frac{1}{2}\begin{Bmatrix} 2, & J \text{ even} \\ 0, & J \text{ odd} \end{Bmatrix}$$

$$= (2J+1)\begin{Bmatrix} 1, & J \text{ even} \\ 0, & J \text{ odd} \end{Bmatrix}$$

Therefore, only the *even J* states are populated! *Odd J* states do not exist for CO_2 in its ground electronic state.

Consider C_2H_2

$$Q'_{\text{rot}} = \frac{1}{2}\frac{T}{\theta_r}(2I_{\text{C}}+1)^2(2I_{\text{H}}+1)^2 = 2\frac{T}{\theta_r}$$

There is one fermion on either side of the axis of symmetry, so Fermi statistics apply.

$$g'_{\text{rot}} = (2J+1)\frac{1}{2}\left[(2I_{\text{C}}+1)^2(2I_{\text{H}}+1)^2\right.$$

$$\left.\pm(2I_{\text{C}}+1)(2I_{\text{H}}+1)\right]\begin{Bmatrix} -, & J \text{ even} \\ +, & J \text{ odd} \end{Bmatrix}$$

$$= (2J+1)\frac{1}{2}\begin{Bmatrix} 2, & J \text{ even} \\ 6, & J \text{ odd} \end{Bmatrix}$$

$$= (2J+1)\begin{Bmatrix} 1, & J \text{ even} \\ 3, & J \text{ odd} \end{Bmatrix}$$

5.4 Case II: Nonlinear Molecules

5.4.1 Asymmetric Rotor (e.g., CHFClBr and N_2H_4)

Recall that asymmetric rotor molecules have three non-zero and unequal moments of inertia, $I_A \neq I_B \neq I_C$. The general expression for the rotational partition function, including contributions from all three axes of rotation and nuclear spin is

$$Q'_{\text{rot}} = \frac{1}{\sigma}\sqrt{\pi\left(\frac{T}{\theta_A}\right)\left(\frac{T}{\theta_B}\right)\left(\frac{T}{\theta_C}\right)}\prod_{n=1}^{L}(2I_n+1) \tag{5.17}$$

The degeneracy is

$$g'_{\text{rot}} = \frac{1}{\sigma}(2J+1)\prod_{n=1}^{L}(2I_n+1) \tag{5.18}$$

Again, the nuclear spin terms can be omitted since there is no nuclear spin-rotation coupling, and the nuclear terms cancel in expressions for the population fraction.

Consider $^1H^2HO$ (HDO)

$$\begin{aligned} Q'_{\text{rot}} &= \sqrt{\frac{\pi T^3}{\theta_A\theta_B\theta_C}}(2I_{\text{O}}+1)(2I_{\text{H}}+1)(2I_{\text{D}}+1) \\ &= 6\sqrt{\frac{\pi T^3}{\theta_A\theta_B\theta_C}} \\ g'_{\text{rot}} &= 6(2J+1) \end{aligned} \tag{5.19}$$

5.4.2 Symmetric Top

Recall that symmetric top molecules have two equal moments of inertia, $I_A \neq I_B = I_C$. The general expression for the rotational partition function, including contributions from all three axes of rotation and nuclear spin is

$$Q'_{\text{rot}} = \frac{\sqrt{\pi}}{\sigma}\sqrt{\frac{T}{\theta_A}}\left(\frac{T}{\theta_B}\right)\prod_{n=1}^{L}(2I_n+1) \tag{5.20}$$

where σ depends on the group symmetry.

(a) C_{3v} **Group Symmetry (e.g. NH$_3$ and CH$_3$Cl)**

$$Q'_{\text{rot}} = \frac{\sqrt{\pi}}{3}\sqrt{\frac{T}{\theta_A}}\left(\frac{T}{\theta_B}\right)\prod_{n=1}^{L}(2I_n + 1) \tag{5.21}$$

The symmetry factor for C_{3v} group symmetry is $\sigma = 3$ due to the threefold rotational symmetry about the A-axis. The degeneracy is dependent on whether the quantum number K is divisible by 3.

For K divisible by 3 (including $K = 0$)

$$g'_{\text{rot}} = (2J + 1)\left[\prod_{c=1}^{C}(2I_C + 1)\right]\frac{2I_{NC} + 1}{3}(4I_{NC}^2 + 4I_{NC} + 3) \tag{5.22}$$

For K not divisible by 3

$$g'_{\text{rot}} = (2J + 1)\left[\prod_{c=1}^{C}(2I_C + 1)\right]\frac{2I_{NC} + 1}{3}(4I_{NC}^2 + 4I_{NC}) \tag{5.23}$$

where I_C is the nuclear spin of the cth nucleus on the threefold symmetry axis and I_{NC} is the nuclear spin of one of the nuclei off the threefold symmetry axis of the molecule (e.g., H in NH$_3$). The degeneracy is independent of the statistics of the nuclei. Thus, for a given J, the K degeneracies vary like $\text{x} : \text{x} : \text{x}' : \text{x} : \text{x} : \text{x}' : \ldots$, with $\text{x} < \text{x}'$.

Consider CH$_3$Cl as an example of a molecule with C_{3v} symmetry.

$$\begin{aligned} Q'_{\text{rot}} &= \frac{\sqrt{\pi}}{3}\sqrt{\frac{T}{\theta_A}}\left(\frac{T}{\theta_B}\right)(2I_{\text{C}} + 1)(2I_{\text{Cl}} + 1)(2I_{\text{H}} + 1)^3 \\ &= \frac{32\sqrt{\pi}}{3}\sqrt{\frac{T}{\theta_A}}\left(\frac{T}{\theta_B}\right) \end{aligned}$$

$$\begin{aligned} g'_{\text{rot}} &= (2J + 1)\frac{1}{3}(2I_{\text{C}} + 1)(2I_{\text{Cl}} + 1) \\ &\quad (2I_{\text{H}} + 1)\left\{\begin{array}{r} 4I_{\text{H}}^2 + 4I_{\text{H}} + 3,\ K \text{ div. by } 3 \\ 4I_{\text{H}}^2 + 4I_{\text{H}},\ K \text{ not div. by } 3 \end{array}\right\} \\ &= (2J + 1)\frac{8}{3}\left\{\begin{array}{l} 6,\ K \text{ div. by } 3 \\ 3,\ K \text{ not div. by } 3 \end{array}\right\} \\ &= (2J + 1)8\left\{\begin{array}{l} 2,\ K \text{ div. by } 3 \\ 1,\ K \text{ not div. by } 3 \end{array}\right\} \end{aligned}$$

(b) **D_{3h} Group Symmetry (e.g., BCl_3)**
The effective partition function is the same as given for the C_{3_v} case above,

$$Q'_{\text{rot}} = \frac{\sqrt{\pi}}{3}\sqrt{\frac{T}{\theta_A}}\left(\frac{T}{\theta_B}\right)\prod_{n=1}^{L}(2I_n+1) \tag{5.24}$$

For levels with $K = 0$, there is now an alternation in nuclear degeneracy as J (or N) increases. Depending on the appropriate nuclear statistics and rotational quantum number, the degeneracy is given by:
For integral I_{NC} and even N *or* half-integral I_{NC} and odd N:

$$g'_{\text{rot}} = (2J+1)\left[\prod_{c=1}^{C}(2I_C+1)\right]\frac{2I_{NC}+1}{3}(2I_{NC}+3)(I_{NC}+1) \tag{5.25}$$

For integral I_{NC} and odd N *or* half-integral I_{NC} and even N:

$$g'_{\text{rot}} = (2J+1)\left[\prod_{c=1}^{C}(2I_C+1)\right]\frac{2I_{NC}+1}{3}(2I_{NC}-1)I_{NC} \tag{5.26}$$

For levels with $K \neq 0$, Eqs. (5.22) and (5.23) apply for K divisible by 3 and not divisible by 3, respectively.

Consider BCl_3 as an example of a molecule with D_{3h} symmetry.

$$\begin{aligned} Q'_{\text{rot}} &= \frac{\sqrt{\pi}}{3}\sqrt{\frac{T}{\theta_A}}\left(\frac{T}{\theta_B}\right)(2I_{\text{B}}+1)(2I_{\text{Cl}}+1)^3 \\ &= 128\sqrt{\pi}\sqrt{\frac{T}{\theta_A}}\left(\frac{T}{\theta_B}\right) \end{aligned} \tag{5.27}$$

Since $I_{NC} \equiv I_{\text{Cl}} = 3/2$, the system follows Fermi statistics.

$$g'_{\text{rot}} = (2J+1)(2I_{\text{B}}+1)\frac{1}{3}$$

$$(2I_{\text{Cl}}+1)\left\{\begin{array}{c} (2I_{\text{Cl}}-1)I_{\text{Cl}},\ K=0, J \text{ even} \\ (2I_{\text{Cl}}+3)(I_{\text{Cl}}+1),\ K=0, J \text{ odd} \\ 4I_{\text{Cl}}^2+4I_{\text{Cl}}+3,\ K\neq 0,\ K \text{ div. by } 3 \\ 4I_{\text{Cl}}^2+4I_{\text{Cl}},\ K\neq 0,\ K \text{ not div. by } 3 \end{array}\right\}$$

$$= 8(2J+1)\left\{\begin{array}{l} 3,\ K=0, J \text{ even} \\ 15,\ K=0, J \text{ odd} \\ 18,\ K\neq 0, K \text{ div. by } 3 \\ 15,\ K\neq 0, K \text{ not div. by } 3 \end{array}\right\}$$

$$= 24(2J+1)\left\{\begin{array}{l} 1,\ K=0, J \text{ even} \\ 5,\ K=0, J \text{ odd} \\ 6,\ K\neq 0, K \text{ div. by } 3 \\ 5,\ K\neq 0, K \text{ not div. by } 3 \end{array}\right\}$$

5.4.3 Others (e.g., C_6H_6, CH_4, and P_4)

These molecules must be considered by symmetry group. See, for example, Herzberg [1, vol. 2, Chap. 2] and [2, vol. 3, Chap. 1].

5.5 Exercises

1. Ammonia, NH_3, is a symmetric top molecule with C_{3v} symmetry, which means $\sigma = 3$, and the molecule can be rotated into itself three different ways.
   (a) Evaluate the relative strengths of the microwave absorption lines for $J'' = 1$ and $J'' = 2$ at 300 K assuming that these relative strengths follow Boltzmann statistics, i.e., evaluate $(N_{J=1}/N_{J=2})$. Use $\theta_A = 9$ K and $\theta_B = 14$ K in your calculations. Hint: don't forget to sum over the allowed values of K.
   (b) Calculate the effective (including nuclear spin) rotational partition function for ammonia at 300 K.

References

1. G. Herzberg, *Molecular Spectra and Molecular Structure, volume II. Infrared and Raman Spectra of Polyatomic Molecules*, 2nd edn. (Krieger Publishing Company, Malabar, FL, 1945)
2. G. Herzberg, *Molecular Spectra and Molecular Structure, volume III. Electronic Spectra and Electronic Structure of Polyatomic Molecules* (Van Nostrand Renhold Company, New York, NY, 1966)

Rayleigh and Raman Spectra 6

Unlike absorption and emission, Rayleigh and Raman spectroscopy are based on how a molecule scatters photons. One key difference to keep in mind as we explore scattering processes further is that while absorption requires that the molecule have energy-level spacings (for allowed transitions) corresponding to the energy of the interacting photon, scattering can occur with an incident photon of almost any energy. That is, for a molecule to absorb light, the light must generally be at specific wavelengths or frequencies. Scattering, on the other hand, can occur at almost any wavelength.

6.1 Light Scattering

For the light scattering experimental schematic in Fig. 6.1, the incident laser power, P_i, becomes scattered (Rayleigh or Raman) when it interacts with molecules inside the volume element $\delta V = \delta A \times L$. The scattered laser power, P_s, is determined by the product of incident laser power, number density of molecules inside the volume element, length L, scattering cross-section, and collection angle of the optics.

Incident Laser Power $P_i = \dot{N}_p h\nu$ [W]

$\dot{N}_p =$ incident photons/s

Scattered Laser Power $P_s = \dot{N}_s h\nu$ [W]

$\dot{N}_s =$ scattered photons/s into collection angle

$$\dot{N}_s = \underbrace{\frac{\dot{N}_p}{\delta A}}_{\substack{\text{photons}\\ \text{per area}\\ \text{per second}}} \times \underbrace{N\frac{\partial\sigma}{\partial\Omega}\Omega}_{\substack{\text{projected}\\ \text{scattering}\\ \text{area, cm}^2}}$$

R.K. Hanson et al., *Spectroscopy and Optical Diagnostics for Gases*,
DOI 10.1007/978-3-319-23252-2_6

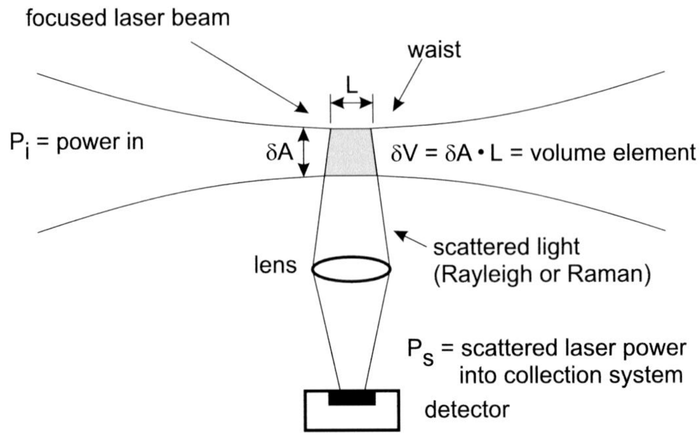

Fig. 6.1 Schematic of a light scattering experiment

$N = n \cdot \delta A \cdot L =$ the number of molecules in δV; n is the molecular density [molec/cm^3]

$\dfrac{\partial \sigma}{\partial \Omega}$ = differential scattering cross-section [cm^2/molec Sr]

Ω = solid angle of collection [Steradians, Sr]

so that

$$\boxed{P_s = P_i n L \frac{\partial \sigma}{\partial \Omega} \Omega} \tag{6.1}$$

Since scattering is essentially instantaneous, the above relation applies at any instant, when the incident power varies with time. For pulsed light, often employed for Rayleigh and Raman scattering, P_i and P_s can be reinterpreted as the total number of incident and scattered (collected) photons.

6.1.1 Cross-Sections

Rayleigh Cross-Sections

The differential Rayleigh cross-section is a property of a molecule that describes its propensity to scatter light. In general, this cross-section depends on the angle between the detection beam and the incident beam as well as the polarization. One can derive the differential Rayleigh cross-section as a function of laser and scattering geometry by modelling a molecule as an infinitesimally small oscillating dipole interacting with an imposed electric field. While a thorough discussion can be found in Miles et al. [1] or in Banwell [2], some key results are given here.

A spherically symmetric molecule's differential cross-section at 90° relative to the plane of incident polarization (formed by the E-field vector of the propagating wave and the direction of propagation) is

$$\left(\frac{\partial \sigma_{SS}}{\partial \Omega}\right)_{\text{Rayleigh}} \cong \frac{4\pi^2 (n_i - 1)^2}{N^2 \lambda^4} \tag{6.2}$$

where n_i is the refractive index and N is the number density of the scattering molecules (written here as a capital N to distinguish number density from the refractive index). Note the strong inverse dependence on wavelength. Also, since the quantity $(n_i - 1)$ of a gas is proportional to its density, the quantity $(n_i - 1)^2/N^2$ is independent of density, and hence the Rayleigh cross-section (per molecule) is an intrinsic property (actually, this property is proportional to something called the polarizability of the molecule). The total cross-section is the differential cross-section integrated over the surface of a sphere enclosing the scatterer. It is given by

$$\sigma_{SS} \cong \frac{32\pi^3}{3\lambda^4}\left(\frac{n_i - 1}{N}\right)^2 = \frac{8\pi}{3}\left(\frac{\partial \sigma_{SS}}{\partial \Omega}\right)_{\text{Rayleigh}} \tag{6.3}$$

This formula will generally allow one to calculate the scattering cross-section to within a few percent for most molecules. If more accuracy is required, particles with some spherical asymmetry (e.g., diatomics) also have a King factor, F_K, that describes their anisotropy:

$$\sigma \cong F_K \sigma_{SS} \tag{6.4}$$

Typical values of the total cross-section in the visible region are on the order of 10^{-27} cm^2/molec for small molecules. A few examples are given in Table 6.1.

Table 6.1 Rayleigh scattering parameters for selected molecules, taken from [1] and [3]

Molccule	λ [nm]	F_K	σ [cm^2/molec]
Ar	250	1	9.77×10^{-26}
	500		5.85×10^{-27}
	1000		3.50×10^{-28}
N_2	250	1.039	1.14×10^{-25}
	500	1.035	6.81×10^{-27}
	1000	1.034	4.08×10^{-28}
Air	250	1.051	1.26×10^{-25}
	500	1.049	6.66×10^{-27}
	1000	1.047	4.01×10^{-28}
SF_6	250	1	7.12×10^{-25}
	500		4.32×10^{-26}
	1000		2.63×10^{-27}

Table 6.2 Vibrational Raman cross-sections for common molecules at 532 nm (in units of $10^{-30}\,\text{cm}^2/\text{molec/Sr}$)

Species	Cross-section
O_2	0.65
N_2	0.46
H_2	0.943
NH_3	1.3
$CO_2(\nu_1)$	0.6
$CH_4(\nu_1)$	2.6

Raman Cross-Sections

The cross-section for Raman scattering is much smaller than for Rayleigh scattering (Table 6.2).

$$\left(\frac{\partial\sigma}{\partial\Omega}\right)_{\text{Raman}} \approx 10^{-3}\left(\frac{\partial\sigma}{\partial\Omega}\right)_{\text{Rayleigh}}$$

Example: Rayleigh and Raman Scattering by N_2 at STP

For a laser source with $P_i = 1\,\text{W}$ at $\lambda = 500\,\text{nm}$, the incident photon rate is $\dot{N}_p = 2.5 \times 10^{18}$ photons/s. The approximate differential scattering cross-section for Rayleigh scattering is

$$\left(\frac{\partial\sigma}{\partial\Omega}\right)_{\text{Rayleigh}} \approx 8.1 \times 10^{-28}\ [\text{cm}^2/\text{molec Sr}]$$

For a system with 4 W of incident laser power at this wavelength, a measurement length $L = 1$ mm and $\Omega \approx 10^{-2}$ Sr,

$$\dot{N}_s = \underbrace{\left(10^{19}\,\frac{\text{photons}}{\text{s}}\right)}_{\dot{N}_p}\underbrace{\left(2.7 \times 10^{19}\,\frac{\text{molec}}{\text{cc}}\right)}_{n=n_L \text{ at STP}}\underbrace{(0.1\,\text{cm})}_{L}\underbrace{\left(8.1 \times 10^{-28}\,\frac{\text{cm}^2}{\text{molec Sr}}\right)}_{\partial\sigma/\partial\Omega}\underbrace{(10^{-2}\,\text{Sr})}_{\Omega}$$

$$\dot{N}_s = 22 \times 10^7 \text{ photons/s}$$

$$P_s \approx 88\,\text{pW}$$

For Raman scattering, however,

$$\dot{N}_s = 22 \times 10^4 \text{ photons/s, and}$$

$$P_s \approx 0.088\,\text{pW} = 88\,\text{fW (small!)}$$

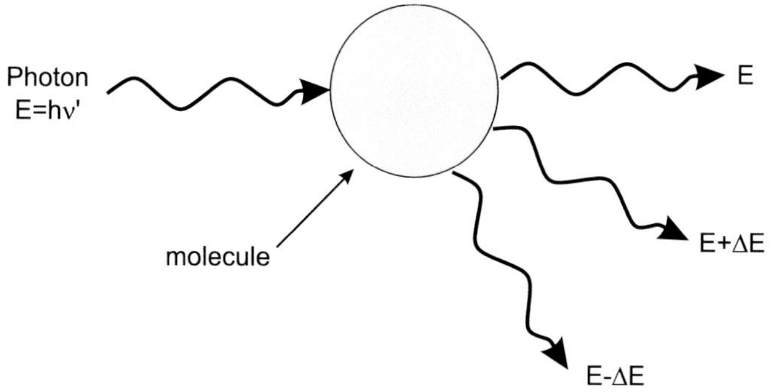

Fig. 6.2 Photon scattering due to interaction with a molecule

6.2 Quantum Model

Consider radiation to be composed of discrete photons that exchange energy with molecules (see Fig. 6.2). There are three different scattering possibilities: elastic (incident and scattered energy/wavelength are the same), and inelastic with either increased or decreased energy. Rayleigh scattering corresponds to elastic interactions, and Raman scattering corresponds to the two inelastic interaction possibilities.

Rayleigh	E (elastic); $\nu = \nu'$
Anti-Stokes Raman	$E + \Delta E$ (inelastic); $\nu = \nu' + \Delta E/h$
Stokes Raman	$E - \Delta E$ (inelastic); $\nu = \nu' - \Delta E/h$

where

$$\Delta E = \Delta E_{\text{rot}} \text{ or } \Delta E_{\text{rot,vib}},$$

with new selection rules for allowed transitions. Scattering is a weak process, and thus requires a sensitive apparatus for making measurements.

6.3 Classical Theory

Classical theory for scattering is based on the polarizability of molecules. Whereas absorption and emission require an *intrinsic* dipole to be resonant with the interacting photon, scattering occurs when an externally applied electric field (the photon in classical terms) *induces* a dipole in the molecule. The induced dipole strength, μ, is given by

$$\mu = \alpha E \tag{6.5}$$

where α is the polarizability and E is the E-field of the light wave. The E-field is oscillatory in nature (e.g., $E = E_o \sin(2\pi \nu t)$) and induces an oscillation in the molecule. The induced oscillation can occur:

1. at the same frequency of the incident field, resulting in "emission" at ν (Rayleigh), or
2. at beat frequencies based on the interaction of the incident field and the molecule's rotational/vibrational frequencies, i.e. with "emission" at $\nu \pm \nu_{\mathrm{vib/rot}}$

These beat frequencies are the Raman sidebands. This occurs because the polarizability *may* oscillate with the rotational and/or vibrational frequencies of the molecule, e.g.

$$\alpha = \alpha_o + \beta \sin(2\pi \nu_{\mathrm{vib}} t),$$

where β is the rate of change of polarizability with the rotation/vibration. From Eq. (6.5), the induced dipole strength becomes

$$\mu = \underbrace{\alpha_o E_o \sin(2\pi \nu t)}_{\text{Rayleigh}} + \frac{1}{2}\beta E_o [\underbrace{\cos(2\pi(\nu - \nu_{\mathrm{vib}})t)}_{\text{Stokes}} - \underbrace{\cos(2\pi(\nu + \nu_{\mathrm{vib}})t)}_{\text{Anti-Stokes}}]$$

Therefore, for molecules with $\beta \neq 0$, molecular motion leads to sidebands. That is, in order to be Raman active a molecular rotation or vibration must cause a *change* in a component of the molecular polarizability. The electron motion and response is a primary factor in polarizability, as nuclei might not move much in the short times of applied field oscillations.

6.4 Rotational Raman Spectra

For now, consider only pure rotational interactions for Raman scattering. Two pertinent cases are linear and symmetric top molecules.

6.4.1 Linear Molecules

The energy for a linear molecule is

$$F(J) = BJ(J+1) - \underbrace{DJ^2(J+1)^2}_{\text{neglect for Raman}} \tag{6.6}$$

The scattering rotational selection rules for linear rigid rotor molecules are

$$\textbf{Selection Rules} \quad \Delta J = 0,\ +2 \tag{6.7}$$

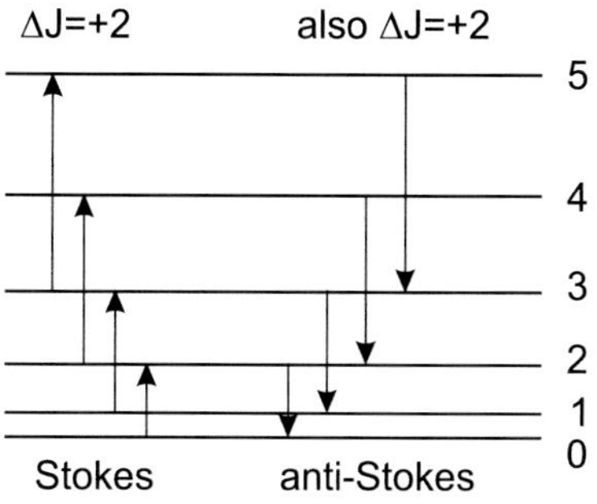

Fig. 6.3 Sample energy levels for rotational Raman transitions

$\Delta J = 0$	corresponds to Rayleigh scattering
$\Delta J = +2$	corresponds for pure rotations to the S branch; pure rotations cannot have $\Delta J = -2$ since, by definition, $\Delta J = J' - J''$

Why do the selection rules allow transitions with $\Delta J = +2$ instead of $+1$? This difference emerges from the symmetry arguments of polarizability (see Banwell [2, p. 105]). The branch associated with increments of 2 in the rotational quantum number J is the S branch (Fig. 6.3). The expression for the transition energy in the S branch as a function of the lower-level value of J (i.e., J'') is given below.

$$\begin{aligned} \Delta E(J) &= E_{J'} - E_{J''} \\ &= B[(J+2)(J+3) - J(J+1)] \\ &= B(4J+6) \\ S(J) &= B(4J+6), \quad J = 0, 1, 2, \ldots \end{aligned} \tag{6.8}$$

Therefore,

$$\boxed{\overline{\nu} = \overline{\nu}_{\text{ex}} \pm B(4J+6)}, \tag{6.9}$$

where J refers to J''. For anti-Stokes Raman transitions, the photon gains energy (the molecule loses energy), and Eq. (6.9) becomes

$$\overline{\nu} = \overline{\nu}_{\text{ex}} + B(4J+6) \tag{6.10}$$

Similarly, the photon loses energy for Stokes Raman transitions, producing

$$\overline{\nu} = \overline{\nu}_{\text{ex}} - B(4J+6) \tag{6.11}$$

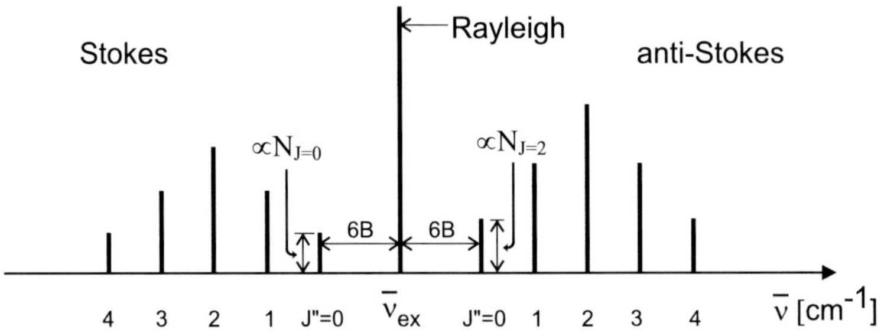

Fig. 6.4 Two branches of Raman spectra. Note that the signal strength is proportional to the population of molecules in the initial state ($N_{J\text{-initial}}$)

Figure 6.4 illustrates the equal transition spacings and unequal transition strengths associated with the Stokes and anti-Stokes branches in rotational Raman spectra. Rayleigh scattering, which occurs at $\overline{\nu} = \overline{\nu}_{ex}$, is shown in the plot as a strong transition halfway between the two branches. Note that IR and μwave *inactive* molecules, such as N_2 and O_2, have Raman spectra.

Nuclear spin effects alter the rotational Raman spectra in a similar fashion to the rotational absorption spectra discussed earlier.

O_2	even J lines are missing
CO_2	odd J lines are missing
N_2	alternating line intensities 2:1 (even J: odd J)

6.4.2 Symmetric Top Molecules

Recall that the rotational energy for a symmetric top molecule is

$$E_{J,K}, \text{cm}^{-1} = F(J,K) = BJ(J+1) + (A-B)K^2 \tag{6.12}$$

The scattering selection rules are:

$$\Delta K = 0$$

$$\Delta J = 0, +1, +2$$

except,

$$\Delta J = +2 \text{ when } K = 0$$

Molecules with $K = 0$ are essentially linear, so the selection rule for a linear molecule applies, namely $\Delta J = +2$. The R branch is for transitions in which $\Delta J = 1$ and the S branch is for transitions with $\Delta J = 2$. The transition spacing as a function of J for the two branches is given by the following equations.

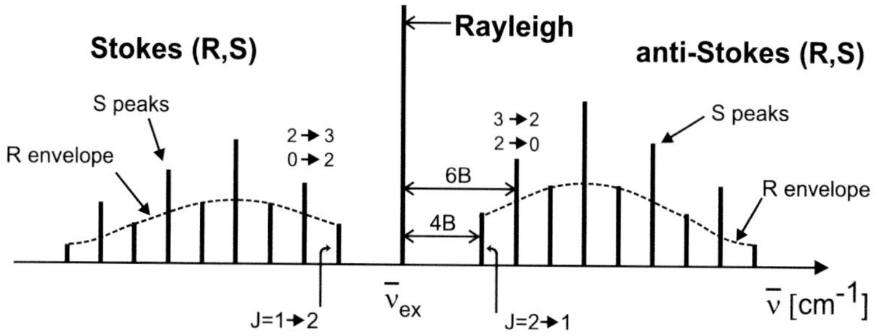

Fig. 6.5 Sample energy levels for symmetric top rotational Raman transitions

R branch: $\Delta J = 1$, spacing $= 2B$

$$\Delta E_R = R(J) = 2B(J+1) \quad J = 1, 2, 3, \ldots \tag{6.13}$$

S branch: $\Delta J = 2$, spacing $= 4B$

$$\Delta E_S = S(J) = B(4J+6) \quad J = 0, 1, 2, \ldots \tag{6.14}$$

Applying these two branches for the Stokes and anti-Stokes conditions gives

$$\overline{\nu}_R = \overline{\nu}_{\text{ex}} \pm 2B(J+1) \tag{6.15}$$

$$\overline{\nu}_S = \overline{\nu}_{\text{ex}} \pm B(4J+6) \tag{6.16}$$

where the $+$ sign applies to the anti-Stokes lines and the $-$ sign is for the Stokes lines.

Figure 6.5 shows the alternating intensities in the Stokes and anti-Stokes regimes due to the additive and overlapping nature of the *R* and *S* branches. These alternating intensities are evidence of the two branches, and thus, symmetric top structure. The spacing of $4B$ instead of $2B$ from the null gap is a clue that the spectra are from Raman transitions, rather than emission or absorption. For spherical tops, all rotational modes are Raman *inactive*. For asymmetric rotors, all rotational modes are Raman *active*.

6.5 Vibrational Raman Spectra

The Raman-activity of a molecule's vibrational modes generally follows a pattern. For asymmetric molecules, all vibrational modes are usually Raman-active. For symmetric molecules:

symmetric stretch vibrations:	very strong
asymmetric stretch vibrations:	usually weak (sometimes unobservable)
bending vibrations:	usually weak

The reader may notice that Raman-activity tends to be strong when infrared-activity is weak or non-existent. Because of this, infrared and Raman spectroscopy are often considered complementary. The relationship between infrared- and Raman-activity can be summarized by the Rule of Mutual Exclusion.

Rule of Mutual Exclusion

> If a molecule has a center of symmetry, then Raman-active modes are IR-inactive, and vice versa. If no center of symmetry exists, then some modes may be both IR- and Raman-active.

For a definition and discussion of the center of symmetry, see Appendix E.

6.5.1 Polarization

Polarization information of Raman-scattered light can help in interpreting spectra and molecular structure.

1. symmetric vibration → partially/fully polarized scattering
2. asymmetric vibration → depolarized scattering

The usefulness of combining IR and Raman techniques is illustrated in the following example.

Example: N_2O
IR and Raman spectra can be used to investigate the symmetry and linearity of N_2O's molecular structure. For example, using Table 6.3 below can we answer the following questions:

Linearity	Linear or nonlinear?
Symmetry	N–N–O or N–O–N?

- some bands with only *P*, *R* branches → hence a linear molecule
- Raman and IR in the same band → hence no center of symmetry, must be N–N–O

Table 6.3 Raman and IR spectra information for N_2O

ν [cm^{-1}]	IR	Raman
589	Strong; *P*, *Q*, *R*	–
1285	VS; *P*, *R*	VS; polarized
2224	VS; *P*, *R*	S; depolarized

- symmetric modes are polarized → hence the 1285 cm^{-1} band is the symmetric stretch (ν_1)
- asymmetric modes are depolarized → hence the 2224 cm^{-1} band is the asymmetric stretch vibrational mode (ν_3)
- the 589 cm^{-1} band has P, Q, and R lines → hence it must be the perpendicular bending mode (ν_2)

6.5.2 Selection Rules

The selection rules for vibrational Raman spectra permit the following vibrational quantum number change for a harmonic oscillator:

$$\Delta v = v' - v'' = +1$$

and for an anharmonic oscillator:

$$\Delta v = v' - v'' = +1, +2, +3, \ldots$$

Note that $\Delta v = 0$ corresponds to pure rotational Raman, and that transitions corresponding to changes greater than 1 in the vibrational quantum number v are generally much weaker than for $\Delta v = 1$.

6.5.3 Diatomics

The selection rules for a diatomic harmonic oscillator are

$$\Delta v = v' - v'' = +1$$
$$\Delta J = J' - J'' = 0, \pm 2$$

The changes in rotational quantum number J are what produce the different branches (Fig. 6.6).

$$\Delta J = \begin{cases} 0 & \text{corresponds to the } Q \text{ branch} \\ +2 & \text{corresponds to the } S \text{ branch} \\ -2 & \text{corresponds to the } O \text{ branch} \end{cases}$$

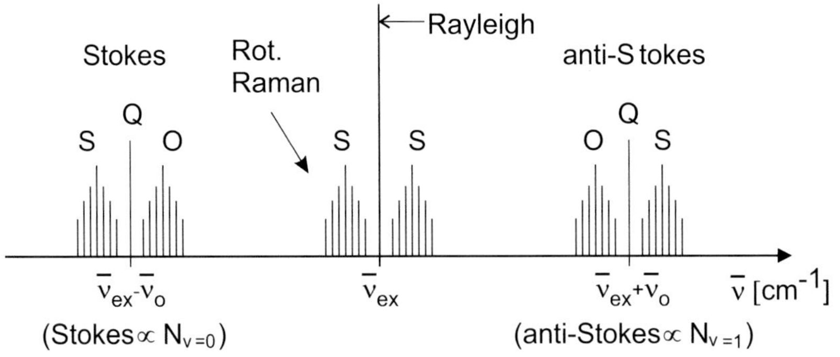

Fig. 6.6 S, Q, and O branches for Stokes and anti-Stokes vibrational and rotational Raman spectra. Note that the signal strength is dependent on the population of molecules in the initial state ($N_{\upsilon\text{-initial}}$)

The general expressions for the transition frequencies in the three branches are

$$\overline{\nu}_Q = \overline{\nu}_{\text{ex}} \pm \overline{\nu}_0 \qquad \overline{\nu}_0 = \omega_e(1 - 2x_e) \tag{6.17}$$

$$\overline{\nu}_S = \overline{\nu}_{\text{ex}} \pm (\overline{\nu}_0 + B(4J + 6)) \qquad J = 0, 1, 2, \ldots \tag{6.18}$$

$$\overline{\nu}_O = \overline{\nu}_{\text{ex}} \pm (\overline{\nu}_0 - B(4J - 2)) \qquad J = 2, 3, 4, \ldots \tag{6.19}$$

where J denotes the rotational state in the *lower* vibrational level. Thus, the anti-Stokes lines are at

$$\overline{\nu}_Q = \overline{\nu}_{\text{ex}} + \overline{\nu}_0 \tag{6.20}$$

$$\overline{\nu}_S = \overline{\nu}_{\text{ex}} + \overline{\nu}_0 + B(4J + 6) \tag{6.21}$$

$$\overline{\nu}_O = \overline{\nu}_{\text{ex}} + \overline{\nu}_0 - B(4J - 2) \tag{6.22}$$

and the Stokes lines are at

$$\overline{\nu}_Q = \overline{\nu}_{\text{ex}} - \overline{\nu}_0 \tag{6.23}$$

$$\overline{\nu}_S = \overline{\nu}_{\text{ex}} - [\overline{\nu}_0 + B(4J + 6)] \tag{6.24}$$

$$\overline{\nu}_O = \overline{\nu}_{\text{ex}} - [\overline{\nu}_0 - B(4J - 2)] \tag{6.25}$$

6.5.4 Temperature

The ratio of anti-Stokes and Stokes signals can be used for temperature measurements. The anti-Stokes signals are proportional to the initial population in the $\upsilon = 1$ state, while the Stokes signals are proportional to the initial population in the $\upsilon = 0$

vibrational state. This ratio can be related to temperature through the Boltzmann relation for vibrational levels.

$$\frac{\text{anti-Stokes}}{\text{Stokes}} \propto \frac{N_{v=1}}{N_{v=0}} \rightarrow T$$

6.5.5 Typical Raman Shift

A typical excitation wavelength is the second harmonic of the output of an Nd:YAG laser,

$$\lambda_{\text{ex}} = 532\,\text{nm} \rightarrow \overline{\nu}_{\text{ex}} = 18{,}797\,\text{cm}^{-1}$$

A representative vibrational frequency of $1880\,\text{cm}^{-1}$ would thus correspond to a Raman shift of 10 % in frequency, shifting the Stokes scattered signal to a longer wavelength, $\lambda = 591$ nm, where the light can be readily separated from that at the excitation wavelength. Also, the efficiency of photo-detection at $\lambda = 591$ nm can be larger than that at $\lambda = 532$ nm.

6.6 Summary of Rayleigh and Raman Scattering

1. Raman is complementary to IR and μwave absorption and emission
2. Techniques are linear (signal is proportional to I_{illum})
3. Observable signals made possible by the availability of intense laser sources (high photons/cm^2 s)
4. Rayleigh $\gg$ Raman and $\sigma_{\text{abs}} \gg \sigma_{\text{Rayleigh}} \gg \sigma_{\text{Raman}}$
5. Scattering is instantaneous; fluorescence takes time
6. Raman spectrum can be observed in the visible (where detector responsivities are generally high)
7. Rayleigh/Raman cross-sections generally scale with $1/\lambda^4$

6.7 Exercises

1. (a) Which type of spectroscopy would one observe the pure rotational spectrum of H_2?
   (b) If the characteristic rotational temperature of H_2, θ_r, is 87.59 K, and it is a rigid rotor, what is the spacing of the lines in the pure rotational spectrum?
   (c) The spin of the hydrogen nucleus is 1/2. How would that affect your answer to part (b)?
2. You are asked to measure the temperature at a point along the centerline of a high-temperature stream of N_2 gas. You elect to infer the temperature from the ratio of the anti-Stokes and Stokes branches of the vibrational Raman spectrum

of N_2. A pulsed laser at 500 nm is used to generate the Raman spectrum. Spectral filters are used to separate the anti-Stokes and Stokes signals so that they can be recorded on separate detectors. The individual features of the anti-Stokes and Stokes are not resolved, just the total signal of each branch is recorded. Assume $\omega_e = 2354\,\text{cm}^{-1}$ and $\omega_e x_e = 0$.

(a) Calculate the temperature of the gas if the ratio of the anti-Stokes and Stokes signals is $1/e$ (i.e., 0.368).

(b) What is the center wavelength of the anti-Stokes signal?

3. (a) A cell contains an unknown gas. Probing of the infrared region has revealed absorption features centered at 712 (VS), 1415 (VW), 2097 (W), and 3311 (S) cm^{-1}, as shown in the sketch below. The first feature can be resolved into three distinct, but unresolved branches, while the others contain only two branches, also unresolved. In addition, Raman scattering from a HeNe laser (632.8 nm) has been measured at 662.7 and 729.8 nm (see sketch). Determine the origin of each of these features and the geometry of the molecule contained in the cell. Try to identify the molecule.

(b) Upon further examination, the separation between the tallest peaks of the two outer branches of the $712\,\text{cm}^{-1}$ IR feature has been found to be $93.6\,\text{cm}^{-1}$ at 1000 K. Use this fact to determine the shift in the purely rotational Raman spectrum for the 18th stokes line and the 14th anti-stokes line.

(c) A second cell containing another gaseous molecule has qualitatively similar IR and Raman spectra, except the IR absorption bands are now centered at 569, 2630, and $1925\,\text{cm}^{-1}$ and the Raman scattering of the HeNe laser is now found to be at 656.4 and 720.6 nm. Suggest the identity of this new molecule.

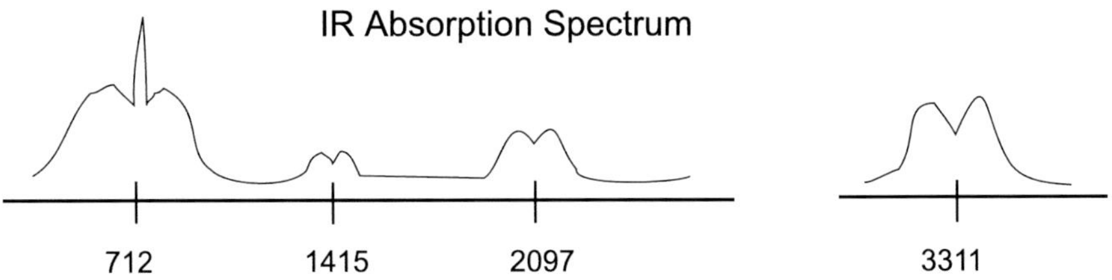

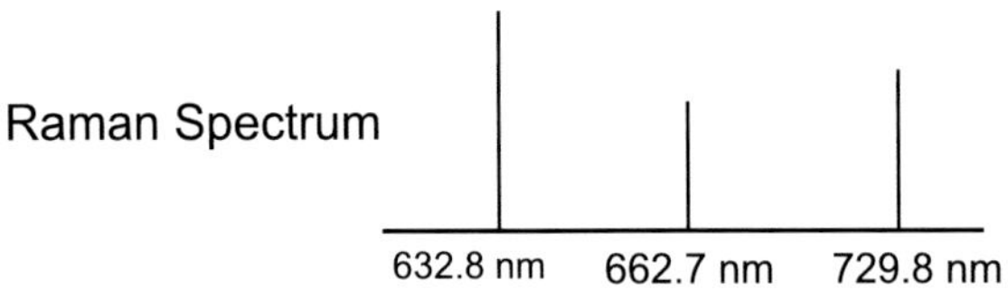

4. (a) A 500 nm laser is used to record the Raman spectrum of CO. Use the following constants for CO: $\omega_e = 2000\,\text{cm}^{-1}$, $\omega_e x_e = 0\,\text{cm}^{-1}$, $B_e = 2\,\text{cm}^{-1}$, and $\alpha_e = 0\,\text{cm}^{-1}$.

(i) What is the frequency of the Rayleigh scattering, in cm^{-1}?

(ii) What is the frequency (in cm^{-1}) of the Q branch of the anti-Stokes ($\Delta v = 1$) Raman spectrum?

(iii) What is the frequency (in cm^{-1}) of the first S branch transition in the anti-Stokes ($\Delta v = 1$) Raman spectrum?
(iv) Calculate the frequency of the $O(2)$ line in the same band.
(v) What is the frequency (in cm^{-1}) of the Q branch of the anti-Stokes ($\Delta v = 2$) Raman scattering?

(b) The above experiment is repeated for O_2 which has a ground state configuration of $X^3\Sigma_g^-$ and the following spectroscopic parameters: $\omega_e = 1500\,cm^{-1}$, $\omega_e x_e = 15\,cm^{-1}$, $B_e = 1.5\,cm^{-1}$, and $\alpha_e = 0\,cm^{-1}$.
(i) What is the frequency (in cm^{-1}) of the first S branch transition in the anti-Stokes ($\Delta v = 1$) Raman spectrum?
(ii) What is the frequency (in cm^{-1}) of the Q branch of the anti-Stokes ($\Delta v = 2$) Raman scattering?

References

1. R.B. Miles, W.R. Lempert, J.N. Forkey, Laser Rayleigh scattering. Meas. Sci. Technol. **12**(5), R33–R51 (2001)
2. C.N. Banwell, E.M. McCash, *Fundamentals of Molecular Spectroscopy*, 4th edn. (McGraw-Hill International (UK) Limited, London, 1994)
3. H. Naus, W. Ubachs, Experimental verification of rayleigh scattering cross sections. Opt. Lett. **25**(5), 347–349 (2000)

7 Quantitative Emission and Absorption

We are ready to begin discussion of quantitative emission and absorption spectroscopy, with the goal of being able to specify emission and absorption as a function of wavelength. Two steps are involved in this treatment. In the first step, a simple form of the equation of radiative transfer will be used to identify a new parameter, known as the spectral absorption coefficient, which will be seen as the governing quantity which characterizes emission and absorption of light, as a function of wavelength. In the second step, the Einstein theory of radiation is employed to show that the spectral absorption coefficient is given simply by the product of the line strength and a lineshape function; the relationship of the line strength to fundamental quantities known as Einstein coefficients is also shown. With these relationships in hand, it will become evident how spectrally resolved absorption (or emission) can be used as a nonintrusive means of measuring a variety of gasdynamic parameters, including: species concentration, pressure, temperature, density, and even flow velocity.

7.1 Spectral Absorption Coefficient

We begin with an overview of possibilities when collimated light at frequency ν enters a gas sample of differential length dx: there are four separate possibilities, with probabilities summing to 1 (100 %).

$$1 = \underbrace{\text{absorption}}_{\alpha_\nu} + \underbrace{\text{reflection}}_{=0} + \underbrace{\text{scattering}}_{=0} + \underbrace{\text{transmission}}_{T_\nu} \tag{7.1}$$

Therefore, in the common case where reflection and scattering are negligible,

$$\alpha_\nu + T_\nu = 1 \tag{7.2}$$

R.K. Hanson et al., *Spectroscopy and Optical Diagnostics for Gases*,
DOI 10.1007/978-3-319-23252-2_7

where α_ν and T_ν are known as the spectral absorptivity and transmissivity, respectively. The result in Eq. (7.2) follows from conservation of energy.

We now express α_ν, the fraction of incident light I_ν for the frequency range $\nu \to \nu + d\nu$ that is absorbed, in terms of an absorption coefficient per unit length, k_ν, i.e.

$$\alpha_\nu \equiv k_\nu dx = \frac{-dI_\nu}{I_\nu} \text{ [no units]} \tag{7.3}$$

k_ν is the *spectral absorption coefficient* (the fraction of incident light I_ν over frequency range $\nu \to \nu + d\nu$ that is absorbed per unit length dx). Thus

$$\boxed{k_\nu \equiv -\frac{(dI_\nu/dx)}{I_\nu} \text{ [cm}^{-1}\text{]}} \tag{7.4}$$

where I_ν may have units of power per unit area per unit spectral interval at frequency ν (i.e., power per unit area over the spectral range $\nu \to \nu + d\nu$), or can be substituted with $I(\nu)$, which denotes power at frequency ν or power per unit area at frequency ν. In the former case, which is most common for our purposes, I_ν is known as the spectral intensity and has units

$$\left(\frac{\text{W/cm}^2}{\text{cm}^{-1}}\right) \text{ or } \left(\frac{\text{W/cm}^2}{\text{Hz}}\right).$$

The spectral intensity I_ν can be integrated over frequency to obtain the total radiant intensity, I:

$$I \text{ [W/cm}^2\text{]} = \int_\nu I_\nu d\nu \tag{7.5}$$

In general, the equations that relate spectral intensity, spectral radiancy, and total radiancy to other parameters can use I_ν or $I(\nu)$ interchangeably. The exceptions are integral relations such as Eqs. (7.5) and (7.13); they require the differential form of spectral intensity, I_ν. Thus, Eqs. (7.3) and (7.4) also define α_ν, the fraction of incident light absorbed, and k_ν, the fraction of incident light absorbed *per unit length*, each *at* frequency ν.

Hence, α_ν and k_ν are the spectral absorptivity and spectral absorption coefficient *at* frequency ν **or** *over the frequency range* $\nu \to \nu + d\nu$.

7.2 Equation of Radiative Transfer: Classical Approach

We wish to perform a simple one-dimensional radiation energy balance on a thin slab of gas. To do this, we must first introduce the spectral emissivity, which is the way that we account for the "emission" from the gas slab. (If a gas sample can absorb light, it follows that it must be allowed to emit, in order to satisfy detailed balance arguments for equilibrium.) The spectral emissivity is conventionally defined as the radiation emitted by the gas sample (I_ν^{em}) relative to that of a blackbody (an "equilibrium" radiator which sets the upper bound on the emission for a specified temperature):

$$\varepsilon_\nu = \frac{I_\nu^{\text{em}}}{I_\nu^{\text{bb}}} \text{ [no units]} \tag{7.6}$$

$$= \frac{I^{\text{em}}(\nu)}{I^{\text{bb}}(\nu)} \text{ [no units]} \tag{7.7}$$

where I_ν^{bb} is the blackbody spectral radiancy. At this point, we employ Kirchhoff's law, which states that "emissivity equals absorptivity," so that

$$\varepsilon_\nu = \alpha_\nu. \tag{7.8}$$

(This law also follows from equilibrium arguments.) Now consider the radiation energy balance at frequency ν for a gas slab of thickness dx; for simplicity, we consider only collimated light (Fig. 7.1).

$$\text{emission} = \varepsilon_\nu I_\nu^{\text{bb}}$$

$$\text{absorption} = \alpha_\nu I_\nu$$

$$dI_\nu = \text{emission} - \text{absorption}$$

$$= \varepsilon_\nu I_\nu^{\text{bb}} - \alpha_\nu I_\nu \tag{7.9}$$

$$= \alpha_\nu (I_\nu^{\text{bb}} - I_\nu) \tag{7.10}$$

Therefore,

$$\boxed{dI_\nu = k_\nu dx (I_\nu^{\text{bb}} - I_\nu).} \tag{7.11}$$

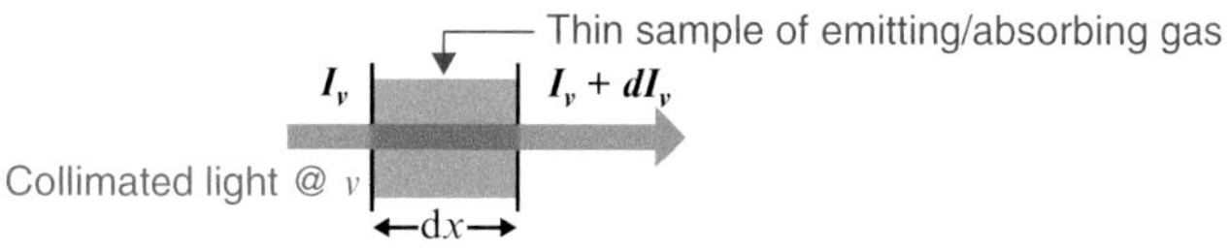

Fig. 7.1 Radiation energy balance across a slab of gas

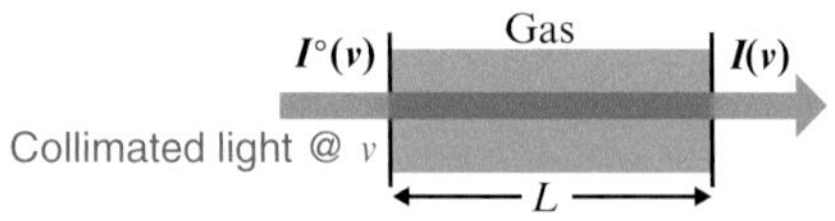

Fig. 7.2 Radiation energy across a slab of gas of width L

This is the *differential form* of the equation of radiative transfer. Integrating over a distance L, for a uniform sample (i.e., constant k_ν, see Fig. 7.2), with an incident intensity I_ν^0 at $x = 0$ (the boundary condition):

$$\boxed{I_\nu(L) = I_\nu^0 \exp(-k_\nu L) + I_\nu^{\mathrm{bb}}[1 - \exp(-k_\nu L)]} \tag{7.12}$$

This important result is the *integrated form* of the equation of radiative transfer. The quantity $k_\nu L$ is known as the "optical depth" (and also as the absorbance at frequency ν). Note that the gas sample is not required to radiate as a blackbody, but we can relate the actual emission to the blackbody level.

Consider the two important cases: Emission and Absorption.

7.2.1 Case 1: Emission Experiments ($I_\nu^0 = 0$)

For emission experiments, the incident radiation intensity $I_\nu^0 = 0$ (i.e., there is no radiation source such as a laser involved).

Spectral radiancy: $I_\nu(L) = I_\nu^{\mathrm{bb}}[1 - \exp(-k_\nu L)]$

Spectral emissivity: $\varepsilon_\nu(k_\nu, L) = \frac{I_\nu(L)}{I_\nu^{\mathrm{bb}}} = 1 - \exp(-k_\nu L)$

We can integrate these relations over frequency to obtain results for the total radiancy:

$$I(L) = \int_0^\infty I_\nu(L)d\nu = \int_0^\infty I_\nu^{\mathrm{bb}}[1 - \exp(-k_\nu L)]d\nu \tag{7.13}$$

$$\varepsilon(L) = \frac{I(L)}{I^{\mathrm{bb}}} = \frac{1}{\sigma T^4}\int_0^\infty I_\nu^{\mathrm{bb}}[1 - \exp(-k_\nu L)]d\nu \tag{7.14}$$

Note:

$$I^{\mathrm{bb}} = \int_0^\infty I_\nu^{\mathrm{bb}} d\nu = \sigma T^4,$$

where

$$\sigma = 5.67 \times 10^{-5}\ [\mathrm{erg\,cm^{-2}\,s^{-1}\,K^{-4}}]$$

is the Stefan–Boltzmann constant.

Emission Types

The emission for the formulas above may be of any type, including single line, multiple line, single or multiple bands, or continuum.

Optical Depth

The optical depth, $k_\nu L$, is a key parameter. When $k_\nu L \gg 1$, the system is *optically thick* and the spectral radiance approaches that of a blackbody. When $k_\nu L \ll 1$, the system is *optically thin* and the spectral radiance approaches $(k_\nu L) I_\nu^{\mathrm{bb}}$.

$$\begin{aligned} &\text{optically thick: } k_\nu L \gg 1, \quad I_\nu(L) \to I_\nu^{\mathrm{bb}} \\ &\text{optically thin: } k_\nu L \ll 1, \quad I_\nu(L) \to (k_\nu L) I_\nu^{\mathrm{bb}} \end{aligned}$$

7.2.2 Case 2: Absorption Experiments ($I_\nu^0 \gg I_\nu^{\mathrm{bb}}$)

For absorption experiments, the incident radiant intensity is much greater than the blackbody radiation intensity.

$$\text{Absorption:} \quad I_\nu^0 \gg I_\nu^{\mathrm{bb}}$$

The equation of radiative transfer, Eq. (7.12), then becomes

$$\boxed{I_\nu(L) = I_\nu^0 \exp(-k_\nu L)} \tag{7.15}$$

This relation, known as Beer's Law or the Beer–Lambert Law, may be *the most important relation in absorption spectroscopy*. Alternate forms in terms of the fractional transmission or "transmissivity," T_ν, are

$$T_\nu = \left(\frac{I}{I^0}\right)_\nu \tag{7.16}$$

$$= \exp(-k_\nu L) \tag{7.17}$$

$$= \frac{I(\nu)}{I^0(\nu)} \tag{7.18}$$

We make two observations:

1. The same equation would apply to the transmission of a pulse of laser excitation, with energy E_ν [$\mathrm{J/cm^2/cm^{-1}}$], i.e. $T_\nu = E_\nu/E_\nu^0$.
2. The fundamental parameter controlling absorption over length L is the spectral absorption coefficient, k_ν.

Our next step is to establish a relationship between k_ν and the fundamental molecular parameters that govern the "strengths" and "shapes" of absorption lines, namely the *Einstein coefficients* and *line-broadening coefficients*.

7.3 Einstein Theory of Radiation

We begin with a simplified theory, without regard to lineshape or structure (sometimes termed the Milne Theory). Consider two states of an atom (or molecule) which are radiatively coupled (i.e., have radiative transitions that are "allowed"), with $E_2 - E_1 = h\nu$.

The total transition rates [molec/s] are N_2A_{21}, $N_1B_{12}\rho(\nu)$, and $N_2B_{21}\rho(\nu)$, where N_i is the total number of molecules in state i. Alternatively the transition rate per unit volume [molec/cm^3/s] can be expressed using the number density n_i.

7.3.1 Einstein Coefficients

A_{21}, B_{12}, and B_{21} in Fig. 7.3 are the Einstein coefficients of radiation.

$B_{12}\rho(\nu)$ — the probability/s that a molecule in state 1 exposed to radiation of spectral density $\rho(\nu)$ [J/(cm^3 Hz)][1] will absorb a quantum $h\nu$ and pass to state 2. The Einstein B-coefficient thus carries units of cm^3 Hz/(J s).

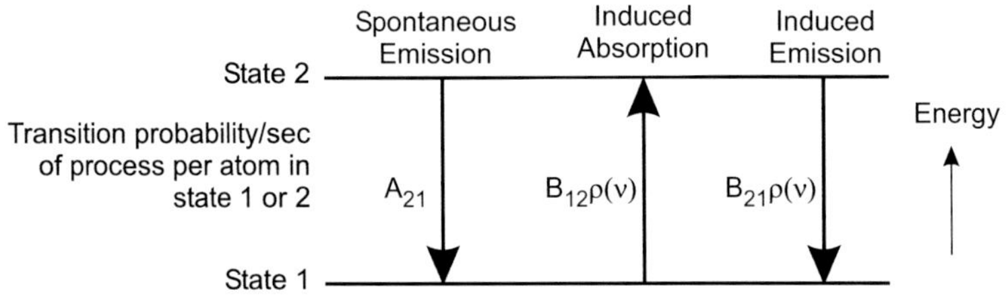

Fig. 7.3 Transition probabilities between states 1 and 2

[1]The spectral density is the energy density per unit frequency contained in an electric field.

$B_{21}\rho(\nu)$ the probability/s that a molecule in state 2 exposed to radiation of spectral density $\rho(\nu)$ will emit a quantum $h\nu$ and pass to state 1.[2]

A_{21} the probability/s of spontaneous transfer from state 2 to 1 with release of photon of energy $h\nu$ (without regard to the presence of $\rho(\nu)$)

Similar expressions apply when isotropic radiation intensity or parallel beam intensity (e.g., a laser) are involved.

7.3.2 Equilibrium

At equilibrium, the net rate of change of molecules in any molecular state is zero. Consider a detailed balance of the optical processes at equilibrium for the population change in state 2:

$$\left(\dot{N}_2\right)_{\text{rad}} = \underbrace{N_1 B_{12}\rho(\nu)}_{\text{molec/s entering state 2}} \underbrace{- N_2(A_{21} + B_{21}\rho(\nu))}_{\text{molec/s leaving state 2}} = 0 \tag{7.19}$$

The molecule balance in Eq. (7.19) is simply the difference between the rate of molecules entering state 2 and the rate of molecules leaving state 2. Equilibrium requires that all transitions from state 1 to 2 (induced absorption) are balanced by transitions from state 2 to 1 (induced and spontaneous emission). Another way to express the equilibrium condition is with the Boltzmann fraction from statistical mechanics:

$$\frac{N_2}{N_1} = \underbrace{\frac{B_{12}\rho(\nu)_{\text{eq}}}{A_{21} + B_{21}\rho(\nu)_{\text{eq}}}}_{\text{rad. equil.}} = \underbrace{\frac{g_2}{g_1}\exp(-h\nu/kT)}_{\text{statistical equil.}} \tag{7.20}$$

Solving for $\rho(\nu)_{\text{eq}}$ in Eq. (7.20) and equating it to the known result for $\rho(\nu)_{\text{eq}}$, i.e. Planck's blackbody distribution,

$$\rho(\nu)_{\text{eq}} = \frac{(8\pi h\nu^3/c^3)}{\exp(+h\nu/kT) - 1} \tag{7.21}$$

gives

$$\rho(\nu)_{\text{eq}} = \frac{(A_{21}/B_{21})}{\frac{g_1}{g_2}\frac{B_{12}}{B_{21}}\exp(h\nu/kT) - 1} = \underbrace{\frac{(8\pi h\nu^3/c^3)}{\exp(h\nu/kT) - 1}}_{\text{Planck's distribution}} \tag{7.22}$$

[2]This induced emission occurs in phase with and in the same direction as the incident beam. Hence, for collimated incident light (e.g., a collimated laser beam) the induced emission appears as gain in the exciting beam.

Equation (7.22) must hold for all ν and T, producing the two following important conclusions [1]:

$$g_1 B_{12} = g_2 B_{21} \tag{7.23}$$

$$A_{21} = \left(\frac{8\pi h\nu^3}{c^3}\right) B_{21} \tag{7.24}$$

$$\equiv 1/\tau_{21} \tag{7.25}$$

where τ_{21}, the inverse of A_{21}, is the "radiative lifetime" in state 2. We must note here that even though Eq. (7.24) was derived from thermodynamic equilibrium arguments and Planck's blackbody distribution, the relationship between A and B holds even for systems not in equilibrium, and it does not depend on $\rho(\nu)$. A_{21}, and hence B_{12} and B_{21} are theoretically calculable from quantum mechanics, but in practice, τ_{21} and/or B_{12} are often measured.

Note: For collimated light (as in the case for most absorption experiments):
$\rho(\nu)_{eq} = n_p \cdot h\nu$ [J/cm^3 s^{-1}] (n_p is the number of photons/cm^3 s^{-1})
$I_\nu = n_p \cdot h\nu \cdot c$ [W/cm^2 s^{-1}] (power per unit area per unit frequency)
Therefore,

$$\rho(\nu) = I_\nu / c \tag{7.26}$$

Where is the link to k_ν ? Find this next.

7.3.3 What is k_ν?

We proceed now to find the relationship between the spectral absorption coefficient, k_ν, and the Einstein coefficients, for the case of a structureless absorption line of width $\delta\nu$. Recall that Beer's Law is

$$T_\nu = \left(\frac{I}{I^0}\right)_\nu = \exp(-k_\nu L) \tag{7.27}$$

where I_ν may be either the spectral intensity [W/cm^2 s^{-1}] or intensity [W/cm^2] or power [W] at frequency ν, and that

$$k_\nu \equiv -\frac{dI_\nu}{I_\nu dx} \tag{7.28}$$

Figure 7.4 plots T_ν and k_ν versus frequency for this case. Note that $T_\nu = 1$ everywhere but in the region of the absorption line.

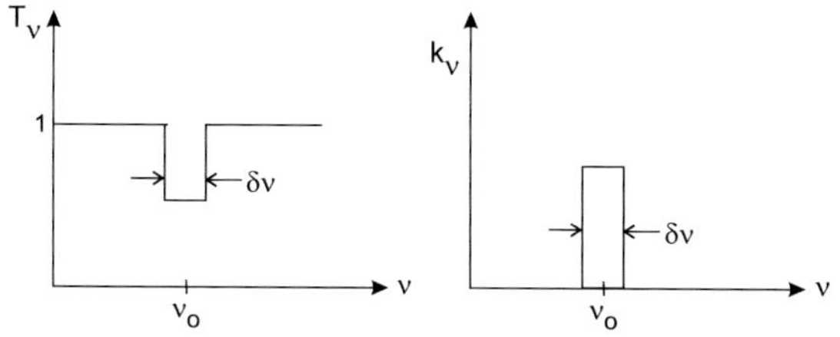

Fig. 7.4 T_ν and k_ν versus frequency for a structureless absorption line of width $\delta\nu$

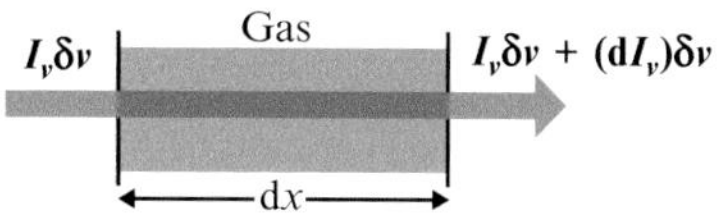

Fig. 7.5 Transmission of laser intensity across a gas slab of depth dx

Imagine that a spectrally broad light source of uniform spectral intensity I_ν^0 is used to illuminate a sample gas with the spectral absorption coefficient shown in Fig. 7.4; the length of the sample is L. What is the absorbed power in W/cm^2?

$$P_{\text{abs}} = (\text{incident power over } \delta\nu) \times (\text{fraction absorbed}) \quad [\text{W/cm}^2]$$

$$= (I_\nu^0\, \delta\nu) \times (1 - T_\nu) \tag{7.29}$$

$$= (I_\nu^0\, \delta\nu)(1 - \exp(-k_\nu L)) \tag{7.30}$$

where I_ν^0 has units of $\text{W/cm}^2\,\text{s}^{-1}$ and $\delta\nu$ has units of s^{-1}. The product $I_\nu^0\delta\nu$ represents the incident power per unit area (W/cm^2) contained in the spectral interval $\delta\nu$ over which absorption may occur.

What happens to Eq. (7.30) for a small (incremental) width dx, such that $k_\nu dx \ll 1$? For small optical thickness, the exponential term can be linearized, leading to the simple result

$$P_{\text{abs}} = (I_\nu^0\, \delta\nu)(k_\nu dx) \tag{7.31}$$

or

$$\frac{P_{\text{abs}}}{I_\nu^0\, \delta\nu} = \text{fraction absorbed} = k_\nu dx. \tag{7.32}$$

This is known as the optically thin limit.

We can use a simple energy balance on an incremental slab of gas to find k_ν (Fig. 7.5).

In this model, we recognize that the *change* in intensity, i.e. $(dI_\nu)\delta\nu$, is equal to the net combined effects of emission and absorption,

$$(dI_\nu)\delta\nu = (\text{induced emission} + \underbrace{\text{spontaneous emission}}_{=0}) - \text{induced absorption} \tag{7.33}$$

where the spontaneous emission is approximately zero for collimated light and

$$\text{induced emission} = \underbrace{(n_2 dx)}_{\substack{\text{molec/cm}^2\\ \text{in state 2}}} \times \underbrace{B_{21}\rho(\nu)}_{\substack{\text{prob/s of}\\ \text{emission}}} \times \underbrace{h\nu}_{\substack{\text{energy per}\\ \text{photon}}} \tag{7.34}$$

$$\text{induced absorption} = \underbrace{(n_1 dx)}_{\substack{\text{molec/cm}^2\\ \text{in state 1}}} \times \underbrace{B_{12}\rho(\nu)}_{\substack{\text{prob/s of}\\ \text{absorption}}} \times \underbrace{h\nu}_{\substack{\text{energy per}\\ \text{photon}}} \tag{7.35}$$

Recalling Eq. (7.26), Eq. (7.33) becomes

$$(dI_\nu)\delta\nu = [n_2 B_{21} - n_1 B_{12}]\frac{h\nu}{c} I_\nu dx \tag{7.36}$$

Therefore,

$$\frac{dI_\nu}{I_\nu dx} \equiv -k_\nu = \frac{h\nu}{c}\frac{1}{\delta\nu}[n_2 B_{21} - n_1 B_{12}] \tag{7.37}$$

which may be simplified further to give

$$\boxed{k_\nu\ [\text{cm}^{-1}] = \frac{h\nu}{c}\frac{1}{\delta\nu} n_1 B_{12}\,(1 - \exp(-h\nu/kT))} \tag{7.38}$$

While this result is immediately helpful in understanding the fundamental coupling between k_ν and B_{12}, n_1, ν, and T, we can see now that the shape and width of absorption lines (evident in Eq. (7.38) with the term $\delta\nu$) are also relevant. How would the use of a more realistic lineshape model affect Eq. (7.38)?

7.4 Revised Treatment of Einstein Theory (with Lineshape)

We now repeat the derivation for k_ν using an improved lineshape model that includes the structure of absorption and emission lines. Compared with the uniform and structureless feature in Fig. 7.4, real spectra have shapes. These realistic shapes for the spectral transmission, T_ν, spectral absorption, k_ν, and normalized lineshape function, ϕ, of a single absorption line are shown in Fig. 7.6.

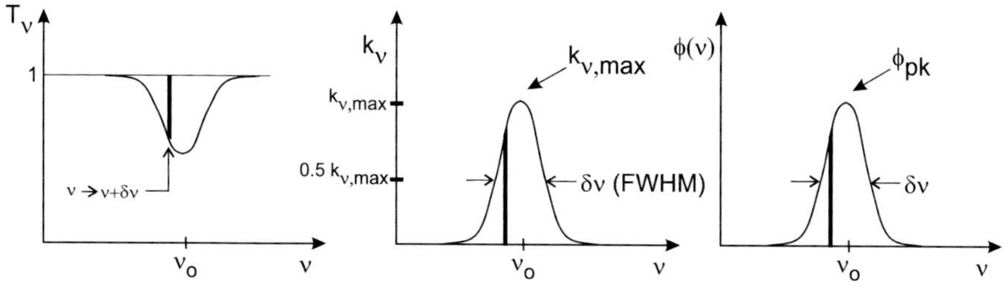

Fig. 7.6 T_ν, k_ν, and ϕ versus frequency for an absorption line with typical structure

Recall that the general form for Beer's Law is

$$T_\nu = \left(\frac{I}{I^0}\right)_\nu = \exp(-k_\nu L) \tag{7.39}$$

Solving for k_ν yields

$$k_\nu = -\frac{1}{L}\ln T_\nu \tag{7.40}$$

In addition, we define a new parameter, ϕ, as the *normalized* lineshape function

$$\phi \equiv \frac{k_\nu}{\int\limits_{\text{line}} k_\nu d\nu} \quad [\text{cm}] \text{ or } [\text{s}] \tag{7.41}$$

so that

$$\int\limits_{\text{line}} \phi d\nu = 1 \tag{7.42}$$

The units for ϕ are inverse frequency (and hence may have units of either centimeters or seconds). The lineshape function, whose integral over the line is 1, provides a useful way of characterizing the shape of a line.

Note: Since

$$\int k_\nu d\nu \approx k_{\nu,\max}\delta\nu$$

where $\delta\nu$ is an average width, then

$$\phi_{pk} = \frac{k_{\nu,\max}}{\int k_\nu d\nu} \approx \frac{1}{\delta\nu}$$

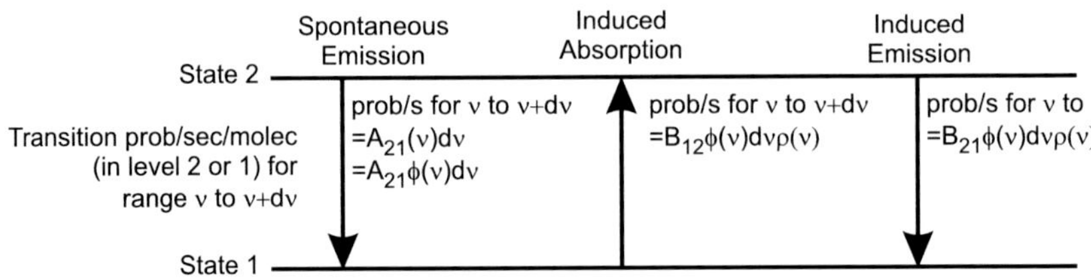

Fig. 7.7 Transition probabilities per second per molecule in level 2 or 1

Thus we should not be surprised if we find that the term $1/\delta\nu$ in Eq. (7.38) is simply replaced by ϕ in the revised formulation, with ϕ varying smoothly from zero in the "wings" of an absorption line to a peak value proportional to $1/\delta\nu$ (see Fig. 7.6 above).

Consider the small frequency interval, $\nu \rightarrow \nu + d\nu$, in Fig. 7.6. What are the relevant transition probabilities for this spectral interval? We recognize that these probabilities should have the same spectral dependence (shape) as k_ν and $\phi(\nu)$, i.e. having a peak value at line center and falling to zero away from the line. We achieve this shape with a simple modification to our previous model (with the constant probabilities A_{21}, B_{12}, B_{21}) by multiplying the Einstein coefficients by $\phi(\nu)d\nu$ (see Fig. 7.7).

$A_{21}\phi(\nu)d\nu$	the probability/s of a molecule undergoing spontaneous emission, in the range $\nu \rightarrow \nu + d\nu$ [Note that the integral of this quantity over the range of allowed ν is just A_{21} [s^{-1}], i.e. $\int A_{21}\phi(\nu)(d\nu) = A_{21}$.]
$B_{12}\phi(\nu)d\nu\rho(\nu)$	the probability/s of a molecule undergoing a transition from $1 \rightarrow 2$, in the range $\nu \rightarrow \nu + d\nu$
$B_{21}\phi(\nu)d\nu\rho(\nu)$	the probability/s of a molecule undergoing a transition from $2 \rightarrow 1$, in the range $\nu \rightarrow \nu + d\nu$

Recall: $\rho(\nu) = I_\nu/c$, where $\rho(\nu)$ is the spectral energy density [J/cm^3 s^{-1}] and I_ν is the spectral intensity [W/cm^2 s^{-1}] of collimated laser light.

We are now ready to do a simple energy/power balance on an incremental gas slab of width dx, for the frequency range $\nu \rightarrow \nu + d\nu$ (Fig. 7.8).

The energy balance for the slab requires that the incremental gain in intensity, $dI_\nu d\nu$, is equal to the difference between the emission and absorption over the frequency interval $d\nu$ in the gas slab:

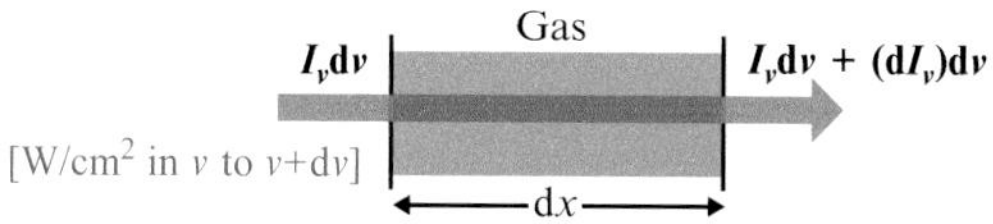

Fig. 7.8 Energy/power balance on an incremental gas slab

$$dI_\nu d\nu = \text{emission in } d\nu - \text{absorption in } d\nu$$

$$= \overbrace{\underbrace{n_2}_{}}^{\#/\text{cc}} \underbrace{\;dx}_{\text{molec/cm}^2} \times \overbrace{[B_{21}\phi(\nu)d\nu I_\nu/c]}^{\text{prob/s}\cdot\text{molec for } d\nu} \times \underbrace{h\,\nu_0}_{\text{energy/photon}} - n_1 dx[B_{12}\phi(\nu)d\nu I_\nu/c]h\nu_0$$

Therefore,

$$-\frac{dI_\nu}{I_\nu dx} \equiv k_\nu = \frac{h\nu}{c}[n_1 B_{12} - n_2 B_{21}]\phi(\nu) \tag{7.43}$$

so

$$\boxed{k_\nu = \frac{h\nu}{c} n_1 B_{12}\,(1 - \exp(-h\nu/kT))\,\phi(\nu)} \tag{7.44}$$

which is, as suggested, equal to our earlier result for k_ν aside from the substitution of $\phi(\nu)$ for $1/\delta\nu$! Integrating k_ν over the absorption line yields

$$S_{12} \equiv \int_{\text{line}} k_\nu d\nu \quad [\text{cm}^{-1}\,\text{s}^{-1}] \tag{7.45}$$

$$\boxed{S_{12} = \frac{h\nu}{c} n_1 B_{12}(1 - \exp(-h\nu/kT))} \tag{7.46}$$

S_{12} is an important quantity known as the "integrated absorption" for the absorption transition $1 \rightarrow 2$[1]. It is also often called the "line strength." Note that the quantity does not depend on lineshape and is simply a function of n_1, T, and B_{12}. Changes in lineshape, e.g. owing to pressure-broadening, thus do not affect S_{12}. This quantity, like Einstein coefficients, is thus fundamental in nature. In view of the interrelationships of A_{21}, B_{12}, and B_{21}, we may also write

$$S_{12} = \frac{\lambda^2}{8\pi} n_1 A_{21} \frac{g_2}{g_1} (1 - \exp(-h\nu/kT)) \quad [\mathrm{cm^{-1}\,s^{-1}}] \tag{7.47}$$

$$S_{12} = \left(\frac{\pi e^2}{m_e c}\right) n_1 f_{12} (1 - \exp(-h\nu/kT)) \quad [\mathrm{cm^{-1}\,s^{-1}}] \tag{7.48}$$

(Note that the units of S_{12} shown here as $\mathrm{cm^{-1}\,s^{-1}}$ depend on the choice of units for frequency, typically either $\mathrm{s^{-1}}$ or $\mathrm{cm^{-1}}$, so that S_{12} may have units of $\mathrm{cm^{-1}\,s^{-1}}$ or $\mathrm{cm^{-2}}$. Other variations also appear in the literature.)

Equation (7.47) makes use of the relationship between A_{21} and B_{12}. Equation (7.48) utilizes the *oscillator strength*, f:

$$f_{12} = \frac{S_{12,\mathrm{actual}}}{S_{12,\mathrm{classical}}(1 - \exp(-h\nu/kT))} \tag{7.49}$$

The oscillator strength of a transition (or group of transitions like a vibrational band or electronic system) compares the integrated strength of the transition with the classical electromagnetic model of an oscillating electron in a magnetic field. $S_{\mathrm{classical}}$ is given by[2]:

$$S_{\mathrm{classical}} = \left(\frac{\pi e^2}{m_e c}\right) n_1 \tag{7.50}$$

where

$$\left(\frac{\pi e^2}{m_e c}\right) = 0.0265\,\mathrm{cm^2\,Hz}.$$

so that

$$S_{12,\mathrm{actual}} = (0.0265\,\mathrm{cm^2\,Hz}) n_1 f_{12} (1 - \exp(-h\nu/kT)) \tag{7.51}$$

Since n_1 can be written in terms of pressure, i.e.,

$$n_1 = \frac{p_1}{kT} \tag{7.52}$$

then at STP, if all the absorbing atoms are in the ground state ($n_1 = n = 2.7 \times 10^{19}\,\mathrm{cm^{-3}}$) and $\exp(-h\nu_{12}/kT) \ll 1$, the line strength S_{12} is simply $7.17 \times 10^{17} f_{12}$ [$\mathrm{cm^{-1}\,Hz/atm}$], or equivalently

$$S_{12}[\mathrm{cm^{-2}/atm}] = 2.380 \times 10^7 f_{12} \tag{7.53}$$

From Eqs. (7.47) and (7.48), the reader may note that the oscillator strength is also directly related to the Einstein coefficients for a given transition, and like the coefficients for stimulated emission,

$$f_{21} = \left(\frac{g_1}{g_2}\right) f_{12} \tag{7.54}$$

We now make two important observations:

1. From the original definition of k_ν and S_{12} we have

$$\boxed{k_\nu = S_{12}\phi(\nu)}$$

2. When

$$h\nu/kT \gg 1,$$

as is common for electronic state transitions,

$$S_{12} \quad [\text{cm}^{-1}\,\text{Hz}] = \left(\frac{\pi e^2}{m_e c}\right) n_1 f_{12} \tag{7.55}$$

$$= (0.0265\,\text{cm}^2\,\text{Hz}) n_1 f_{12} \tag{7.56}$$

$$= \frac{\lambda^2}{8\pi} n_1 A_{21} \frac{g_2}{g_1} \tag{7.57}$$

and, by comparison of right-hand sides

$$f_{12}/A_{21} = f_{12}\tau_{21} = 1.51 \frac{g_2}{g_1} (\lambda\ [\text{cm}])^2 \tag{7.58}$$

where $\tau_{21} = 1/A_{21}$ is the *radiative lifetime* of the $2 \rightarrow 1$ transition.

Example: "Resonance Transition"
A resonance transition is one that couples the ground state to the first excited state. Let's look at a case for an electronic transition of a sodium atom:

$$\text{Na}\ (\overbrace{3^2S_{1/2}}^{\text{lower(L)}} - \overbrace{3^2P_{1/2}}^{\text{upper(U)}}),$$

$$\frac{g_2}{g_1} = 1,$$

$$\lambda = 589\,\text{nm} = 5.89 \times 10^{-5}\,\text{cm} \tag{7.59}$$

Table 7.1 Oscillator strengths of selected sodium transitions, abstracted from [1]

Transition	f_{21}	λ[nm]
$3^2S_{1/2} - 3^2P_{1/2}$	0.33	589.6
$3^2S_{1/2} - 3^2P_{3/2}$	0.67	589.0
$3^2S - 4^2P$	0.04	330.2

Conventions:

atoms: (L – U)

molecules: (U $\leftrightarrow$ L), arrow denotes absorption or emission

f_{ij}: i denotes initial state, j denotes final

From Eq. (7.58),

$$f_{12}\tau_{589\,\text{nm}} = 5.24 \times 10^{-9}\,\text{s},$$

where the radiative lifetime is

$$\tau = 16.1 \times 10^{-9}\,\text{s (measured, corresponds to } A = 0.62 \times 10^8\,\text{s}^{-1})$$

Thus, $f \approx 0.325$ (strong atomic transition; single electron). Values of f for molecular transitions tend to be much smaller, $\sim 10^{-2}$–10^{-4}, owing to the relatively strong coupling between the multiple electrons and protons in a molecule and can be very much smaller for highly forbidden transitions. The "term symbols" for electronic states in atoms will be discussed in Chap. 9.

Table 7.1 lists the emission (f_{21}) oscillator strengths and spectral locations for a few transitions of the sodium atom.

These values can be compared with electronic and vibrational oscillator strengths in a few molecules (Table 7.2).

7.5 Radiative Lifetime

The concept of radiative lifetime merits further discussion. If we write a rate equation for radiative decay from an upper level u, accounting for all allowed spontaneous decay paths to lower states l, then

Table 7.2 Absorption oscillator strengths of selected vibrational and vibronic bands of a few molecules

Molecule	$v' \leftarrow v''$	Electronic transition	Band center [cm^{-1}]	f_{12}
CO	$1 \leftarrow 0$	–	2143	1.09×10^{-5}
	$2 \leftarrow 0$	–	4260	7.5×10^{-8}
OH	$1 \leftarrow 0$	–	3568	4.0×10^{-6}
	$0 \leftarrow 0$	$^2\Sigma \leftarrow {}^2\Pi$	32,600	1.2×10^{-3}
CN	$0 \leftarrow 0$	$^2\Pi \leftarrow {}^2\Sigma$	9117	2.0×10^{-2}

$$\frac{dn_u}{dt} = -n_u \sum_l A_{u \to l} \quad \text{(radiation only)} \tag{7.60}$$

Thus, for an initial number density $n_u(0)$, the time-dependent density, allowing for radiative decay only, is:

$$n_u(t) = n_u(0) \exp\left[-t \sum_l A_{u \to l}\right] \tag{7.61}$$

so

$$\boxed{\tau_r = \text{radiative lifetime} = \frac{1}{\sum_l A_{u \to l}}} \tag{7.62}$$

Of course, collisions and radiative excitation into the upper levels will also be present and will maintain a non-zero population in the upper level. τ_r is also sometimes described as the *zero-pressure lifetime*.

The decay $u \rightarrow l$ may also occur through non-radiative (i.e., collisional) processes; corresponding energy may be given to neighboring molecules as translational energy or, quite commonly, as internal energy. In the case of non-radiative decay, the rate of decays per unit volume can be written in terms of a rate parameter $k_{\mathrm{nr}}(s^{-1})$

$$\left(\frac{dn_u}{dt}\right)_{\mathrm{nr}} = -k_{\mathrm{nr}} n_u = -\frac{n_u}{\tau_{\mathrm{nr}}} \tag{7.63}$$

where τ_{nr} is the *non-radiative decay time*. This parameter depends on the transition considered and on the surrounding molecules.

With the simultaneous presence of both radiative and non-radiative transitions, the time variation of the upper level u population can be written as

$$\frac{dn_u}{dt} = -\frac{n_u}{\tau_r} - \frac{n_u}{\tau_{\mathrm{nr}}} = -\frac{n_u}{\tau} \tag{7.64}$$

where $\tau^{-1} = \tau_r^{-1} + \tau_{\mathrm{nr}}^{-1}$ is called the *lifetime* of level u.

7.6 Alternate Forms

There are many alternate forms for the linestrength and Beer's Law, each with its own units and notation.

7.6.1 Line Strengths

Alternate forms for linestrength notation and units are as follows:

1.

$$k_\omega\ [\mathrm{cm}^{-1}] = S_{12}\ [\mathrm{cm}^{-2}]\ \phi_\omega\ [\mathrm{cm}],\ \text{or}$$
$$k_\nu\ [\mathrm{cm}^{-1}] = S_{12}\ [\mathrm{cm}^{-2}]\ \phi_\nu\ [\mathrm{cm}]$$

where

$$\nu\ [\mathrm{cm}^{-1}]\ \text{or}\ \omega\ [\mathrm{cm}^{-1}] \equiv 1/\lambda\ [\mathrm{cm}],$$
$$\phi_\nu\ [\mathrm{cm}]\ \text{or}\ \phi_\omega\ [\mathrm{cm}] = c\ [\mathrm{cm/s}]\ \phi_\nu\ [\mathrm{s}]$$

and

$$d\nu\ [\mathrm{cm}^{-1}]\ \text{or}\ d\omega\ [\mathrm{cm}^{-1}] = (1/c)d\nu\ [\mathrm{s}^{-1}]$$

2.

$$S_{12}\ [\mathrm{cm}^{-2}] = (1/c)S_{12}\ [\mathrm{cm}^{-1}/\mathrm{s}]$$

3.

$$\begin{aligned} S_{12}\ [\mathrm{cm}^{-2}/\mathrm{atm}] &= S_{12}\ [\mathrm{cm}^{-2}]/P_i\ [\mathrm{atm}] \\ &= \left(\frac{n_1}{P_i\ [\mathrm{atm}]}\right)\left(\frac{c}{8\pi\nu^2}\right)A_{21}\frac{g_2}{g_1}\left(1 - \exp(-h\nu/kT)\right) \end{aligned}$$

where n_1 is the number density of the absorbing species i in state 1.

4. The typical units for linestrengths include a per-unit-pressure version, S [$cm^{-2}\,atm^{-1}$], and a per-unit-number-density version used by HITRAN96, S^* [$cm^{-1}/(molecule\,cm^{-2})$]. The conversion between S and S^* is

$$S\,[cm^{-2}\,atm^{-1}] = \frac{S^*\,[cm^{-1}/(molecule\,cm^{-2})] \times n\,[molecules/cc]}{P\,[atm]}, \tag{7.65}$$

where n is the number density of the absorbing species in [molecules/cc] and P is the corresponding partial pressure in [atm]. Using the ideal gas law and converting pressure units from [$dynes/cm^2$] to [atm] yields the following relation

$$S\,[cm^{-2}\,atm^{-1}] = \frac{S^*\,[cm^{-1}/(molecule\,cm^{-2})] \times 1013250\,[dynes/(cm^2\,atm)]}{kT}, \tag{7.66}$$

where $k = 1.38054 \times 10^{-16}$ erg/K is the Boltzmann constant and T [K] is the temperature at which the conversion is being performed and the linestrength is known. Equation (7.66) reduces to

$$S = \frac{S^* \times (7.34 \times 10^{21})}{T}\,[cm^{-2}\,atm^{-1}]. \tag{7.67}$$

For converting room-temperature linestrength ($T = 296$ K), the conversion is

$$S = S^* \times (2.4797 \times 10^{19})\ [cm^{-2}\,atm^{-1}]. \tag{7.68}$$

7.6.2 Beer's Law

It follows from the alternate forms for the linestrength that multiple expressions for Beer's Law also exist, e.g.

$$\left(\frac{I}{I^0}\right)_{\nu,\omega,\lambda} = \exp\left(-k_\nu L\right) \tag{7.69}$$

$$= \exp\left(-n\sigma_\nu L\right) \tag{7.70}$$

$$= \exp\left(-\beta_\omega P_i L\right) \tag{7.71}$$

$$= \exp(-S\phi_\nu P_i L) \tag{7.72}$$

where

n = number density of the absorbing species [$molecules/cm^3$]

σ_ν = absorption cross-section [$cm^2/molecule$]

$$S = \text{“linestrength” } [\text{cm}^{-2}\,\text{atm}^{-1}] \text{ or } [\text{cm}^{-1}\,\text{s}^{-1}/\text{atm}]$$

$$\beta_\omega = \text{frequency-dependent absorption coefficient } [\text{cm}^{-1}/\text{atm}] \tag{7.73}$$

$$P_i = \text{partial pressure of species } i \text{ [atm]}$$

$$\phi_\nu = \text{frequency-dependent lineshape function [cm] or [s]}$$

In the IR, it is common to use atmosphere and wavenumber units, i.e. $\beta_\omega = k_\nu / P_i$ = absorption coefficient per atmosphere of pressure. Thus,

$$\begin{aligned} S_{12}\,[\text{cm}^{-2}/\text{atm}] &= \int \beta_\omega d\omega \\ &= \frac{S_{12}\,[\text{cm}^{-1}\,\text{s}^{-1}]}{cP_i\,[\text{atm}]} \\ &= 8.82 \times 10^{-13} \frac{n_1}{P_i\,[\text{atm}]} f_{12}\,(1 - \exp(-h\nu/kT)) \\ &= \frac{c}{8\pi\nu^2} \frac{n_1}{P_i} A_{21} \frac{g_2}{g_1}\,(1 - \exp(-h\nu/kT)) \end{aligned}$$

7.7 Temperature-Dependent Linestrengths

As Eq. (7.46) demonstrates, linestrengths are directly dependent on the number density, n_1, and exponentially dependent on the temperature, T. Using the Boltzmann fraction to relate n_1 at various temperatures and combining with Eq. (7.46) yields an expression for the linestrength as a function of temperature.

The linestrength $S_i(T)$ for a particular transition i at some temperature T can be determined from the molecule's reference temperature linestrength $S_i(T_0)$; the absorbing molecule's partition function $Q(T)$; the frequency of the transition, $\nu_{0,i}$; and the lower-state energy of the transition, E_i''. This relationship is given by

$$\begin{aligned} S_i(T) = S_i(T_0) &\frac{Q(T_0)}{Q(T)} \left(\frac{T_0}{T}\right) \exp\left[-\frac{hcE_i''}{k}\left(\frac{1}{T} - \frac{1}{T_0}\right)\right] \\ &\times \left[1 - \exp\left(\frac{-hc\nu_{0,i}}{kT}\right)\right]\left[1 - \exp\left(\frac{-hc\nu_{0,i}}{kT_0}\right)\right]^{-1}, \end{aligned} \tag{7.74}$$

where S is in units of $[\text{cm}^{-2}\,\text{atm}^{-1}]$. For units of $[\text{cm}^{-1}/(\text{molecule}\,\text{cm}^{-2})]$, the following temperature scaling can be used

$$S_i^*(T) = S_i^*(T_0)\frac{Q(T_0)}{Q(T)}\exp\left[-\frac{hcE_i''}{k}\left(\frac{1}{T}-\frac{1}{T_0}\right)\right]$$
$$\times\left[1-\exp\left(\frac{-hc\nu_{0,i}}{kT}\right)\right]\left[1-\exp\left(\frac{-hc\nu_{0,i}}{kT_0}\right)\right]^{-1}. \quad (7.75)$$

Thus, a ratio of linestrengths with the different units can be calculated as follows:

$$\frac{S(T)}{S(T_0)} = \frac{S_i^*(T)}{S_i^*(T_0)} \times \frac{T_0}{T}. \quad (7.76)$$

7.8 Concept of Band Strength

The concept of band strength is common in the IR. Recall that a band is a group of lines for different upper and lower vibrational quantum numbers. The band strength is determined by the number and strength of individual lines, and is expressed as a sum of the linestrengths.

$$S_{\text{band}} = \sum_{\text{band}} S_{\text{lines}} \quad (7.77)$$

Example: Heteronuclear Diatomic Band Strength

For the $1 \leftarrow 0$ band of a heteronuclear diatomic molecule, the band strength is

$$S^{1\leftarrow 0} = \sum_{J''}^{v'=1\leftarrow v''=0}\left[S_{J'\leftarrow J''}^{1\leftarrow 0}(P) + S_{J'\leftarrow J''}^{1\leftarrow 0}(R)\right]$$

where

$$S_{J'J''}^{10}(R) = \frac{c}{8\pi\nu^2}\overbrace{\frac{n_{J''}}{n_i kT/1.013\times 10^6}}^{\frac{n_{J''}}{P_{i,\ \text{atm}}}}\overbrace{\left[\frac{g_{J'}}{g_{J''}} = \frac{2J'+1}{2J''+1}\right]}^{\approx 1}\overbrace{\left[A_R^{10} \approx \frac{J''+1}{2J''+1}A^{10}\right]}^{A_P^{10}\approx\frac{J''}{2J''+1}A^{10}}$$
$$\times\left(1-\exp(-h\nu/kT)\right).$$

Note: The approximations shown for A_R^{10} and A_P^{10} are based on the normalized Hönl–London factors, to be discussed in Chap. 10.

Then

$$S^{10}(R) = \frac{(1.013\times 10^6)\,c}{8\pi\nu^2 kT}A^{10}\sum_{J''}\left[\frac{n_{J''}}{n_i}\frac{J''+1}{2J''+1}\right],$$

and similarly

$$S^{10}(P) = \frac{(1.013 \times 10^6)\,c}{8\pi \nu^2 kT} A^{10} \sum_{J''} \left[\frac{n_{J''}}{n_i} \frac{J''}{2J'' + 1} \right]$$

Therefore, since $\sum_{J''}(n_{J''}/n_i) = 1$,

$$\boxed{S^{10}(T) = \frac{(1.013 \times 10^6)\,c A^{10}}{8\pi \nu^2 kT}}$$

Example: Band Strength of CO

The measured band strength of CO at 273 K, as measured at Stanford, is

$$S^{10}_{\mathrm{CO}}(273\,\mathrm{K}) = \frac{3.2 \times 10^{28} A^{10}}{\nu^2} \approx 280\,\mathrm{cm}^{-2}/\mathrm{atm}$$

But,

$$\omega \approx 2150\,\mathrm{cm}^{-1}$$

and

$$\nu \approx 6.4 \times 10^{13}\,\mathrm{s}^{-1},$$

yielding

$$A^{10} = 36\,\mathrm{s}^{-1},$$

or, equivalently,

$$\tau^{10} = 0.028\,\mathrm{s}$$

Compare the value for τ^{10} with the previous example of $\tau_{\mathrm{Na}} \approx 16\,\mathrm{ns}$. CO requires 28 ms to decay by radiation! Thus, IR transitions, due to their smaller changes in dipole moment, have much lower values of A and longer radiative lifetime τ than UV/Visible transitions.

7.9 Exercises

1. Light is transmitted through an optically thin (but absorbing) medium. If the path length (L) is doubled, what happens to the fractional absorption?
2. The fractional transmission of monochromatic light through a uniform absorbing medium of length L is 0.75. What is the fractional transmission if the path length is doubled?
3. A discrete electronic transition of a monatomic gas at high temperature ($T = 5000$ K) and low pressure ($P = 0.01$ atm) has a measured linecenter spectral absorption coefficient k_{ν_o} of 0.1 cm^{-1} at a wavelength of 500 nm. Determine the linecenter spectral emissivity of the gas if the gas sample is 10 cm thick and $\phi(500\,nm) = 6$ cm.
4. The Einstein-A coefficient for a particular rovibrational transition of CO_2 is 220 s^{-1}. In the absence of collisions, what is the characteristic lifetime of the upper state? Compare this with the Na transition near 589.6 nm which has an Einstein-A coefficient of $6.14 \times 10^7\ s^{-1}$.
5. The partial pressure of H_2O is 0.10 atm and an absorption transition with a linestrength of $7.58 \times 10^{-22}\ cm^{-1}$/molecule cm^2 is excited by a 10 mW laser near 1392 nm that is resonant with the transition linecenter. How much power is absorbed by the 10 cm gas sample if $\phi(1392\,nm) = 5$ cm?
6. Given the Einstein-A coefficient of a transition, what additional information do you need to calculate the temperature-dependent linestrength of this transition?

References

1. S.S. Penner, *Quantitative Molecular Spectroscopy and Gas Emissivities* (Addison-Wesley Publishing Company, Reading, MA, 1959)
2. A.C.G. Mitchell, M.W. Zemansky, *Resonance Radiation and Excited Atoms* (Cambridge University Press, London, 1971)

Spectral Lineshapes

8

The lineshape function $\phi(\nu)$ characterizes the relative variation in the spectral absorption coefficient with frequency and appears directly in Beer's Law:

$$\left(\frac{I}{I^0}\right)_\nu = \exp(-S\phi_\nu P_i L) \tag{8.1}$$

This variation with frequency is caused by broadening mechanisms in the medium. An understanding of these mechanisms allows accurate predictions of the lineshape function. Likewise, a measurement of the lineshape function, and the center frequency, ν_0, can be used to determine properties of the medium such as temperature, pressure, and velocity.

8.1 Lineshape Introduction

The lineshape function has been defined so that its integral over frequency is unity,

$$\phi(\nu) = \frac{k_\nu}{\int\limits_{\text{line}} k_\nu d\nu}, \tag{8.2}$$

so that

$$\int_{-\infty}^{+\infty} \phi(\nu)\, d\nu = 1. \tag{8.3}$$

A typical lineshape of an isolated absorption line centered at ν_0 is shown in Fig. 8.1 as a function of frequency.

The lineshape has a maximum value $\phi(\nu_0)$ at the center frequency ν_0. The width of the feature, $\Delta\nu$, is defined by the width at half the maximum value (the fullwidth

R.K. Hanson et al., *Spectroscopy and Optical Diagnostics for Gases*,
DOI 10.1007/978-3-319-23252-2_8

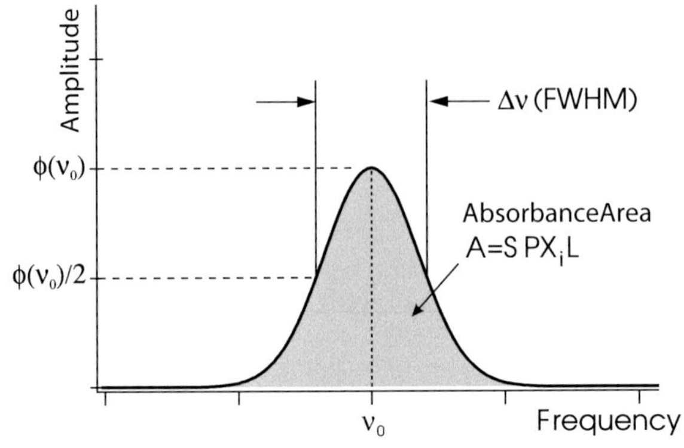

Fig. 8.1 Sample lineshape as a function of frequency

at half maximum, or, FWHM). Note that the halfwidth at half maximum, HWHM, is also used in many texts. The integral in Eq. (8.3) is defined to have no dimensions. Since the units of $d\nu$ are typically either cm^{-1} or s^{-1}, $\phi(\nu)$ will have units of cm or s, respectively.

It is often convenient to investigate the lineshape and other absorption parameters by converting transmission or absorption into *absorbance*, where absorbance is defined as

$$\alpha(\nu) = -\ln\left(\frac{I}{I^0}\right)_\nu = S\phi_\nu P_i L = S\phi_\nu P X_i L, \tag{8.4}$$

where X_i = mole fraction of the absorbing species i. Note that absorbance has no units.

8.2 Line Broadening Mechanisms

Broadening of an absorption feature occurs due to physical processes within the medium that perturb the transition's energy levels or the way in which individual atoms and molecules interact with light. If this effect is the same for all atoms of the species, the broadening is said to be *homogeneous*. If, however, there are separate classes or subgroups for which the interaction varies, the broadening is said to be *inhomogeneous* or heterogeneous. This subtle difference will become more clear when examining the types of broadening mechanisms.

The energy of an optical transition is equal to the energy difference between two states. The Heisenberg Uncertainty Principle relates the uncertainty of these energy levels to their lifetimes [1]. The uncertainty in the energy of level i is limited by

$$\Delta E_i \geq \frac{h}{2\pi\tau_i} \tag{8.5}$$

where τ_i is the lifetime of level i. The lifetimes of the upper and lower states, τ' and τ'', respectively, can be combined to give the total energy uncertainty ΔE of a transition. Thus, there is a range of energies over which the transition has a non-zero probability of being measured. Since $\Delta E = h\Delta\nu$, that range is characterized by a linewidth (FWHM) given (in units of frequency) by:

$$\Delta\nu = \frac{1}{2\pi}\left(\frac{1}{\tau'} + \frac{1}{\tau''}\right) \tag{8.6}$$

Since the Uncertainty Principle applies to all atoms in the same way, lifetime broadening is *homogeneous*. The lineshape function can be derived by modelling the atomic system as a damped oscillator. It takes the form of a Lorentzian function:

$$\phi_L(\nu) = \frac{1}{2\pi}\frac{\Delta\nu}{(\nu-\nu_0)^2 + \left(\frac{\Delta\nu}{2}\right)^2} \tag{8.7}$$

The Lorentzian lineshape function at linecenter is

$$\phi_L(\nu_0) = \frac{2}{\pi\Delta\nu}. \tag{8.8}$$

Another form of Eq. (8.7) is

$$\phi_L(\nu) = \phi_L(\nu_0)\frac{1}{1 + 4(\frac{\nu-\nu_0}{\Delta\nu})^2}. \tag{8.9}$$

There are different mechanisms that lead to broadening of spectroscopic transitions. Some of these mechanisms are described below.

8.2.1 Natural Broadening

In the absence of interactions with other atoms (e.g., via collisions), the lifetime of an atom in a particular level is determined by spontaneous emission [2]:

$$\Delta\nu_N = \frac{1}{2\pi}\left(\frac{1}{\tau_i} + \frac{1}{\tau_j}\right) \tag{8.10}$$

$$= \frac{1}{2\pi}\left(\sum_k A_{ik} + \sum_k A_{jk}\right) \tag{8.11}$$

This is the sum of the Einstein A coefficients for all downward transitions from the two levels of the transitions i and j. Thus, the sum is over all k such that $E_k < E_i$ or E_j. Note that if the lower level, level j, is the ground state, there can be no radiative decay and its contribution to the halfwidth is zero. The subscript N indicates that the FWHM is due to natural broadening which is described by the Lorentzian lineshape function in Eq. (8.7).

8.2.2 Collisional Broadening (Pressure Broadening)

Molecules (or atoms) interact with each other via collisions. During a collision, energy may be transferred among various energy modes within and between molecules, thus changing the molecular state each molecule exists in. As a result, collisions can reduce the average lifetime of a molecule in a state below that of the natural lifetime. According to Eq. (8.6), shortening the lifetime τ_i of a molecule in level i leads to greater uncertainty, and thus a broader absorption lineshape. Hence, the more frequently collisions occur, the more lifetimes are shortened and transitions are broadened. This relationship between collision frequency and transition breadth is the fundamental mechanism underlying collisional (or pressure) broadening.

A collision can be modelled as occurring when two molecules approach each other within a certain distance called the optical collision diameter. If the diameters of molecules A and B are defined as σ_A and σ_B, respectively, the optical collision diameter is given by

$$\sigma_{AB} = \frac{1}{2}(\sigma_A + \sigma_B) \tag{8.12}$$

The associated optical cross-section area is $\pi(\sigma_{AB})^2$. Note that this cross-section may be significantly different than the cross-sections for collisions between A and B that give rise to other phenomena (e.g., the elastic collision cross-section). The number of collisions per second *of a single B* with all A is given by

$$Z_{AB} = n_A \pi \, \sigma_{AB}^2 \, \bar{c} \tag{8.13}$$

where n_A is the number density of species A and $\bar{c}$ is the mean relative speed of the molecules given by

$$\bar{c} = \left(\frac{8kT}{\pi \, \mu_{AB}} \right)^{1/2} \tag{8.14}$$

The reduced mass, μ_{AB}, is a function of the masses of A and B

$$\mu_{AB} = \frac{m_A m_B}{m_A + m_B} \tag{8.15}$$

The total collision frequency (*of a single B*) for a variety of different collision partners is obtained by summing over the different species:

$$Z_B = \sum_A n_A \pi\, \sigma_{AB}^2 \left(\frac{8kT}{\pi\, \mu_{AB}}\right)^{1/2} \tag{8.16}$$

Note that for a pure gas, i.e. all B, Eq. (8.16) simplifies to the expression $Z_B = n_B \pi \sigma_{BB}^2 (8kT/\pi \mu_{BB})^{1/2}$. By introducing the ideal gas law, $P = nkT$, we can rewrite Eq. (8.16) for the single-molecule collision frequency as

$$Z_B = P \sum_A X_A \pi\, \sigma_{AB}^2 \left(\frac{8}{\pi\, \mu_{AB} kT}\right)^{1/2} \tag{8.17}$$

where P is the total pressure and X_A is the mole fraction of species A. Using Eq. (8.6) with $1/\tau' = 1/\tau'' = Z_B$, it follows that $\Delta\nu_C$, the FWHM due to collisional broadening, is given by

$$\Delta\nu_C = \frac{Z_B}{\pi}. \tag{8.18}$$

The frequency uncertainty due to collisional effects, $\Delta\nu_C$, is the lineshape's halfwidth (FWHM). This net uncertainty for the interrogated species, B, is often modelled as the product of the system pressure and sum of the mole fraction for each perturbing species A multiplied with its process-dependent collisional broadening coefficient $2\gamma_{B-A}$

$$\Delta\nu_C = P \sum_A X_A 2\gamma_{B-A}. \tag{8.19}$$

Thus, the collisional width of an absorption transition is directly proportional to pressure. The standard notation for the different broadening coefficients is $2\gamma_{B-A}$, where B is the molecule whose lineshape is being studied and A is the collision partner (or perturber) that interacts with species B to broaden its absorption line. Thus, $2\gamma_{CO-CO}$ is the broadening coefficient for CO–CO collisions (self-broadening, $2\gamma_{self}$), and $2\gamma_{CO-N_2}$ is the coefficient for CO lineshape broadening due to collisions with N_2 (nitrogen-broadening). Mass-averaged O_2 and N_2 coefficients are contained in air-broadening coefficients, $2\gamma_{CO-Air}$. Values of 2γ are published for specific transitions and specific collision partners.

If P has units of atmospheres, and $\Delta\nu_C$ has units of s^{-1} then $2\gamma_B$ is defined by

$$2\gamma_B = 1.013 \times 10^6\, \sigma_{AB}^2 \left(\frac{8}{\pi\, \mu_{AB} kT}\right)^{1/2} \tag{8.20}$$

where all parameters are in *cgs* units. The optical cross-section $\pi\sigma_{AB}^2$ and reduced mass μ_{AB} depend on the molecule of interest and its various collision partners, though typical values for small molecules are $2\gamma \approx 0.1\,\mathrm{cm}^{-1}$/atm, at room temperature.

Example: Pressure Broadening of CO

Calculate the collisional halfwidth (FWHM) for the $R(9)$ line of CO's second overtone for 50 ppm of CO in air at 300 K and 1.0 atm pressure. The standard species populations in air are 77 % N_2, 21 % O_2, 2 % H_2O (for 85 % standard humidity), and 380 ppm CO_2. The broadening coefficients and species mole fractions are listed in the following table. According to Eq. (8.19), the collisional halfwidth for this CO transition is:

Species, A	Mole fraction, X_A	$2\gamma_{CO-A}(300\,K)\,cm^{-1}/atm$
N_2	0.77	0.116
H_2O	0.02	0.232
CO	50e−6	0.128
CO_2	380e−6	0.146
O_2	0.21	0.102

$$\begin{aligned}\Delta\nu_C &= P(X_{N_2}\cdot 2\gamma_{CO-N_2} + X_{H_2O}\cdot 2\gamma_{CO-H_2O} + X_{CO}\cdot 2\gamma_{CO-CO} \\ &\quad + X_{CO_2}\cdot 2\gamma_{CO-CO_2} + X_{O_2}\cdot 2\gamma_{CO-O_2}) \\ &= 0.114\,cm^{-1}\end{aligned}$$

The variation of the broadening coefficient 2γ with temperature is often modeled using the following approximation

$$2\gamma(T) = 2\gamma(T_0)\left(\frac{T_0}{T}\right)^N \tag{8.21}$$

where T_0 is the reference temperature, typically 296 or 300 K, $2\gamma(T_0)$ is the broadening coefficient at the reference temperature, and N is the temperature coefficient, which is generally less than 1 and typically 0.5–0.8. Inspection of Eq. (8.20) reveals that $N = 0.5$ when σ_{AB} is constant, independent of T, but in fact σ_{AB} is typically a weak function of T.

In the absence of actual data, useful approximations are that $2\gamma(300) \approx 0.1\,cm^{-1}/atm$ for molecules (atoms tend to have larger values) and $N = 0.5$. An exception, however, is NO, whose broadening coefficients for electronic transitions are $2\gamma(300) \approx 0.5\,cm^{-1}/atm$. Another exception is H_2O for which 2γ and N vary widely, N can even be < 0 for high-J transitions (see Chap. 14).

Table 8.1 Some collisional broadening coefficients 2γ [cm^{-1}/atm] in Argon and Nitrogen at 300 K

Species	Wavelength [nm]	Ar	N_2
Na	589	0.70	0.49
K	770	1.01	0.82
Rb	421	2.21	1.51
OH	306	0.09	0.10
NH	335	0.038	
NO	225	0.50	0.58
NO	5300	0.09	0.12
CO	4700	0.09	0.11
HCN	3000	0.12	0.24

Table 8.2 Some collisional broadening coefficients 2γ [cm^{-1}/atm] in Argon and Nitrogen at 2000 K

Species	Wavelength [nm]	Ar	N_2
NO	225	0.14	0.14
OH	306	0.034	0.04
NH	335	0.038	

Tables 8.1 and 8.2 contain broadening coefficients for a variety of species at 300 and 2000 K, respectively. Note that there are some radicals listed here that are not stable at 300 K. However, it is common to extrapolate values of 2γ down to 300 K to facilitate comparisons. Some molecules, like OH, have a strong J (rotational quantum number) dependency on 2γ, while others such as NO do not.

Collisional broadening follows the Lorentzian spectral distribution in Eq. (8.7). The total halfwidth for the Lorentzian lineshape can be determined by adding the natural and collisional halfwidths, $\Delta\nu = \Delta\nu_N + \Delta\nu_C$. However, $\Delta\nu_C$ is usually much greater than the natural broadening, so $\Delta\nu_N$ can often be neglected.

8.2.3 Doppler Broadening

When a molecule has a velocity component in the same direction as the propagation of a beam of light, there will be a shift in the frequency at which it will absorb a photon. This effect is called the Doppler shift. The molecules of any gas are in constant motion and the distribution of their random velocities (achieved in the absence of collisions, see 8.5.1) is described by the Maxwellian velocity distribution function. We can consider each group of molecules with the same velocity component to be part of a velocity class. The Maxwellian velocity distribution function tells us what portion of the molecules are in each class. Each velocity class will have its own

Doppler shift. Thus the distribution function leads directly to an *inhomogeneous* (meaning it varies with frequency) lineshape function with a Gaussian form:

$$\phi_D(\nu) = \frac{2}{\Delta\nu_D}\left(\frac{\ln 2}{\pi}\right)^{1/2} \exp\left\{-4\ln 2\left(\frac{\nu-\nu_0}{\Delta\nu_D}\right)^2\right\} \tag{8.22}$$

The Gaussian lineshape function at linecenter is

$$\phi_D(\nu_0) = \frac{2}{\Delta\nu_D}\left(\frac{\ln 2}{\pi}\right)^{1/2} \tag{8.23}$$

The Doppler halfwidth (FWHM) $\Delta\nu_D$ is given by

$$\Delta\nu_D = \nu_0\left(\frac{8kT\ln 2}{mc^2}\right)^{1/2}, \tag{8.24}$$

for which a more convenient form is

$$\Delta\nu_D = \nu_0(7.1623\times 10^{-7})\left(\frac{T}{M}\right)^{1/2}, \tag{8.25}$$

where T is in Kelvins and M is the molecular weight in grams/mole. The Doppler halfwidth $\Delta\nu_D$ is to be used with Eq. (8.22) and never with (8.7).

The Gaussian lineshape can also be described in normalized terms as an exponential relationship,

$$K_D(x) = \exp(-x^2), \tag{8.26}$$

where

$$x = \frac{(\nu-\nu_0)}{\Delta\nu_{D,1/e}} \tag{8.27}$$

is the normalized frequency detuning relative to the linecenter frequency, ν_0, and $\Delta\nu_{D,1/e}$ is the $1/e$ halfwidth,

$$\Delta\nu_{D,1/e} = \nu_0\sqrt{\frac{2kT}{mc^2}}. \tag{8.28}$$

Figure 8.2 compares Gaussian (Doppler; inhomogeneous) and Lorentzian (collisional and natural; homogeneous) lineshapes which have the same FWHM and by definition [Eq. (8.3)] the same area. The Gaussian lineshape has a peak value which is about 50 % higher than the Lorentzian, but it drops off much faster in the wings. The molecules of each velocity class will also have some finite lifetime and thus some Lorentzian halfwidth. If the Lorentzian halfwidth is much smaller than the Gaussian halfwidth, the Lorentzian component can be ignored, and vice versa.

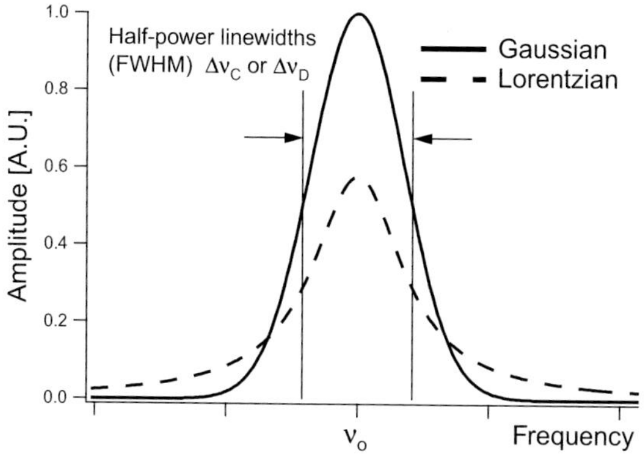

Fig. 8.2 Comparison of Gaussian and Lorentzian lineshapes with the same FWHM

8.2.4 Stark Broadening

Plasmas are characterized by the presence of charged particles (ions and electrons) which exhibit long range Coulomb forces. These forces can perturb the energy levels causing significant broadening, particularly in atomic hydrogen and other light atoms. Both numerical and experimental results are published for specific transitions as a function of electron number density. See the book by Griem [3] for a detailed description.

8.2.5 Artifactual/Instrument Broadening

A measured lineshape can be broader than expected based on the conditions found in a medium. Broadening can be caused by the way in which a measurement is made and thus is an artifact of the experiment. One common example is where an instrument such as a monochromator has insufficient resolution to measure an actual lineshape. This low resolution causes the measured profile to look broader than it really is. Another example is power broadening where a laser is strong enough to perturb the populations of the molecular states (this effect is often referred to as saturation) and thereby modify the frequency distribution of the absorption coefficient. A third example is transit-time broadening where the interaction between the molecules and the light is limited to a time comparable to (or less than) the lifetime determined by collisions or natural emission. As an example, the transit time for a gas moving at velocity V across a laser beam of diameter D is D/V; this leads to an apparent broadening for absorption lines of $\Delta\nu_{\text{transit}}(\text{FWHM}) \approx V/D$ in the case of a top hat intensity distribution [2].

8.3 Voigt Profile

8.3.1 Analytical Expressions

In the common case in which both Doppler and collisional broadening are significant and neither can be neglected, the appropriate lineshape will be a combination of the two. If we assume that the effects of Doppler and collisional broadening are decoupled, we can view each velocity class to be collisionally broadened. This leads to a lineshape that is a convolution of Doppler and collisional broadening [4] (assuming natural broadening is much less than collisional broadening, $\Delta\nu_N \ll \Delta\nu_C$).

$$\phi_V(\nu) = \int_{-\infty}^{+\infty} \phi_D(u)\,\phi_C(\nu - u)\,du \tag{8.29}$$

This convolution is called the Voigt function. Before inserting relations for ϕ_D and ϕ_C, we define some useful parameters:

(1) Voigt "a" parameter

$$a = \frac{\sqrt{\ln 2}\,\Delta\nu_C}{\Delta\nu_D} \tag{8.30}$$

(2) nondimensional line position, w

$$w = \frac{2\sqrt{\ln 2}\,(\nu - \nu_0)}{\Delta\nu_D} \tag{8.31}$$

(3) linecenter magnitude, $\phi_D(\nu_0)$, the magnitude of Eq. (8.22) at ν_0

$$\phi_D(\nu_0) = \frac{2}{\Delta\nu_D}\sqrt{\frac{\ln 2}{\pi}} \tag{8.32}$$

(4) integral variable, y

$$y = \frac{2u\sqrt{\ln 2}}{\Delta\nu_D} \tag{8.33}$$

With these parameters, Eq. (8.29) becomes

$$\phi_V(\nu) = \phi_D(\nu_0)\,\frac{a}{\pi}\int_{-\infty}^{+\infty} \frac{\exp(-y^2)\,dy}{a^2 + (w - y)^2} \tag{8.34}$$

$$= \phi_D(\nu_0)\,V(a, w) \tag{8.35}$$

where $V(a, w)$ is the famous "Voigt function." The a parameter indicates the relative significance of Doppler and collisional broadening, with a increasing as the effects of collisional broadening increase. The w parameter is a measure of the distance from linecenter. The integral can be calculated using standard mathematical routines, but it is common to make use of existing tables for this function.

Note that at the linecenter, where $w = 0$, $V(a, w)$ reduces to

$$V(a, 0) = \exp(a^2)\mathrm{erfc}(a) \tag{8.36}$$

$$= \exp(a^2)[1 - \mathrm{erf}(a)] \tag{8.37}$$

where erf(a) is the error function.

8.3.2 Numerical Approximations

There are many different numerical approximations for the Voigt lineshape that have been published. One of the most accurate and quickest to calculate is the algorithm published by Humlíček (*J. Quant. Spectrosc. Radiat. Transfer*, Vol. 27, No. 4, pp. 437–444, 1982). The algorithm can be processed using standard programs such as Igor Pro, Matlab, and the C language. Appendix C contains an example Matlab *.m file for calculating the normalized Voigt profile.

8.4 Line Shifting Mechanisms

Just as there are physical mechanisms that broaden absorption lines, there are processes by which the lineshapes are shifted in frequency. Two of these shift mechanisms, similar to broadening, are pressure shift and Doppler shift.

8.4.1 Pressure Shift of Absorption Lines

Interactions between two collision partners can have a perturbing effect on the intermolecular potential of the molecule whose lineshape is being studied. Changes in the intermolecular potential lead to differences in the energy level spacings, and hence the linecenter frequencies of the different transitions. These differences from the equilibrium linecenter values that result from collisions are known as pressure shift. Just as in Eq. (8.19), where the collisional halfwidth, $\Delta\nu_C$, is proportional to the pressure and a broadening coefficient, 2γ, the expression for pressure shift depends directly on pressure but contains the shift coefficient δ instead. Both 2γ and δ have units of cm^{-1} per atm.

$$\Delta\nu_S = P \sum_A X_A \delta_A \tag{8.38}$$

The shift coefficient also scales from a reference temperature similarly to the broadening coefficient, but with a different temperature exponent, M.

$$\delta_A(T) = \delta_A(T_0)\left(\frac{T_0}{T}\right)^M \tag{8.39}$$

Note that while $2\gamma > 0$, the pressure shift can be either negative or positive. Average values for near-infrared H_2O spectra are $\delta \approx -0.017\,\mathrm{cm}^{-1}/\mathrm{atm}$ and $M \approx 0.96$.

8.4.2 Doppler Shift Measurements of Velocity

For gases that have a bulk velocity relative to an incident laser beam, the entire Voigt profile is shifted by

$$\frac{\delta\nu}{\nu_0} = \frac{u}{c} \tag{8.40}$$

where u is the bulk or mean speed in the laser direction and c is the speed of light. Measurements of this frequency shift provide a non-intrusive means of measuring gas velocities. This capacity is especially attractive for high speeds and low densities where conventional laser Doppler anemometry (with particles) is not feasible.

8.5 Lineshapes Beyond the Voigt Profile

The Voigt profile assumes that Doppler and collisional broadening are independent/uncorrelated and does not account for velocity-changing collisions. In some cases, these assumptions can lead to relatively large errors (typically 1–10 % of peak values) in modeling the lineshape of measured spectra. This section will briefly discuss the underlying physics responsible for these errors and introduce a few advanced lineshape profiles that address these processes. See Sect. 14.3 for a working example.

8.5.1 Line Narrowing Mechanisms

While collisional broadening is usually more significant, collisions can also narrow lineshapes via two primary mechanisms: (1) Dicke narrowing [5] and (2) speed-dependent (i.e., heterogeneous) collisional broadening. These processes, more generally referred to as "collisional narrowing," lead to larger peak absorbances and smaller FWHM. Collisional narrowing is most significant for molecules with large rotational-energy spacing (e.g., H_2O, HF, HCN) and at temperatures where the thermal energy kT is small compared to the spacing between rotational energy levels ($\sim 2BJ''$). In this case only strong collisions are rotationally inelastic, and therefore, state-changing collisions are less common (i.e., γ is small).

Dicke narrowing describes a collision-induced reduction of the Doppler width [compared to that calculated using Eq. (8.25)] that results from velocity-changing collisions reducing the average thermal velocity of the absorber with respect to the photon. Recall, the Maxwellian velocity distribution does not account for collisions. Dicke narrowing is expected to be observed at modest number densities where the mean free-path is comparable to $\lambda/2\pi$ where λ is the wavelength of the transition [6]. Dicke narrowing is insignificant when collisional broadening dominates the lineshape (e.g., at high pressures).

Speed-dependent collisional broadening refers to collisional broadening that depends on the speed of the absorber/emitter (i.e., which velocity class it belongs to) and, therefore, is heterogeneous/inhomogeneous. For example, a molecule moving at 1 m/s is likely to experience a very different type of collision (e.g., head-on vs. glancing) than a molecule traveling at 1000 m/s. Despite its name, speed-dependent broadening leads to a narrower lineshape compared to lineshape models that assume homogeneous collision broadening (e.g., Lorentzian, Voigt).

8.5.2 Rautian and Galatry Profiles

The Rautian [7] and Galatry [8] profiles address Doppler and collisional broadening in the same manner as the Voigt profile; however, they also address Dicke narrowing through either hard- or soft-collision models, respectively. The hard-collision model assumes that the velocity of the radiator after a collision is uncorrelated with its velocity prior to the collision while the soft-collision model assumes many collisions are needed to significantly alter the velocity of the radiator. Both lineshape models address Dicke narrowing through one additional (compared to the Voigt) parameter β, commonly expressed in units of cm^{-1} or $cm^{-1}/$ atm, that represents the frequency of velocity-changing collisions. β is lineshape-model specific, and thus, cannot be used interchangeably between the Rautian and Galatry profiles. If $\beta = 0$, both the Rautian and Galatry profiles reduce to the Voigt profile.

8.5.3 Speed-Dependent Voigt Profile

In reality, Doppler and collisional broadening are not independent processes and, in fact, collisional broadening depends on which velocity class an absorber/emitter belongs to. The speed-dependent Voigt profile [9] acknowledges this by introducing a speed-dependent broadening coefficient γ_2, and γ represents the speed-averaged collisional-broadening coefficient. The speed-dependent Voigt also models the speed dependence of the pressure-shift by introducing a speed-dependent pressure shift coefficient δ_2. If speed-dependent effects are negligible (i.e., $\gamma_2 = \delta_2 = 0$), the speed-dependent Voigt profile reduces to the standard Voigt profile.

8.6 Quantitative Lineshape Measurements

The size and shapes of different transitions can be used to make quantitative measurements of species concentration, pressure, temperature, and flow velocity.

8.6.1 Species Concentration and Pressure

A measurement of the integrated absorbance,

$$A_i = \int_{-\infty}^{+\infty} \alpha(\nu) d\nu \tag{8.41}$$

or area under the absorption transition, can be used to calculate the species molefraction, X_j, and the pressure, P. The integral in Eq. (8.41) removes the normalized contribution of the lineshape, reducing Eq. (8.4) to

$$A_i = S_i P X_j L \tag{8.42}$$

where S_i is the line strength of the transition, X_j is the molefraction of the absorbing species, P is the pressure, and L is the pathlength. When $S_i PL$ is known, the species concentration is then given by

$$X_j = \frac{A_i}{S_i PL} \tag{8.43}$$

Similarly, if $S_i X_j L$ is known, the pressure is determined by

$$P = \frac{A_i}{S_i X_j L} \tag{8.44}$$

8.6.2 Temperature

There are two lineshape techniques available for measuring the temperature. The first can be used in Doppler-limited applications, where the pressure broadening is negligible. For these applications, the width of the absorption lineshape, or FWHM, can be related directly to the temperature of the absorbing species via Eq. (8.24).

For measurement regimes in which pressure broadening cannot be ignored, a two-line technique can be used. Earlier, it was demonstrated that a line strength for a particular transition has a temperature dependency [Eq. (7.74)]. Taking the ratio of the integrated absorbance area for two transitions removes the pressure and species concentration parameters, leaving only the temperature and lower-state energies for the respective transitions. Since the lower-state energies are typically known (either by quantum mechanical calculation or, more commonly, from tabulations), the temperature can be inferred. The ratio of the integrated areas, R, is related to the temperature by

$$R = \frac{S(T,\nu_1)}{S(T,\nu_2)} = \frac{S(T_0,\nu_1)}{S(T_0,\nu_2)} \exp\left[-\left(\frac{hc}{k}\right)(E_1'' - E_2'')\left(\frac{1}{T} - \frac{1}{T_0}\right)\right] \tag{8.45}$$

where ν_1, ν_2, E_1'', and E_2'' are the linecenter frequencies and lower-state energies for the two transitions, $S(T_0, \nu_1)$ and $S(T_0, \nu_2)$ are the strengths of the two lines at the reference temperature T_0, and the induced emission terms have been neglected. Thus, the temperature is

$$T = \frac{\frac{hc}{k}(E_2'' - E_1'')}{\ln R + \ln \frac{S_2(T_0)}{S_1(T_0)} + \frac{hc}{k}\frac{(E_2''-E_1'')}{T_0}} \tag{8.46}$$

Differentiating Eq. (8.45) yields the temperature sensitivity for a given line pair at a specific temperature in percent per Kelvin:

$$\frac{1}{R}\frac{dR}{dT}\ [\%/\mathrm{K}] = \left(\frac{hc}{k}\right)\frac{(E_1'' - E_2'')}{T^2} \times 100 \tag{8.47}$$

This shows that a large $\Delta E''$ is desirable, to enhance sensitivity, regardless of the temperature. There are additional factors, however, which must be considered. The absorbance of each transition (α_1 and α_2) must be properly bounded. A good rule of thumb is $0.1 < \alpha < 2.3$. This means that for a given partial pressure, path length, and lineshape, the two linestrengths, $S_1(T)$ and $S_2(T)$, will be bounded. Generally, a large difference between E_1'' and E_2'' results in a large difference between $S_1(T)$ and $S_2(T)$. Thus, we see that a tradeoff exists between acceptable absorbance levels and temperature sensitivity.

8.6.3 Examples

Nitric Oxide, NO

Nitric oxide, NO, is a common species found in combustion flows. It has an electronic transition ($A^2\Sigma - X^2\Pi$) near 225 nm (44,000 cm^{-1}) in the UV region. Some typical values (in cm^{-1}) for the halfwidths are shown below for a pressure of 1 atm and two different temperatures.

T [K]	$\Delta\nu_N$	$\Delta\nu_C$	$\Delta\nu_D$
300	$\sim 5\times10^{-5}$	0.58	0.10
2000		0.14	0.26

Since neither Doppler nor collisional broadening is dominant, Voigt functions are necessary at both temperatures to accurately describe the lineshape function. It is clear that Doppler broadening becomes more significant with increasing temperature (scales with $T^{1/2}$). Likewise, collisional broadening becomes more significant with increasing pressure (scales approximately with $P/T^{1/2}$). Note that the collisional broadening coefficient for NO's electronic transition is relatively high compared to other molecules.

Because Doppler broadening is proportional to the transition frequency, it becomes less significant for IR spectroscopy. NO has a vibrational band at 5.3 μm. Values (in cm^{-1}) for the FWHM are shown below for a pressure of 1 atm. While Doppler broadening may be neglected at 300 K, it clearly cannot be neglected at 2000 K.

T [K]	$\Delta\nu_C$	$\Delta\nu_D$
300	0.12	0.004
2000	0.05	0.01

Sodium, Na

A narrowband laser is tuned to the linecenter of the 589 nm sodium transition. It is used to infer the partial pressure of sodium which is seeded into an atmospheric pressure flame. The temperature has been determined to be 1600 K. The partial pressure is obtained from a linecenter absorption measurement via the absorption coefficient. First, solve for k_{ν_0} from the measured fractional transmission at linecenter; next, evaluate the linecenter value of the lineshape function; finally, solve for P_i using the known value for the line strength at 1600 K.

(1) find $k_{\nu_0} = (-1/L)\ln(I/I^0)_{\nu_0}$
(2) find $\phi(\nu_0)$
(3) find P_i using Eq. (8.48)

$$P_i = \frac{k_\nu}{S\phi(\nu_0)} \tag{8.48}$$

To evaluate $\phi(\nu_0)$, first calculate the collisional and Doppler halfwidths. Assume that the collisional broadening is adequately characterized by collisions with N_2, the major species in an air flame:

$$\Delta\nu_C = P\cdot 2\gamma(1600\,\text{K}) = P\cdot 2\gamma(300\,\text{K})\sqrt{\frac{300}{1600}} = 0.21\,\text{cm}^{-1}$$

This example uses the value of 2γ at 300 K (see Table 8.1) and scales it to 1600 K. Next, convert the wavelength to frequency:

$$(589\times 10^{-7}\,\text{cm})^{-1} = 16,978\,\text{cm}^{-1}$$

and calculate the Doppler halfwidth with Eq. (8.24):

$$\Delta\nu_D = (7.17 \times 10^{-7})(16,978\,\text{cm}^{-1})\left(\frac{1600}{23}\right)^{1/2} = 0.10\,\text{cm}^{-1}$$

Next, calculate the Voigt a parameter with Eq. (8.30):

$$a = \frac{\sqrt{\ln 2}\,\Delta\nu_C}{\Delta\nu_D} = \frac{\sqrt{\ln 2}\,(0.21)}{0.10} = 1.75$$

Using the Voigt tables, interpolate between $a = 1.70$ and 1.80 for $w = 0$ (linecenter) to get $V(a, w) = V(1.75, 0) = 0.2852$. Calculate $\phi_D(\nu_o)$ from Eq. (8.32):

$$\phi_D(\nu_o) = \frac{2}{\Delta\nu_D}\sqrt{\frac{\ln 2}{\pi}} = \frac{2}{0.10}\sqrt{\frac{\ln 2}{\pi}} = 9.39\,\text{cm}$$

The lineshape function can now be found from Eq. (8.35):

$$\phi(\nu_o) = \phi_D(\nu_o)\,V(0) = 9.39 \times 0.2852 = 2.68\,\text{cm}$$

Now solve for P_i using Eq. (8.48). Note that $S(T)$ can be calculated from Einstein coefficients or the oscillator strength, if $S(T)$ is not already known (see Chap. 7).

Atomic H Velocity

LIF (Laser Induced Fluorescence) in an arcjet thruster is used to measure the Doppler shift of atomic hydrogen at 656 nm. The laser is directed into the flow in an axial direction (relative to the arcjet), and a shift of $0.70\,\text{cm}^{-1}$ is observed. The corresponding velocity component is found by using Eq. (8.40):

$$u = \frac{c\,\delta\nu}{\nu_0} = \frac{3 \times 10^8\,\text{m/s} \times 0.70\,\text{cm}^{-1}}{15,232\,\text{cm}^{-1}} = 13,800\,\text{m/s}$$

8.7 Exercises

1. Neglecting all broadening mechanisms except collision and Doppler broadening, write a general expression for the Voigt a parameter of a single gas system, in terms of the total pressure, temperature, optical collision diameter (σ), and the transition center frequency.
2. Given a sample cell filled with pure $^{12}C^{16}O$ at atmospheric pressure and $\sigma = 4.0$ Å, determine at what temperature $a = \sqrt{\ln 2}$ for $\overline{\nu}_0 = 2100\,\text{cm}^{-1}$ (an IR transition) and for $\overline{\nu}_0 = 22,170\,\text{cm}^{-1}$ (a visible transition).
3. For $T = 300$ K, $\sigma = 4.0$ Å, and a line strength $S = 10\,\text{cm}^{-2}\,\text{atm}^{-1}$, plot the spectral absorption coefficient at line center, $\beta(\overline{\nu}_0)\,\text{cm}^{-1}$/atm, versus pressure from 0 to 3 atmospheres, for the $2100\,\text{cm}^{-1}$ transition.

4. What atmospheric path length, L (cm), is required at 300 K, 1 atm total pressure, to give 60 % absorption at the center of this transition when CO is present at a level of 10 ppm in the atmosphere? You may neglect the small differences in mass between CO, O_2, and N_2. You may also assume that the optical collision diameters for CO, O_2, and N_2 are equal.
5. A pure gas of molecular weight 25 g/mole is at $T = 300$ K, and $P = 1$ atm. The rovibronic transition at $\lambda = 700$ nm is characterized by a Voigt a parameter $a = 0.5$. Determine the new Voigt a parameter value at $T = 600$ K and $P = 4$ atm. You may assume that the optical collision diameter is invariant with temperature.

   Hints for these problems:

   (a) In order to eliminate interpolation of the Voigt tables you may choose to pick values of P that correspond to convenient values of a. In other words, select a value of a, then calculate the corresponding value of P.

   (b) For $a > 10$, the effect of Doppler broadening is small and purely Lorentzian lineshape may be assumed.

   (c) Use a semi-log plot ($\log \beta$ vs. P) to more clearly display the wide variations in $\beta(\overline{\nu}_0)$.

References

1. S.S. Penner, *Quantitative Molecular Spectroscopy and Gas Emissivities* (Addison-Wesley, Reading, 1959)
2. W. Demtröder, *Laser Spectroscopy: Basic Concepts and Instrumentation*, 2nd enl. edn. (Springer, New York, 1996)
3. H.R. Griem, *Principles of Plasma Spectroscopy* (Cambridge University Press, London, 1997)
4. A.C.G. Mitchell, M.W. Zemansky, *Resonance Radiation and Excited Atoms* (Cambridge University Press, London, 1971)
5. R.H. Dicke, The effect of collisions upon the Doppler width of spectral lines. Phys. Rev. **89**, 472–473 (1953)
6. P.L. Varghese, R.K. Hanson, Collisional narrowing effects on spectral line shapes measured at high resolution. Appl. Opt. **23**(14), 2376–2385 (1984)
7. S.G. Rautian, I.I. Sobel'man, The effect of collisions on the Doppler broadening of spectral lines. Sov. Phys. Usp. **9**, 701–716 (1967)
8. L. Galatry, Simultaneous effect of Doppler and foreign gas broadening on spectral lines. Phys. Rev. **122**, 1218–1223 (1961)
9. F. Rohart, H. Mäder, H.-W. Nicolaisen, Speed dependence of rotational relaxation induced by foreign gas collisions: Studies on CH_3F by millimeter wave coherent transients. J. Chem. Phys. **101**(8), 6475 (1994)

Electronic Spectra of Atoms

9

Electronic spectra of atoms includes the following topics:

1. the role of electron spin and orbital angular momentum
2. the notation employed in describing the electronic state
3. the building-up principle

Key references for the material in this chapter are [1, 2].

9.1 Electron Quantum Numbers

Instead of the vibrational and rotational quantum numbers that have been introduced for molecules, atoms have quantum numbers that derive from the quantum states occupied by their electrons. The individual electrons are characterized by four different quantum numbers:

n	principal quantum number: $1, 2, \ldots$
ℓ	orbital quantum number: $0, \ldots, n-2, n-1$
m or m_ℓ	magnetic quantum number: $0, \ldots, \pm(\ell-1), \pm\ell \quad (2\ell+1$ values)
s or m_s	spin quantum number: $\pm 1/2$

9.2 Electronic Angular Momentum

Thus far we have discussed nuclear motions—now it is time to consider the role of electrons. There are two motions/momenta for electrons: orbital angular momentum and spin angular momentum (Fig. 9.1).

R.K. Hanson et al., *Spectroscopy and Optical Diagnostics for Gases*,
DOI 10.1007/978-3-319-23252-2_9

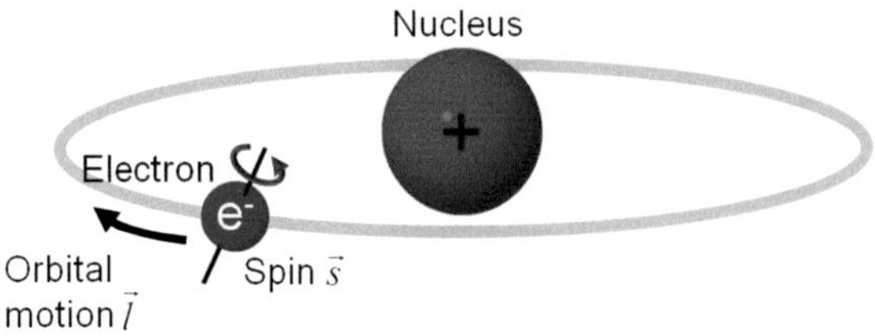

Fig. 9.1 Motions for orbital and spin angular momentum of a single electron atom

9.3 Single Electron Atoms

9.3.1 Orbital Angular Momentum

The orbital angular momentum, $\vec{\ell}$, is a function of the orbital angular momentum quantum number, ℓ [1]. ℓ is an integer in the range $0, 1, 2, \ldots, n-1$, where n is the principal quantum number. n is the primary determinant of energy, roughly related to the electron's distance from the center of the atom.

$$\left|\vec{\ell}\right| = \sqrt{\ell(\ell+1)}\,\hbar \tag{9.1}$$

Orbital angular momentum is a vector quantity with $2\ell+1$ "allowed" directions (each corresponding to a value for m, the magnetic quantum number) with respect to an applied electromagnetic field. Therefore, the degeneracy (i.e., the number of values of m) is $2\ell+1$. Orbital notation utilizes letters to represent specific values of ℓ:

$\ell = 0$, s electron
$\ell = 1$, p electron
$\ell = 2$, d electron

9.3.2 Spin Angular Momentum

The spin angular momentum is

$$\left|\vec{s}\right| = \sqrt{s(s+1)}\,\hbar \tag{9.2}$$

where s is the spin quantum number (1/2). There are two allowed directions: $s_x = \pm 1/2$.

9.3.3 Total Electronic Angular Momentum

The total angular momentum, based on vector addition, is

$$\vec{j} = \vec{s} + \vec{\ell} \tag{9.3}$$

But, owing to quantum constraints, the allowed quantum numbers are limited to

$$\begin{aligned} j &= \ell + s,\ \ell - s \\ &= \ell + 1/2,\ \ell - 1/2 \end{aligned} \tag{9.4}$$

This multiplicity of j values is called spin–orbit coupling.

9.3.4 Term Symbols

Term symbols are used to succinctly summarize the state of the atom. Small letters characterize the "state" of a single electron and large letters characterize the "state" of a whole atom (including multi-electron atoms).

$$\text{Term symbol} = {}^{2S+1}L_J \tag{9.5}$$

where the $2S + 1$ term is the "multiplicity" of state, L is the quantum number for the total orbital angular momentum of the atom, and J is the quantum number for the total angular momentum. For atoms with a single electron in the outer shell, we have the simple result that $S = 1/2$, $L = \ell$, and $J = L \pm 1/2$ (except for $L = 0$ where only $J = 1/2$ is possible, as J cannot be negative). When multiple electrons become involved, only the electrons in unfilled shells contribute. When $L \geq S$, the multiplicity gives the number of possible J values.

$$\textbf{For a single electron:} \quad \vec{j} = \vec{\ell} + \vec{s}$$

$$\textbf{For the atom:} \quad \vec{J} = \vec{L} + \vec{S}$$

or in terms of quantum numbers, $J = L + S, L + S - 1, \ldots, |L - S|$.

There are different couplings possible for atoms. For an atom's four quantum numbers, the selection rules are as follows:

$$\begin{aligned} &\Delta n = \text{any integer} \\ &\Delta L = \pm 1 \\ &\Delta J = 0, \pm 1, (\text{but not } J = 0 \rightarrow 0) \\ &\Delta S = 0\ (\Delta S \neq 0 \text{ is spin-forbidden}) \end{aligned}$$

For example, ${}^2S_{1/2} \rightarrow {}^2P_{1/2}$ or ${}^2P_{3/2}$ are allowed transitions, while ${}^2S_{1/2} \rightarrow {}^4S$ is forbidden.

9.3.5 Example: Hydrogen Atom

Quantum Numbers

$n = 1, 2, 3, \ldots$ principal quantum number
ℓ (and hence L)$= 0, 1, 2, \ldots$ for s, p, d states ($\ell_{max} = n - 1$)
s (and hence S)$= 1/2$

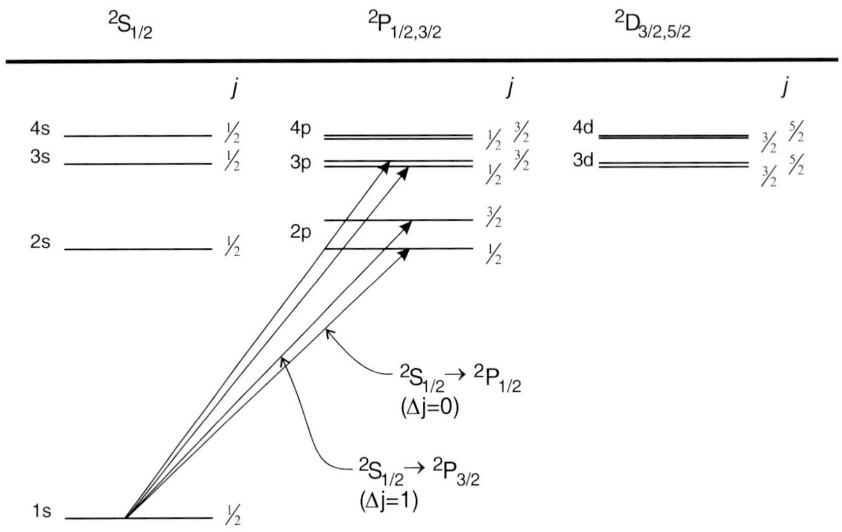

Fig. 9.2 Some of the lower energy levels for the hydrogen atom

States

$^2S_{1/2}, {}^2P_{3/2,1/2}, {}^2D_{5/2,3/2}, \ldots$

Selection Rules

$$\Delta n = \text{any integer}$$
$$\Delta \ell \text{ (and hence } \Delta L) = \pm 1$$
$$\Delta j \text{ (and hence } \Delta J) = 0, \pm 1$$

These selection rules yield "doublets," i.e. a pair of closely spaced lines for each Δn (Fig. 9.2).

9.4 Multi-Electron Atoms

9.4.1 Building-Up Principle

Electron shells fill as the periodic table progresses and obey the Pauli Principle, which states that no two electrons in an atom share all n, ℓ, $\ell_z(m)$, and s_z [1]. Therefore, we fill the $n = 1$ shell with two electrons ($1s$ orbital), and the $n = 2$ shell with eight electrons ($2s$, $2p$ orbitals) (Tables 9.1 and 9.2).

Table 9.1 Building-up principle for the first few quantum numbers

n	ℓ	m	s_z	Number of electrons
1	0	0	$\pm 1/2$	2 electrons
2	0	0	$\pm 1/2$	2 electrons
2	1	1	$\pm 1/2$	6 electrons
		0	$\pm 1/2$	
		-1	$\pm 1/2$	

Table 9.2 Building-up principle for the first ten elements

			$2p$				Atomic term symbol
Element	$1s$	$2s$	p_x	p_y	p_z	Notation	(ground state)
Hydrogen	↑					$1s^1$	$^2S_{1/2}$
Helium	↑↓					$1s^2$	$^1S_0$
Lithium	↑↓	↑				$1s^2 2s^1$	$^2S_{1/2}$
Beryllium	↑↓	↑↓				$1s^2 2s^2$	$^1S_0$
Boron	↑↓	↑↓	↑			$1s^2 2s^2 2p_x^1$	$^2P_{1/2}$
Carbon	↑↓	↑↓	↑	↑		$1s^2 2s^2 2p_x^1 2p_y^1$	$^3P_0$
Nitrogen	↑↓	↑↓	↑	↑	↑	$1s^2 2s^2 2p_x^1 2p_y^1 2p_z^1$	$^4S_{3/2}$
Oxygen	↑↓	↑↓	↑↓	↑	↑	$1s^2 2s^2 2p_x^2 2p_y^1 2p_z^1$	$^3P_2$
Fluorine	↑↓	↑↓	↑↓	↑↓	↑	$1s^2 2s^2 2p_x^2 2p_y^2 2p_z^1$	$^2P_{3/2}$
Neon	↑↓	↑↓	↑↓	↑↓	↑↓	$1s^2 2s^2 2p_x^2 2p_y^2 2p_z^2$	^1S_o

Note:

- The notations p_x, p_y, and p_z correspond to the three allowed values of m for the p-orbital ($m = 1, 0, -1$)
- $2p_x^2 2p_y^2 2p_z^2$ can also be denoted $2p^6$.
- Neon and Helium have closed shells and are thus inert. They have no net electronic (orbital or spin) angular momentum. That is, the spins ($\pm 1/2$) and m ($\pm 1, 0$) quantum numbers of their electrons sum to zero.
- Hund's principle: electrons tend to occupy degenerate orbitals singly with their spins parallel, when possible (e.g., note the progression for B, C, and N).
- Only the electrons in unfilled shells need be considered in establishing the atomic term symbol, e.g., only the $2s$ electron for Lithium.

9.4.2 Examples

Carbon

What is the term symbol for C in its ground (lowest energy) state?

The electron distribution can be represented by $1s^2 2s^2 2p^2$ or $1s^2 2s^2 2p_x^1 2p_y^1$. We have two electrons in an unfilled shell with the same n and ℓ (called equivalent

Table 9.3 Some possible electronic configurations for the carbon atom

1*s*	2*s*		2*p*	
↑↓	↑↓	↑	↑	
↑↓	↑↓	↑		↑
↑↓	↑↓	↑↓		
↑↓	↑↓		↓	↑

Table 9.4 Terms symbols for equivalent electrons

Configuration	Possible term symbols
s^2	^1S
p^2	^1S, ^1D, ^3P
p^3	^2P, ^2D, ^4S
p^4	^1S, ^1D, ^3P
p^5	^2P
p^6	^1S
d^2	^1S, ^1D, ^1G, ^3P, ^3F

electrons), and we must decide how they are arranged. There are, in fact, 15 possible electronic configurations (e.g., see Banwell and McCash [2, pp. 149–151]), of which four are shown in Table 9.3.

There are three net ways for the orbital (angular) momentum to sum, giving a total orbital momentum of $2, 1,$ or 0. The Pauli-allowed spin pairings for each configuration determine the spin multiplicity ($2S + 1$) and the possible values of J (here, the possible values of J simply contribute to the total multiplicity).

For a given number of equivalent electrons in a specified orbital, there are several states of varying energies and thus several possible term symbols. The possibilities are expressed in Table 9.4.

Since carbon has two p electrons, there are three term symbols that describe the available energy states: ^3P, ^1D, and ^1S. Hund's Rules will help us determine which state has the lowest energy.

Hund's Rules:

1. Maximize the value of S consistent with the Pauli Principle. The lowest energy state will be the one with the largest spin multiplicity.
2. Having fixed S, maximize L, i.e., for states of equal multiplicity, the greatest L has lowest energy.

Therefore, for C:

- $1s^2 2s^2 2p^2 (^3P)$ is the ground state, and
- $1s^2 2s^2 2p^2 (^1D)$ and $1s^2 2s^2 2p^2 (^1S)$ are "excited"

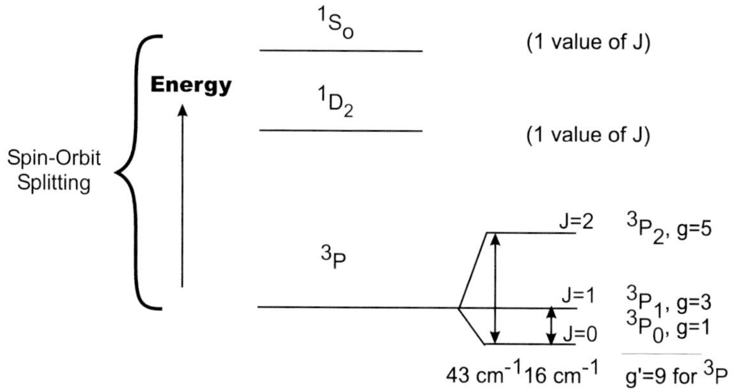

Fig. 9.3 States for a carbon atom

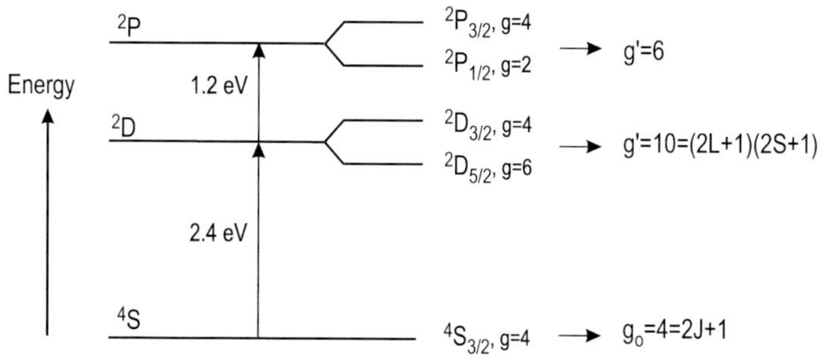

Fig. 9.4 States for a nitrogen atom

Note: 1. degeneracy = $2J + 1$ for a specific J
2. multiplicity = the # of J states when $L \geq S$

The energy spacing of the $J = 0$, 1, and 2 states of the ^3P level increases with the atomic number (i.e., nuclear charge, Z). When energy increases with J, as shown in Fig. 9.3, the multiplet is called "normal"; cases in which energy decreases as J increases are termed "inverted."

Atomic Nitrogen

The electron distribution for N is $1s^2 2s^2 2p^3$. What is its term symbol? There are three "equivalent" $2p$ electrons. From Table 9.4, the possible solutions for term symbols are: ^4S, ^2D, ^2P. Following Hund's rules:

Here g' is the total degeneracy for each level, and $g' = (2L + 1)(2S + 1)$. Note that the ^2D states are *inverted*; i.e., the $^2D_{5/2}$ state has lower energy than $^2D_{3/2}$. Recall from statistical mechanics,

$$Q^N_{\text{elec}} = \sum g_i \exp(-E_i/kT)$$
$$= 4 + 6\exp\left(\frac{-27{,}659}{T}\right) + 4\exp\left(\frac{-27{,}669}{T}\right) + 6\exp\left(\frac{-41{,}495}{T}\right)$$

Thus, except for very high temperatures,

$$\boxed{Q^N_{\text{elec}} \approx g^N_o = 4}$$

owing to large spin–orbit splitting.

9.4.3 Hydrogen-Like Species

After hydrogen, the next simplest "hydrogen-like" species (i.e., those with 1 odd electron and a closed shell) are lithium (Li), sodium (Na), potassium (K), rubidium (Rb), etc., all members of the first column of the periodic table. These atoms are known as the alkali metals.

Lithium (Li)

The ground state for Lithium is $1s^22s^1$ ($^2S_{1/2}$). The excited states are $1s^22p^1$, $1s^23s^1$, $1s^23p^1$, $1s^23d^1$, as shown in Fig. 9.5.

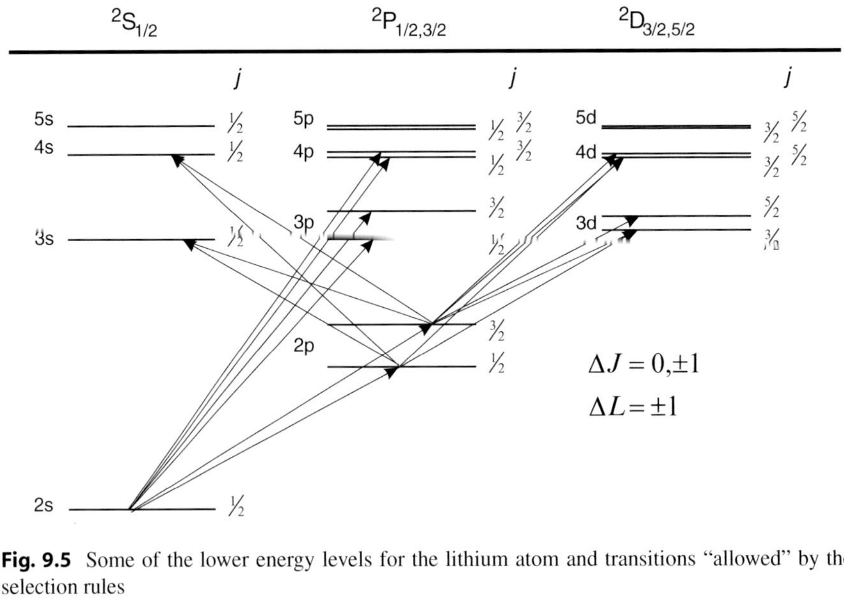

Fig. 9.5 Some of the lower energy levels for the lithium atom and transitions "allowed" by the selection rules

CN Radical

$$-\mathrm{C}\equiv\mathrm{N} \qquad \begin{array}{c} | \\ -\ \mathrm{C}\ - \\ | \end{array}$$

Note that three shared electrons are needed for N-bonding to complete the $n = 2$ ($2s$, $2p$) shell; four for C-atoms. This sharing of electrons to fill outer shells is *covalent bonding*.

9.4.4 Zeeman Effect

We can use Zeeman splitting to infer J, L, g, etc. The basis for the Zeeman effect is that circulating current (moving electrons) can interact with an external magnetic field. The magnetic dipole of an atom is $\vec{\mu}$ and is given by

$$\vec{\mu} = \frac{ge}{2m}\vec{J} \tag{9.6}$$

where g = Landé factor with a value of 0–2, $\vec{J}$ = total angular momentum, e is electron charge and m is the mass of the atom. Interaction with a field of strength B leads to ΔE-splitting between levels of equal J

$$\Delta E = \vec{\mu}\cdot\vec{B} \ \rightarrow\ \mu_z B_z \text{ for a field in the } z\text{-direction} \tag{9.7}$$

Example:

$$\begin{aligned} &\text{for } g = 1,\ \mu \approx 10^{-23}\ \mathrm{J/T} \\ &\text{for } B = 1\ \mathrm{T},\ \Delta E = 10^{-23}\ \mathrm{J} \\ &\Delta \nu \approx 0.5\,\mathrm{cm}^{-1} \end{aligned}$$

Therefore, this effect is observable with a narrow linewidth laser (i.e., a laser with $\Delta\nu \leq 0.5\,\mathrm{cm}^{-1}$ linewidth).

9.4.5 Nuclear Spin

The total angular momentum for an atom also includes the intrinsic momentum of the nucleus; however, we have thus far ignored it because the change in energy due to nuclear spin is small (called hyperfine splitting). Nonetheless, nuclear spin needs to be considered for a thorough analysis of atomic angular momentum. Similar to molecules, the nuclear spin quantum number is I, and has values of zero, integers, or half-integers.

$$I = 0, \text{ integers, half-integers}$$

For an atom, the total angular momentum is F.

$$\vec{F}\ (\text{electronic} + \text{nuclear}) = \sqrt{F(F+1)}\,\hbar \tag{9.8}$$

$$F = J + I, J + I - 1, \ldots, |J - I| \tag{9.9}$$

Therefore, there are $2I + 1$ or $2J + 1$ states (whichever is less). Typical values of energy difference between states (i.e., splitting) are $\approx 10^{-3}$ that of electron spin splitting. Thus, we can conclude that nuclear spin is:

1. often negligible, and
2. observable only with
   (a) narrow linewidth lasers, and
   (b) narrow lineshapes (i.e., Doppler-free)

9.5 Exercises

1. Is the term symbol accurate for a particular state quoted as (a) $^4S_1$, (b) $^2D_{7/2}$, and (c) $^0P_1$? Explain why.
2. The term symbol for a particular atomic state is quoted as $^4\mathrm{D}_{5/2}$. What are the values of L, S, and J for this state? What is the minimum number of electrons which could give rise to this?
3. The moving electrons can interact with an external magnetic field, resulting in the magnetic dipole of an atom [Eq. (9.6)]. For a field in the z-direction,

$$\vec{\mu}_z = \frac{ge}{2m}\vec{J}_z,$$

where $J_z = J, J - 1, \ldots, -J$; g is the Landé splitting factor. This factor depends on the state of the electrons in the atom and is given by:

$$g = \frac{3}{2} + \frac{S(S+1) - L(L+1)}{2J(J+1)}.$$

This expression shows that the application of the field splits the original levels into $(2J + 1)$ different energy levels. Note that the selection rule for the split energy levels is:

$$\Delta J = 0, \pm 1.$$

With this knowledge, try to answer the following questions:

(a) We discussed the electronic energy levels of lithium (hydrogen-like) in class and reproduced some of the lower energy levels in the figure below, showing the doublet lines arising between the ^2S and ^2P states. Write down the corresponding quantum numbers (S, L, J) of these three states and the Landé splitting factor g.

(b) When a magnetic field is applied to this lithium atom, how would its spectrum be affected?

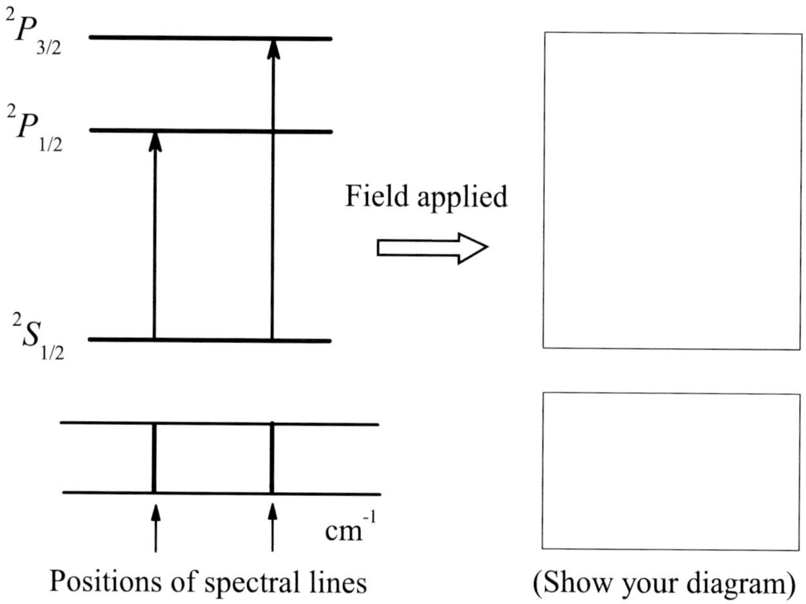

(Part of the energy levels of lithium atom)

References

1. G. Herzberg, *Atomic Spectra and Atomic Structure*, 2nd edn. (Dover, New York, 1944)
2. C.N. Banwell, E.M. McCash, *Fundamentals of Molecular Spectroscopy*, 4th edn. (McGraw-Hill International (UK) Limited, London, 1994)

Electronic Spectra of Diatomic Molecules: Improved Treatment

10

10.1 Term Symbols for Diatomic Molecules

Term symbols, introduced in the previous chapter, are the notation used to characterize key features of electron spin and orbital angular momentum.

For an atom, the term symbol is: ${}^{2S+1}L_J$
For a diatomic, the term symbol is: ${}^{2S+1}\Lambda_\Omega$

Important terms to define are $\vec{S}$, $\vec{\Sigma}$, $\vec{\Lambda}$, and $\vec{\Omega}$:

$\vec{\Lambda}$ projection of orbital angular momentum on the internuclear axis
Its magnitude is $\left|\vec{\Lambda}\right| = \Lambda\hbar$, where Λ is an integer, and the symbols associated with different values of Λ are shown in the table below.

Λ	0	1	2
Symbol	Σ	Π	Δ

$\vec{\mathbf{S}}$ total electronic spin angular momentum (the net sum of electron spin in unfilled shells). The magnitude is $\left|\vec{S}\right| = S\hbar$, where S will have $1/2$-integer values.

$\vec{\Sigma}$ projection of $\vec{S}$ onto the internuclear axis (only defined when $\Lambda \neq 0$). The magnitude of this projection is $\left|\vec{\Sigma}\right| = \Sigma\hbar$, and the allowed values of Σ are

$$\Sigma = S, S-1, \ldots, -S \qquad (2S+1 \text{ values})$$

R.K. Hanson et al., *Spectroscopy and Optical Diagnostics for Gases*,
DOI 10.1007/978-3-319-23252-2_10

$\vec{\Omega}$ sum of projections along the internuclear axis of electron spin and orbital angular momentum

$$\vec{\Omega} = \vec{\Sigma} + \vec{\Lambda}$$

$$\Omega = \Lambda + S, \Lambda + S - 1, \ldots, |\Lambda - S| \quad (2S + 1 \text{ values for } \Lambda \geq S)$$

Examples

NO The ground state for NO is $X^2\Pi$.
$S = 1/2$, $\Lambda = 1$, $\Omega = 3/2, 1/2$
There are two spin-split sub-states: ${}^2\Pi_{1/2}$, ${}^2\Pi_{3/2}$
Separation: $121\,\mathrm{cm}^{-1}$

CO The ground state for CO is $X^1\Sigma^+$.
$S = 0$ and $\Lambda = 0$, therefore Ω is unnecessary. This is a rigid rotor molecule with no influence from electrons. Easiest case!

O_2 The ground state for O_2 is $X^3\Sigma_g^-$.
$S = 1$, $\Lambda = 0$
The $^-$ and $_g$ are notations about symmetric properties of wave functions. This is an example of a molecule that is modelled by Hund's case ***b***, discussed below.

10.2 Common Molecular Models for Diatomics

There are four common molecular models that are used to describe diatomic electronic spectra:

Rigid rotor	$\Lambda = 0, S = 0$
Symmetric top	$\Lambda \neq 0, S = 0$
Hund's ***a***	$\Lambda \neq 0, S \neq 0$
Hund's ***b***	$\Lambda = 0, S \neq 0$

The rigid rotor and symmetric top models have no spin, and thus their multiplicity $(2S + 1)$ is one; these states are called "singlets." For the Hund's cases, the influence of spin on the electronic state structure must be considered through interactions of Λ and Σ.

10.2.1 Rigid Rotor (${}^1\Sigma$)

The simplest model for molecular rotation assumes that electron motions do not contribute to the rotational energy. Rotation of the nuclei occurs about an axis perpendicular to the A-axis (i.e., the B-axis). Recall that $I_A \approx 0$ and $I_B = I_C$ (Fig. 10.1).

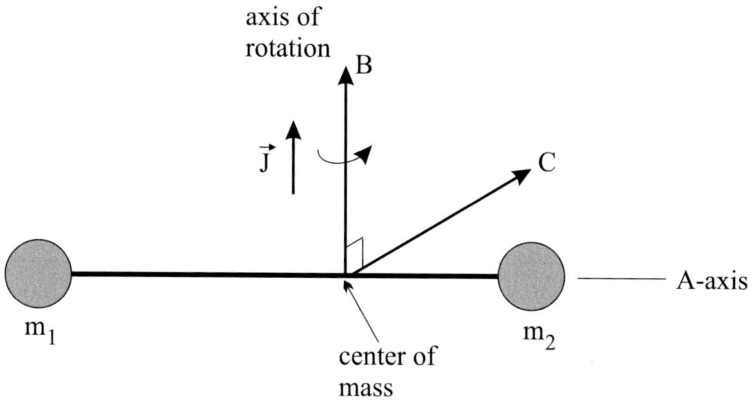

Fig. 10.1 Rigid rotor model for molecular motion

When Λ and $S = 0$, the molecule is $^1\Sigma$ type and Ω is not defined. Note that $\Lambda = 0$ means that the projection of the orbital angular momentum onto the A-axis is zero, and that rotation must thus be around the B-axis.

Rotational Energy

The total rotational energy for a rigid rotor, including centrifugal distortion is, as before,

$$F(J) = B_v J(J+1) - D_v J^2(J+1)^2 \tag{10.1}$$

Total Energy

Rovibronic transitions (those that include elect*ronic*, *vib*rational, and *ro*tational changes in quantum number) have a total energy that includes contributions from each mode, i.e. $\Delta E = \Delta T_e + \Delta G + \Delta F$

$$\text{where} \quad E(T_e, v, J) = T_e + G(v) + F(J). \tag{10.2}$$

Selection Rules

The selection rules for these transitions, as described before, are:

Rotational spectra:	$\Delta J = J' - J'' = +1$
Rovibrational spectra:	$\Delta v = v' - v'' = +1$
	$\Delta J = \pm 1$
Rovibronic spectra:	Δv determined by Franck–Condon factors
	$\Delta J = \pm 1$

Note: An alternate form for the selection rules is used in some texts, i.e. $\Delta\alpha = \alpha_{\text{final}} - \alpha_{\text{initial}}$ where ($\alpha = J$ or v).

Intensity Distribution

Within each band (v', v''), the intensity distribution follows the Boltzmann distribution for J modified by a J-dependent branching ratio (i.e., for the P and R branches), known as the Hönl–London factor (defined later in this chapter). Similarly, the relative intensities among all the vibrational bands originating from a single initial level v_{initial} to all possible final levels v_{final} are given by Franck–Condon factors (a manifestation of the Franck–Condon principle). The relative total emission or absorption from v_{initial} (i.e., to all values of v_{final}) depends directly on the Boltzmann fraction in that level, i.e. $n_{v_{\text{initial}}}/n$, and also on an overall Einstein coefficient or oscillator strength for the specific electronic system (i.e., a specific pair of lower and upper electronic states), as presented in Sect. 10.3.

Examples

Most stable diatomics, including CO, Cl_2, Br_2, N_2, H_2 are Rigid Rotors. Exceptions are NO ($X^2\Pi$) and O_2 ($X^3\Sigma$). (Note that there are no $X\Delta$ states for the diatomics listed in Herzberg—all X states are Σ or Π!) Some linear polyatomics such as CO_2 ($\tilde{X}^1\Sigma_g^+$) and both HCN and N_2O ($\tilde{X}^1\Sigma^+$) are Rigid Rotors with ${}^1\Sigma$ ground states. Remember, however, that nuclear spin can have an impact on the statistics of homonuclear diatomic molecules.

10.2.2 Symmetric Top

Symmetric tops have a non-zero projection of orbital angular momentum on the internuclear axis and zero spin ($\Lambda \neq 0, S = 0$). Thus, its ground states are ${}^1\Pi$, ${}^1\Delta$ (although, as mentioned in the previous section, there are no known $X^1\Delta$ ground states for diatomics) (Fig. 10.2).

The important components are

$\vec{N}$	angular momentum of nuclei
$\vec{\Lambda}$	A-axis projection of electron orbital angular momentum
$\vec{J}$	total angular momentum; $\vec{J} = \vec{N} + \vec{\Lambda}$

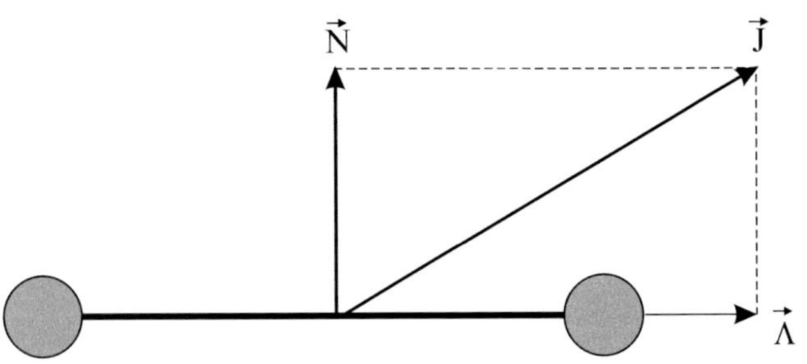

Fig. 10.2 Symmetric top model for molecular motion

Only the axial component of orbital angular momentum is used, because only $\vec{\Lambda}$ is a "good" quantum number, i.e. a constant of the motion.

Rotational Energy

The total rotational energy for a symmetric top is

$$F(J) = BJ(J+1) + (A-B)\Lambda^2 \qquad J = \Lambda, \Lambda+1, \ldots \tag{10.3}$$

The constants A and B are given as before by

$$A, B = \frac{h}{8\pi^2 c I_{A,B}} \tag{10.4}$$

Therefore, the symmetric top energy levels have the same spacing as the Rigid Rotor, but with a constant offset. Note, however, that since I_A is small compared to I_B, A is large compared to B. Lines with $J < \Lambda$ are missing, as $J = \Lambda, \Lambda+1, \ldots$.

Selection Rules

The selection rules for symmetric top electronic spectra are

$$\begin{aligned} \Delta\Lambda &= 0 \qquad \Delta J = \pm 1, 0 \qquad (\Delta J = 0 \text{ may be weak}) \\ \Delta\Lambda &= \pm 1 \qquad \Delta J = \pm 1, 0 \end{aligned}$$

As a result of having a Q branch (i.e., $\Delta J = 0$), the bands for a symmetric top will be double-headed, in contrast to the single-headed character of rigid rotor bands.

Spectra

The spectra for the case where $\Delta\Lambda = 0$ are discussed here. The upper- and lower-state energies are described by the following equations:

$$T' = B'J'(J'+1) + (A'-B')\Lambda^2 + G(v') + T'_e \tag{10.5}$$

$$T'' = B''J''(J''+1) + (A''-B'')\Lambda^2 + G(v'') + T''_e \tag{10.6}$$

$$\bar{\nu}_\infty = \text{upper (for } J' = 0) - \text{lower (for } J'' = 0) = \text{constant} \tag{10.7}$$

$T''_e = 0$ for the ground state. Then, for the three branches, the line positions are as follows:

$$P(J'') = \bar{\nu}_\infty - (B'+B'')J + (B'-B'')J^2 \tag{10.8}$$

$$Q(J'') = \bar{\nu}_\infty + (B'-B'')J + (B'-B'')J^2 \tag{10.9}$$

$$R(J'') = \bar{\nu}_\infty + (B'+B'')(J+1) + (B'-B'')(J+1)^2 \tag{10.10}$$

Using the rotational number m for the three branches,

$$m_P = -J$$

$$m_Q = +J$$

$$m_R = J + 1$$

the line positions for the P and R branches become

$$\overline{\nu} = \bar{\nu}_\infty + am + bm^2, \qquad (10.11)$$

where $a = B' + B''$ and $b = B' - B''$, and the line positions for the Q branch become

$$\overline{\nu} = \bar{\nu}_\infty + bm + bm^2 \qquad (10.12)$$

Note that the three branches now can lead to two bandheads (see the Fortrat parabola plot below for the case of a $^1\Delta \leftarrow {}^1\Delta$ band) (Fig. 10.3).

The Fortrat parabola shows the bandheads in the Q and R branches for the typical case of $B' < B''$. For the $^1\Delta \leftarrow {}^1\Delta$ case, $J_{\min} = 2$ and therefore $m_{\min} = 3$ for the R branch, $m_{\min} = 2$ for the Q branch and $m = -3$ is the first line in the P branch, resulting in multiple missing lines near the origin.

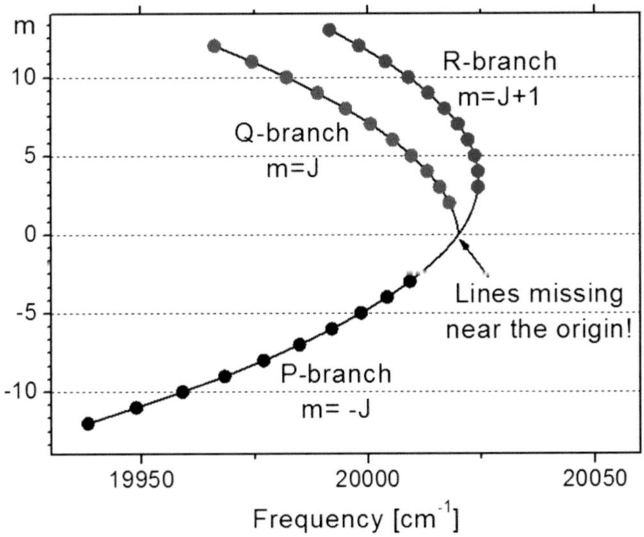

Fig. 10.3 Symmetric top model for molecular motion, $^1\Delta \leftarrow {}^1\Delta$ case

Intensity Distribution

Actual relative intensities depend on n_J/n and "relative intensity factors/line strengths," also known as Hönl–London factors, denoted below by $S_J^{P,Q,R}$. Differences in Hönl–London factors indicate the breakdown of the principle of equal probability [1].

Example: Hönl–London factors for symmetric top (p. 208 of [1])

For $\Delta\Lambda = 0$

$$\left.\begin{aligned} S_J^R &= \tfrac{(J+1+\Lambda)(J+1-\Lambda)}{J+1} \approx J+1 \ \ (J \gg \Lambda) \\ S_J^Q &= \tfrac{(2J+1)\Lambda^2}{J(J+1)} \approx \tfrac{2\Lambda^2}{J} \approx 0 \ \ \text{for large } J \\ S_J^P &= \tfrac{(J+\Lambda)(J-\Lambda)}{J} \approx J \ \ \text{for large } J \end{aligned}\right\} \sum S_J = 2J+1$$

Note:
1. $\Sigma S_J = 2J + 1$, the total degeneracy!
2. The R-branch line, for a specified J, is approximately $(J + 1)/J$ times as strong as the P branch line.
3. For $\Delta\Lambda = \pm 1$, and $J \gg \Lambda$

$$\left.\begin{aligned} S_J^R &\approx \tfrac{(2J+1)}{4} \\ S_J^Q &\approx \tfrac{(2J+1)}{2} \\ S_J^P &= \tfrac{(2J+1)}{4} \end{aligned}\right\} \sum S_J = 2J+1$$

The Q branch lines are thus twice as strong as the P and R lines! Therefore, the $\Delta\Lambda$ value is important in determining the relative line and branch strengths of rovibronic spectra.

Example: $X = {}^1\Pi$

If $X = {}^1\Pi$, then the following transitions are possible for changes in Λ of 0 or ± 1. That is, there are three separate "systems" of bands possible from $X^1\Pi$.

${}^1\Pi \leftarrow {}^1\Pi$	${}^1\Delta \leftarrow {}^1\Pi$	${}^1\Sigma \leftarrow {}^1\Pi$
$\Delta\Lambda = 0$	$\Delta\Lambda = 1$	$\Delta\Lambda = -1$

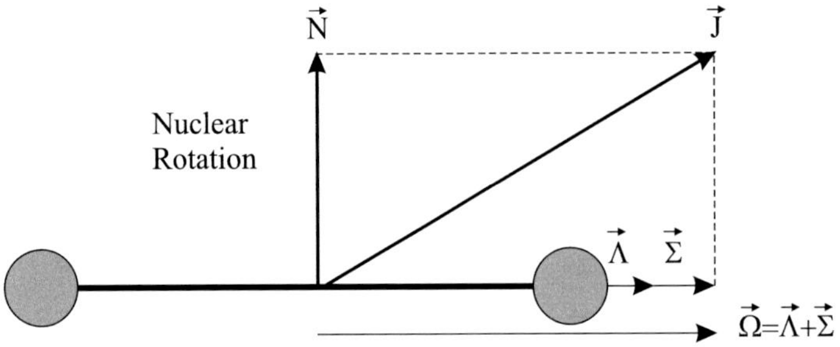

Fig. 10.4 Coupling of Σ and Λ

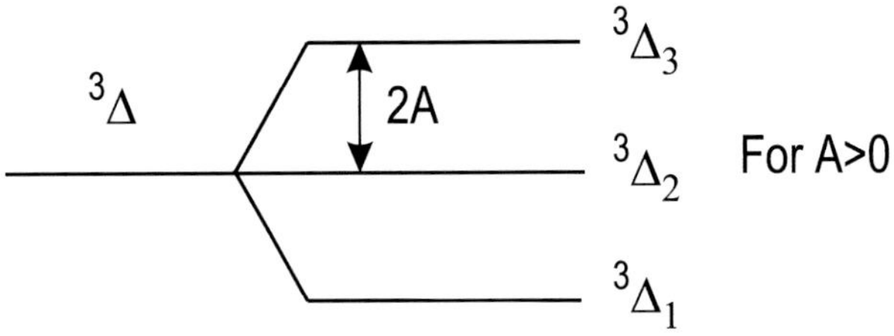

Fig. 10.5 Spin–orbit splitting

10.2.3 Interaction of Λ and Σ

Understanding the interactions and coupling of Λ and Σ is a key to modelling the influence of spin on the electronic state structure. When $\Lambda \neq 0$ and $S \neq 0$, they combine to form a net component of Ω, as shown in Fig. 10.4.

The presence of $\Lambda \neq 0$ implies that there is an associated magnetic field due to net current about the axis. This field interacts with spinning electrons. This effect is known as spin–orbit coupling (or spin-splitting) (Fig. 10.5).

Examples

$^3\Delta$ has three components ($^3\Delta_3$, $^3\Delta_2$, $^3\Delta_1$) corresponding to $S = 1$, $\Lambda = 2$ and $\Omega = 3, 2, 1$ (since $\Sigma = 1, 0, -1$). These states have different electronic energies, which may be represented by

$$T_e = T_0 + A\Lambda\Sigma \tag{10.13}$$

Here, A is the spin–orbit coupling constant (it's not exactly the same A as in the symmetric top, but it is similar; keep in mind that these are *models*). A generally increases with molecular weight and the number of electrons. T_0 is the energy without interaction.

For $^3\Delta$, $\Lambda = 2$, $S = 1$, and $\Sigma = 1, 0, -1$. Therefore,

$$T_e = T_0 + A(2)\begin{pmatrix} 1 \\ 0 \\ -1 \end{pmatrix}$$

Some sample spin–orbit coupling constants are listed below.

$$A_{\text{BeH}} \approx 2\,\text{cm}^{-1} \tag{10.14}$$

$$A_{\text{NO}} \approx 124\,\text{cm}^{-1} \tag{10.15}$$

$$A_{\text{HgH}} \approx 3600\,\text{cm}^{-1} \tag{10.16}$$

$$A_{\text{OH}} \approx -140\,\text{cm}^{-1} \tag{10.17}$$

Note that the spin–orbit coupling constant for OH is negative. (See Herzberg, Vol. I for details; p. 215/216, 232, 558/559, 561.)

Note: 1. $Y = \frac{A}{B_v}$ is the ratio of the spin–orbit constant and B_v
2. Values for A are given in tables in Herzberg, Vol. I

Now, finally, we are ready to consider rotational levels for cases where $S \neq 0$. There are two primary cases: Hund's ***a*** and Hund's ***b***.

10.2.4 Hund's Case *a*

Hund's case ***a*** is for $\Lambda \neq 0$ and $S \neq 0$ with Σ defined as $\Sigma = S, S-1, \ldots, -S$. Replace Λ with Ω in $F(J)$ for the symmetric top to get [1]

$$F(J) = BJ(J+1) + (A-B)\Omega^2 \tag{10.18}$$
$$\Omega = \Lambda + S, \Lambda + S - 1, \ldots, |\Lambda - S|$$
$$J = \Omega, \Omega + 1, \ldots$$

Remember that A for the equation above is $A = h/8\pi^2 I_A c$; it is *not* the spin–orbit constant.

There are P, Q, and R branches for each value of Ω. For example, with $^2\Pi$, we have $\Omega = 3/2$ and $1/2$, i.e. two electronic sub-states, giving a total of $2 \times 3 = 6$ branches.

10.2.5 Hund's Case *b*

Hund's case ***b*** applies when spin is *not* coupled to the A-axis [1] (Fig. 10.6), e.g.:

1. for $\Lambda = 0$ (so $\vec{\Sigma}$ is not defined, and we must use $\vec{S}$)
2. at high J, especially for hydrides, even with $\Lambda \neq 0$

The allowed J are $J = N + S, N + S - 1, \ldots, N - S$, and $J \geq 0$ only. For this case, $\vec{S}$ and $\vec{N}$ couple directly.

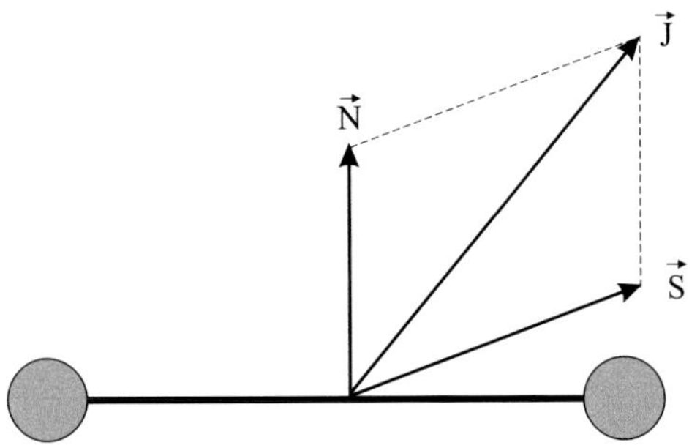

Fig. 10.6 Molecular model for Hund's ***b***

Example: O_2
The ground state $X^3\Sigma$ has three J's for each N except $N = 0$. These energy levels are labelled $F_1(N)$, $F_2(N)$ and $F_3(N)$.

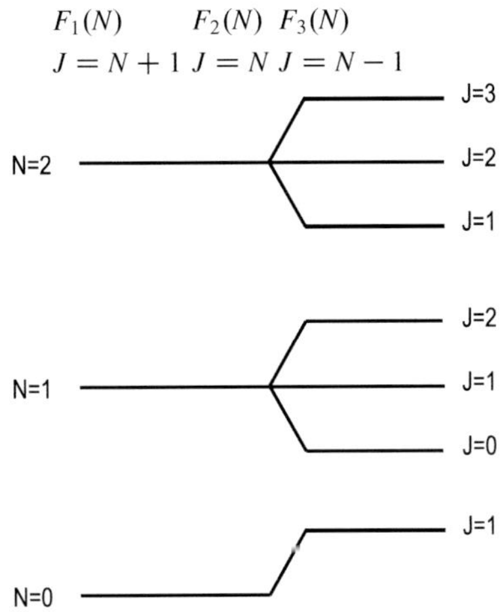

There are split rotational levels for $N > 0$, and each level has a degeneracy of $2J + 1$ and a sum of Hönl–London factors of $2J + 1$. The minimum J is $|N - S|$.

Note: In the $N = 0$ level, since only spin is active and is not equal to zero (i.e., $S = 1$), the minimum value of J is 1.

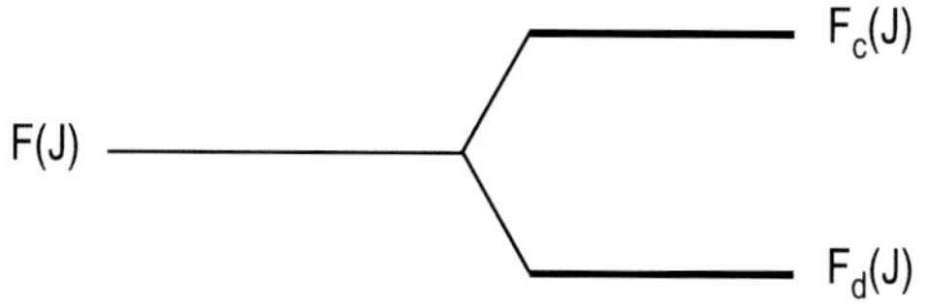

Fig. 10.7 Lambda-doubling results in two different energy levels

10.2.6 Λ-Doubling

There is further complexity in the energy levels resulting from a phenomenon known as Λ-doubling. The two orientations of $\vec{\Lambda}$ ($\pm\Lambda$ along the A-axis) have slightly different energies, owing to different coupling with nuclear rotation (i.e., $\vec{N}$ and $\vec{\Lambda}$ interaction). The result is

$$F(J) \rightarrow F_c(J) \text{ and } F_d(J)$$

That is, there are two components to the energy, $F(J)$ (Fig. 10.7).

By definition, $F_c(J) > F_d(J)$,[1] i.e. the c state is the higher-energy state.

Lambda doubling usually results in a very small change in energy, thus affecting the Boltzmann fraction only slightly (other than adding a state). A more important aspect is found in the change of parity between Λ-doubled states, which reduces the accessible fraction of molecules for a given transition (due to selection rules).

10.2.7 Comment on Models

1. Models are only approximations to real molecules; don't think of them as exact!
2. Coupling may change as J ranges from low to high values.

10.3 Quantitative Absorption

This section is a review of Beer's Law and spectral absorption as interpreted for molecules with multiplet structure.

$$\text{Beer's Law} \quad \left(\frac{I}{I^0}\right)_\nu = \exp(-k_\nu L) \tag{10.19}$$

For a two-level system, we had

$$k_\nu = S_{12}\phi(\nu) = \left(\frac{\pi e^2}{m_e c}\right) n_1 f_{12}\left(1 - \exp(-h\nu/kT)\right)\phi(\nu), \tag{10.20}$$

[1]Note that the subscripts c and d are replaced by e and f, respectively, in some literature.

where S_{12} is the integrated absorption intensity with units of $\mathrm{cm^{-1}\,s^{-1}}$ [see Eq. (7.48)]. How do we evaluate n_1 and f_{12} in a complex, multiple level system? There are two issues:

1. The Boltzmann fraction, and
2. The oscillator strength for a specific transition

10.3.1 Boltzmann Fraction

$$n_1 = n_i \frac{n_1}{n_i} \tag{10.21}$$

$$\begin{aligned} \text{where } n_i &= \text{the total number density of species i and} \\ n_1/n_i &= \text{the fraction of species } i \text{ in state/level 1} \end{aligned}$$

The state/level is specified through the quantum numbers, e.g. n (elec), v (vib), Σ (spin), Λ (orbital), J (total angular momentum), N (nuclear rotation), c or d (Λ-component).

$$\frac{n_1}{n_i} = \frac{N_i(n, v, \Sigma, \Lambda, J, N)}{N_i} \tag{10.22}$$

Λ gives the orientation and distinguishes Λ-doubling.

10.3.2 Oscillator Strength

It is common to denote the oscillator strength of a *specific*, single transition (i.e., from *one* of the J'' substates to a specific J' substate) by $f_{J''J'}$, and to view this transition strength as the product of a system strength, vibrational factor (fraction) and rotational factor (fraction), i.e.,

$$f_{12} = f_{(m,v'',J'')(n,v',J')} \tag{10.23}$$

$$= f_{J''J'} \tag{10.24}$$

$$= \underbrace{f_{\text{el}}}_{\substack{\text{“system” osc.}\\ \text{strength}}} \times \underbrace{q_{v''v'}}_{\substack{\text{Franck–Condon}\\ \text{factor}}} \times \underbrace{\frac{S_{J''J'}}{2J''+1}}_{\substack{\text{normalized H-L}\\ \text{factor or line}\\ \text{strength}}} \tag{10.25}$$

We will show later that the sum of line strengths from a specific substate of J'' (i.e., with a specific N and Λ component (c or d)) to all possible J' is simply $2J'' + 1$; further, $\sum_{v'} q_{v''v'} = 1$. **Thus, the effective total oscillator strength (i.e., for the sum of all radiative transitions) for each of the substates, i.e. with specific values of J'', N'', Λ'' and v'', is f_{el}.**

Note: 1. $\sum_{J'} S_{J''J'} = (2J'' + 1)[(2S + 1)\delta]$
$\sum_{J'} S_{J''J'}$ is the sum over all allowed J' (upper states) for the *combined* lower substates with specific $J = J''$. $\delta = 1$ for $\Sigma - \Sigma$, otherwise $\delta = 2$ (to account for Λ-doubling!). The term $[(2S+1)\delta] = 4$ for OH's $A^2\Sigma \leftarrow X^2\Pi$ system, since there are four substates with a given J''.
2. $\sum_{v',J''} f_{J''J'} = [(2S + 1)\delta]f_{el}$
where $\sum_{v',J''}$ is the sum over v' and all the substates in J''. This sum is f_{el} for a *single* J'' substate.

Remarks

1. It is common to use band oscillator strengths (available in the literature).

$$f_{v''v'} = f_{el} q_{v''v'}$$

e.g., $f_{00} = 0.001$ (OH; $A^2\Sigma \leftarrow X^2\Pi$)

2. Then

$$f_{J''J'} = f_{v''v'}\left(\frac{S_{J''J'}}{2J'' + 1}\right)$$

e.g., if only P and R are allowed,

$$S^P_{J''J'} = J'' \tag{10.26}$$

$$S^R_{J''J'} = J'' + 1 \tag{10.27}$$

Therefore, the intensities for R and P transitions from a specific J'' are similar, except for small J'', but the R-branch transitions are stronger than the P-branch transitions by the ratio $(J'' + 1)/J''$.

3. In some cases, an additional "correction term" $T_{J''J'}$ is used, e.g. in OH.

$$f_{J''J'} = f_{v''v'}\left(\frac{S_{J''J'}}{2J'' + 1}\right)T_{J''J'},$$

where $T_{J''J'}$ is always near 1.

4. In terms of A-coefficients, we may also write

$$f_{v''v'} = \left(\frac{m_e c\lambda^2}{8\pi^2 e^2}\right) A_{v'v''} \left(\frac{g_{e'}}{g_{e''}}\right)$$

5. And recalling Eq. (7.54) we may write

$$f_{v''v'} = \frac{g_{e'}}{g_{e''}} f_{v'v''}$$

We now know enough to tackle a real (and important) molecule, OH; see Chap. 14.

10.4 Exercises

1. A rovibronic transition in nitric oxide (NO) occurs at 226 nm. The oscillator strength for the transition between states 1 and 2 is given by

$$f_{12} = f_{\text{el}} \times q_{v''v'} \times \frac{S_{J''J'}}{2J''+1} = f_{21}\frac{g'}{g''}$$

where f_{el} is the oscillator strength for transitions between the lower and upper electronic states, $q_{v''v'}$ is the Franck–Condon factor and $S_{J''J'}/(2J''+1)$ is the normalized Hönl–London factor. For this transition of NO,

$$f_{\text{el}} = 3.0 \times 10^{-3}$$
$$q_{v''v'} = 1.673 \times 10^{-1}$$
$$\frac{S_{J''J'}}{2J''+1} = 0.5$$

Consider a gas mixture at 1000 K and 2 atm, with an effective collisional broadening coefficient for this NO transition of $2\gamma = 0.09\,\text{cm}^{-1}\,\text{atm}^{-1}$, and a number density in the absorbing state 1 of $n_1 = 3.7 \times 10^{14}\,\text{cm}^{-3}$.

(a) Determine the peak absorption coefficient k_{ν_0} in cm^{-1} at the above temperature and pressure.

(b) Find the fractional absorption at line center and at a detuning of $0.2\,\text{cm}^{-1}$ for path lengths of 1 and 10 cm.

(c) Determine the spontaneous lifetime of the upper state. You may assume the electronic degeneracies of the upper and lower states are both 2.

(d) What is the mole fraction of NO (in ppm) for this gas mixture, assuming an approximate Boltzmann fraction of NO molecules in the absorption state of 4.2 %. (This is a reasonable estimate, if one neglects issues of Λ-doubling and assumes that the absorbing state is in $v'' = 0$ and J'' near the peak of the rotational distribution.) This step requires no spectroscopic calculation, only a simple use of the ideal gas equation of state, but is useful in giving a sense of the species detection sensitivity of spectrally resolved absorption.

Reference

1. G. Herzberg, *Molecular Spectra and Molecular Structure, Volume I. Spectra of Diatomic Molecules*, 2nd edn. (Krieger Publishing Company, Malabar, 1950)

Laser-Induced Fluorescence 11

This chapter discusses the theory and practical application of laser-induced fluorescence (LIF). Fluorescence, generally, is just another name for spontaneous emission. LIF is a two-step process: (1) absorption of the laser photon, followed by (2) emission.

11.1 Introduction

Why is this technique of interest?

1. the LIF signal can be monitored at 90° to the exciting laser beam, thereby giving spatial resolution to the absorption measurement
2. the signal rides on a dark background, rather than being based on signal differences as in absorption
3. LIF is a stronger process for spatially resolved species-specific measurements than Raman scattering

11.1.1 Background

LIF is a multiple step process in which absorption is followed after some delay by spontaneous emission (Fig. 11.1). LIF is NOT instantaneous. The lasers for LIF are either continuous wave (CW) or pulsed and are often tunable, thereby enabling access to specific features in an absorption spectrum.

History of LIF
This is a relatively young field. Key advances have been linked to the evolution of laser sources and detectors.

R.K. Hanson et al., *Spectroscopy and Optical Diagnostics for Gases*,
DOI 10.1007/978-3-319-23252-2_11

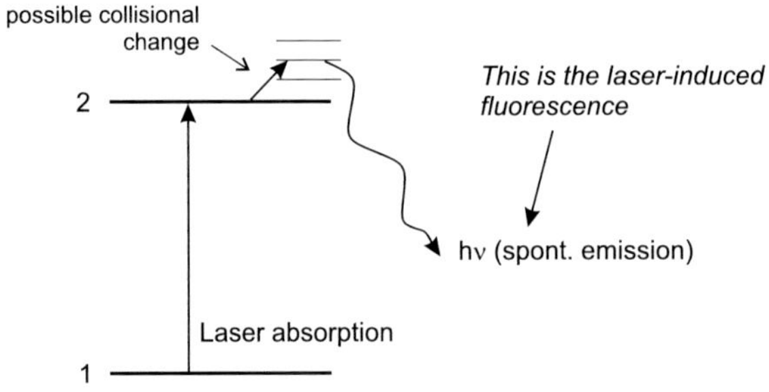

Fig. 11.1 Simple model for laser-induced fluorescence

- Excitation with a flash-lamp pumped tunable dye laser [2]
- Excitation with a CW laser source (1970)
- Excitation with a frequency-doubled CW laser source (around 1980)
- Excitation with Nd:YAG- and excimer-pumped tunable dye lasers (also around 1980)
- 2-D imaging using array detectors, i.e. PLIF, enabling collection of UV/visible fluorescence images (1982)
- Extension of PLIF to infrared excitation and detection, i.e. IR-PLIF (1999)
- Extension of PLIF to high pressures (2004)
- Extension of PLIF to high-speed imaging (2008)

Light detection for LIF can be accomplished with photomultiplier tubes (PMTs) for single-point applications, and array photodetectors for multi-point applications. These arrays can be 1-D or 2-D and may be intensified. Quantum detectors are typically used, owing to their sensitivity. Intensification is achieved by converting the incident photons to electrons, which then undergo multiplication, followed by conversion on a phosphor to an increased number of photons that reach the detector array.

LIF can be used to measure multiple properties, including the following:

$n_i\ (n, v, J)$	number density of species i in a state described by n, v, and J
T	temperature (from the Boltzmann fraction)
X_i	species concentration
$\vec{v}$	velocity (from the Doppler shift of the absorption frequency)
P	pressure (from line broadening)

LIF can also be used to measure different species:

Radicals or atoms:	OH, C_2, CN, NH, ..., O, H, C,...
Stable diatoms:	O_2, NO, CO, I_2, ...
Polyatomics:	NO_2, NCO, CO_2, CH_2O, Acetone (CH_3COCH_3), Biacetyl ($(CH_3CO)_2$), Toluene (C_7H_8), and other carbonyls and aromatic compounds.

11.1.2 Typical Experimental Set-Up

There are several different recording modes available for LIF: (1) *emission intensity* in a fixed wavelength interval or band, (2) *fluorescence spectrum*, or (3) *excitation spectrum*. A recording of emission intensity, $I(t)$, through filters centered at a fixed wavelength λ_{ex} or λ_{det}, can be used to monitor the temporal behavior of LIF. The fluorescence spectrum refers to the emission spectrum, i.e. the intensity as a function of wavelength, $I(\lambda_{em})$, for a fixed excitation wavelength, λ_{ex}. The excitation spectrum resolves the absorption spectrum by varying the excitation wavelength, λ_{ex}, while monitoring the LIF emission at a fixed detection wavelength, λ_{det}, or fixed range of wavelengths (Fig. 11.2).

11.1.3 Measurement Volume

The pathlength, L, for LIF is adjustable, but typically is in the range of $L \approx 0.5 - 5\,\text{mm}$. The solid angle of collection, Ω, is given below (Fig. 11.3)

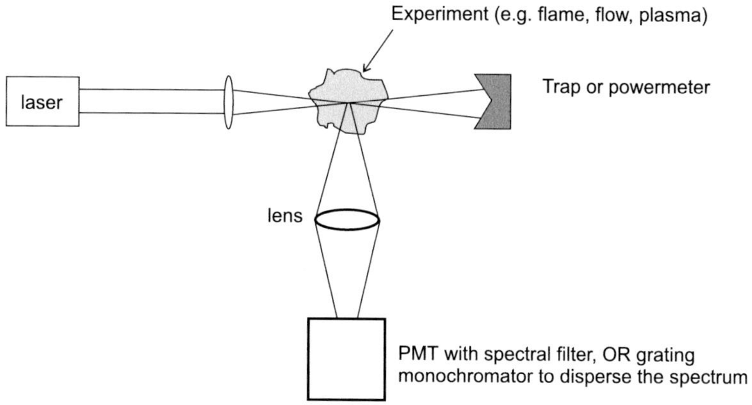

Fig. 11.2 Typical experimental setup for LIF systems

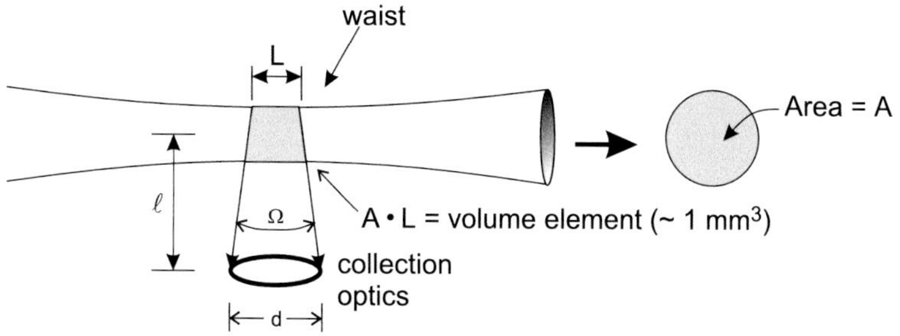

Fig. 11.3 Measurement volume for LIF

$$\Omega \equiv \frac{\text{lens area}}{\ell^2} \tag{11.1}$$

$$= \frac{\pi d^2/4}{\ell^2} \tag{11.2}$$

$$\approx \frac{1}{(f\#)^2} \tag{11.3}$$

The $f\#$ of collection is

$$f\# \equiv \frac{\ell}{d} \tag{11.4}$$

Therefore, small $f\#$'s (also known as fast optics) are required for efficient light collection.

Note: At $f\# = 2.5$, $\Omega/4\pi = 0.01$, or 1 % ! Therefore, the collection process is relatively inefficient, even for fast lenses.

11.1.4 Signal Level (Two-Level System)

See Fig. 11.4

Steady Conditions

For steady conditions, the signal level of fluorescence, S_F, is

$$S_F^{21}\ [\#\ \text{photons/s}] = N_2 \times A_{21} \times \frac{\Omega}{4\pi} \tag{11.5}$$

$$= n_2 \times V \times A_{21} \times \frac{\Omega}{4\pi} \tag{11.6}$$

where

$N_2 =$ number of molecules in the measurement volume in state 2
$n_2 =$ number density of molecules (#/cc) in state 2

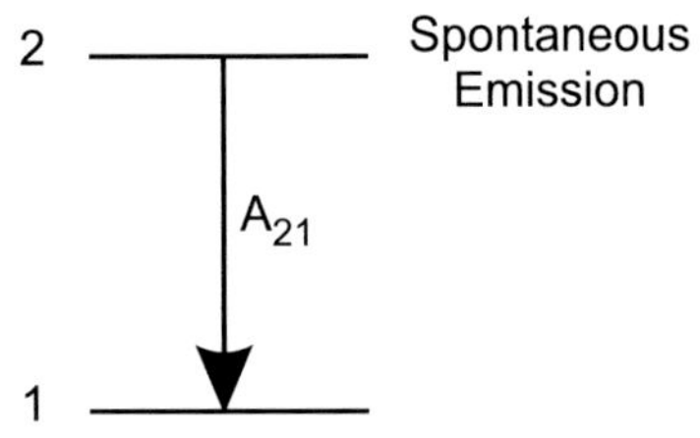

Fig. 11.4 Spontaneous emission corresponds to an atomic (or molecular) transition from level 2 to 1, with release of a photon

$V =$ volume (cc)
$A_{21} =$ probability/s of emission from state 2 to state 1
$\Omega/4\pi =$ collection fraction

The collected signal power is

$$S_F \text{ [power collected]} = (S_F^{21} \text{ [\# photons/s]})\, h\nu \tag{11.7}$$

Pulsed Conditions

For unsteady conditions (e.g., with pulsed excitation), the collected fluorescence can be integrated over time:

$$S_F^{21} \text{ [\# photons]} = \left\{ \int_0^{\tau} n_2(t)dt \right\} \times V \times A_{21} \times \frac{\Omega}{4\pi} \tag{11.8}$$

$n_2(t)$ depends on the laser and the collision process and is further explained in the following sections.

11.2 Two-Level Model

The two-level model provides a good introduction to LIF modelling. As more levels are added to the model, the complexity naturally increases, but the general method of analysis remains the same (Fig. 11.5).

Recall from Einstein theory that the probability per second of a molecule undergoing a transition from state 1 to state 2 due to absorption in the frequency range $\nu \rightarrow \nu + d\nu$ is

$$\text{prob/s}_{1\rightarrow 2}^{\nu\rightarrow\nu+d\nu} = B_{12}{}^{\rho}\rho(\nu)\phi(\nu)d\nu \tag{11.9}$$

where

$$\rho(\nu) = I_\nu / c$$

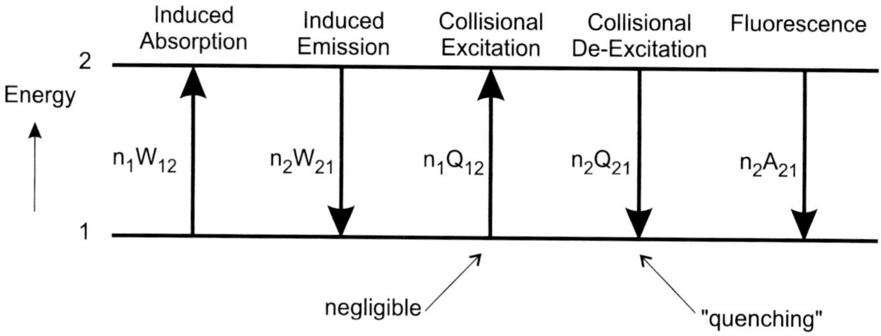

Fig. 11.5 Different transitions that are included in the two-level model

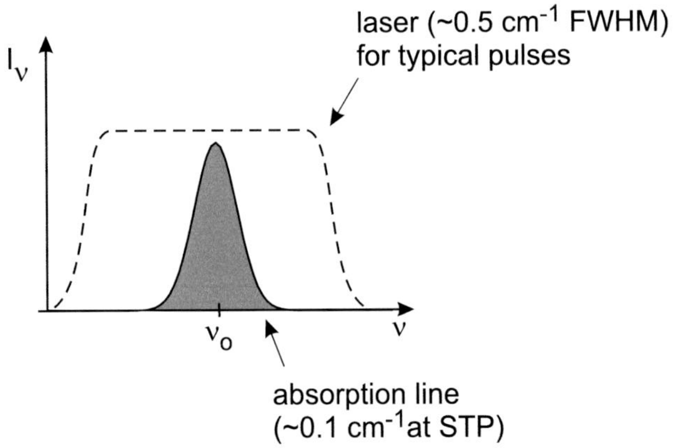

Fig. 11.6 Pulsed lasers are often spectrally broad compared to absorption lines

and

$$\int \phi(\nu)d\nu = 1$$

Most pulsed lasers are spectrally broad compared with absorption lines (see Fig. 11.6), so that $I_\nu \approx$ constant over the spectral width of the absorption line. Hence, the total rate of absorption (i.e., the spectral integral for the $2 \leftarrow 1$ transition) becomes

$$\text{Rate}_{1\to2} = n_1 \int_{\text{line}} B_{12} I_\nu \phi(\nu) d\nu = n_1 B_{12} I_\nu \tag{11.10}$$

$$= n_1 W_{12}. \tag{11.11}$$

Thus W_{12} is the rate (s^{-1}) that individual molecules in state 1 undergo the transition to state 2, and we have introduced $B_{12} = B_{12}^{\rho}/c$ as the probability per second of absorption per unit spectral intensity.

Rate analysis can be performed to determine the rate of change of the population of molecules in level 2

$$\dot{n}_2 = n_1(I_\nu B_{12}) - n_2(I_\nu B_{21} + Q_{21} + A_{21}) \tag{11.12}$$

$$= n_1 W_{12} - n_2(W_{21} + Q_{21} + A_{21}) \tag{11.13}$$

At steady-state, $\dot{n}_2 \approx 0$, thus

$$(n_2)_{SS} = n_1 \frac{W_{12}}{W_{21} + A_{21} + Q_{21}} \tag{11.14}$$

> Two limits emerge from the steady-state analysis for n_2:
>
> - weak excitation ("linear LIF")
> - strong excitation ("saturated LIF")

11.2.1 Weak Excitation Limit

If the induced emission from level 2 is much weaker than the sum of collisional and spontaneous decay processes, i.e. if

$$W_{21} \ll A_{21} + Q_{21}, \tag{11.15}$$

then $n_2 \ll n_1$, and $n_1 \approx n_1^0$ (usually $n_1^0 \approx n^0$ where n^0 is the conserved total number density, $n^0 = n_1 + n_2$). The population fraction in level 2 becomes

$$\boxed{n_2 = n_1^0 \frac{W_{12}}{A_{21} + Q_{21}}} \tag{11.16}$$

and the fluorescence signal is given by

$$S_F = n_2 \times V \times A_{21} \times \frac{\Omega}{4\pi} \quad \text{(photons/s)} \tag{11.17}$$

$$= \underbrace{(n_1^0\, V)(W_{12} = B_{12} I_\nu)}_{\text{photons absorbed/s}} \underbrace{\left(\frac{A_{21}}{A_{21} + Q_{21}}\right)}_{\text{"fluor yield"}} \underbrace{\left(\frac{\Omega}{4\pi}\right)}_{\substack{\text{fraction}\\ \text{collected}}} \tag{11.18}$$

Note:
1. Need to know Q_{21} (the rate of the electronic quenching, a collisional process), as well as A_{21} and I_ν
2. $S_F \propto n_1^0 = n_i f_{v,J}(T)$
   Thus, we can see that the LIF signal is directly proportional to the population density in the lower quantum level. Hence, LIF is typically viewed as a species diagnostic.
3. Alternate view:

$$\begin{aligned} \text{\# of absorptions/s} &= \text{(\# molec in 1)(prob/s of abs)} \\ &= (n_1^0\, V)(B_{12} I_\nu) \end{aligned}$$

11.2.2 Saturation Limit

If the induced emission rate is much larger than the collisional and spontaneous emission rates ($W_{21} \gg A_{21} + Q_{21}$), then

$$n_2 = n_1 \frac{W_{12}}{W_{21}} \tag{11.19}$$

$$= n_1 \frac{B_{12}}{B_{21}} \tag{11.20}$$

$$= n_1 \frac{g_2}{g_1} \tag{11.21}$$

(recall $g_1 B_{12} = g_2 B_{21}$).

Thus, in the "saturation" limit, the population in level 2 is independent of Q and I_ν.

Note: 1. If $g_2 = g_1$, then $n_2 = n_1$ when the transition is "saturated"
2. $n_1^0 = \text{total} = n_2 + n_1$, so that

$$n_2 = \frac{n_1^0}{1 + g_1/g_2}$$

$$= \frac{n_1^0}{2} \quad \text{when } g_2 = g_1$$

Therefore, the LIF signal level (#photons/s reaching the detector) for the saturation limit in the two-level model is

$$\boxed{S_F = n_1^0 \left(\frac{g_2}{g_1 + g_2} \right) \times V \times A_{21} \times \frac{\Omega}{4\pi}} \tag{11.22}$$

The virtue of working in the saturation limit is that the fluorescence signal does not depend on the quenching rate Q_{21}, which may be poorly known, nor does it depend on the laser intensity, which may be difficult to measure accurately. However, there are other complications which arise when one tries to do "saturated LIF." For instance, it may be difficult to reach the saturated limit in all parts of the laser beam (whose intensity varies in space). If the intensity is not sufficiently high, one may reach an intermediate situation between weak excitation and saturation.

11.2.3 Intermediate Result

The solution for n_2, valid at all values of I_ν is found directly from Eq. (11.14), with $n_1 + n_2 = n_1^0$,

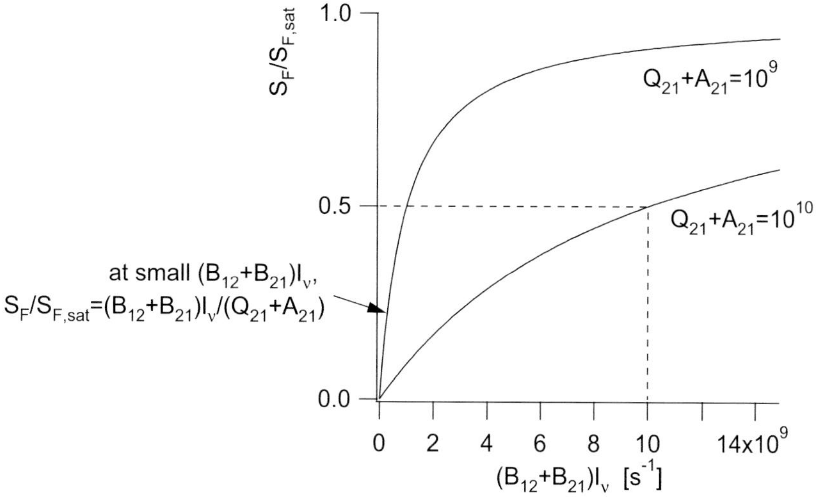

Fig. 11.7 Intermediate fluorescence signal levels

$$n_2 = n_1^0 \frac{B_{12}}{B_{12}+B_{21}} \left\{ 1 + \frac{Q_{21}+A_{21}}{(B_{12}+B_{21})I_\nu} \right\}^{-1} \tag{11.23}$$

Since $g_1 B_{12} = g_2 B_{21}$, then $B_{12}/(B_{12}+B_{21}) = g_2/(g_1+g_2)$ and therefore

$$\frac{S_F}{S_{F,\text{sat}}} = \frac{n_2}{n_{2,\text{sat}}} = \frac{1}{1+\frac{Q_{21}+A_{21}}{(B_{12}+B_{21})I_\nu}} \tag{11.24}$$

See Fig. 11.7.

11.2.4 Typical Values for A and Q in Electronic Transitions

1. $A_{21} \approx 10^5 - 10^8\ \text{s}^{-1}$ ($10^6\ \text{s}^{-1}$ for NO, OH; $10^8\ \text{s}^{-1}$ for Na)
2.

$$\begin{aligned} Q_{21} &\approx Z \text{ (collision frequency)} \\ &= n \times \pi\sigma^2 \times \left(\bar{c} = \sqrt{8kT/\pi\mu}\right) \\ &\propto \sigma^2 P/\sqrt{T} \\ Q_{21} &\approx 10^9 - 10^{10}\ \text{s}^{-1} \text{ at STP} \end{aligned}$$

3. Therefore, the fluorescence yield, $A_{21}/(A_{21}+Q_{21})$, is much less than 1, except at low P [i.e., LIF is an inefficient process!]

4. Note: spontaneous emission for IR transitions is much weaker (smaller A_{21}) than for UV transitions. Therefore LIF is weak in the IR, unless the effective quenching rate Q_{21} is small (as may be the case!)

$$A_{21,\mathrm{IR}}/A_{21,\mathrm{UV}} \ll 1 \tag{11.25}$$

$$Q_{21,\mathrm{IR}}/Q_{21,\mathrm{UV}} = ?$$

5. What is I_ν for $B_{21}I_\nu \gg A_{21} + Q_{21}$? Call this $(I_\nu)_{\mathrm{sat}}$
   (a) For low Q_{21}, $I_{\nu,\mathrm{sat}} \gg A_{21}/B_{21}$
   (b) For high Q_{21}, $I_{\nu,\mathrm{sat}} \gg Q_{21}/B_{21}$ so that $I_{\nu,\mathrm{sat}} \propto (P/\sqrt{T})(1/B_{21})$
   (c) Recall from Eq. (7.24) that $B_{21}^{\rho} = A_{21}(\lambda^3/8h\pi)$

$$\therefore\ B_{21} = B_{21}{}^{I} = \frac{B_{21}{}^{\rho}}{c} = A_{21}\frac{\lambda^3}{8hc\pi}$$

at $\lambda = 500\,\mathrm{nm}$,

$$B_{21} = 25\,A_{21}\ [\mathrm{cm}^2/\mathrm{erg\ s}]$$

   (d) Assume $Q_{21} \approx 10^4 A_{21}$ (e.g., $Q_{21} = 10^{10}\,\mathrm{s}^{-1}$ and $A_{21} = 10^6\,\mathrm{s}^{-1}$), then

$$\begin{aligned} I_{\nu,\mathrm{sat}} &\gg 10^4 A_{21}/25\,A_{21} \\ &= 400\left[\mathrm{ergs/cm}^2 = \frac{\mathrm{erg/s}}{\mathrm{cm}^2\,\mathrm{Hz}}\right] \\ &= 0.4\,\mathrm{J/m}^2 \end{aligned}$$

   (e) But,

$$I_\nu = \frac{\text{Power}}{\text{Area}\cdot\Delta\nu_{\text{laser}}}$$

Take $d = 1$ mm, $\Delta\nu_L = 0.5\,\mathrm{cm}^{-1} = 1.5\times10^{10}\,\mathrm{s}^{-1}$, then

$$\begin{aligned} I_\nu &= \frac{P[\mathrm{W}]\cdot 10^7\ [\mathrm{ergs/J}]}{(0.8\times10^{-2}\,\mathrm{cm}^2)\ (1.5\times10^{10}\,\mathrm{s}^{-1})} \\ &= (P[\mathrm{W}])(0.08) \end{aligned}$$

Therefore, $P_{\mathrm{sat}}\ [\mathrm{W}] \gg \frac{400\ [\mathrm{ergs/cm}^2]}{0.08} = 5\times10^3\ \mathrm{W} = 5\,\mathrm{kW}!$
   (f) Note: Nd:YAG-pumped dye lasers give 1–10 mJ per 10 ns pulse at 225 or 300 nm, giving $10^5 - 10^6$ W!
   (g) Therefore, it's easy to saturate within the limits of a two-level model! In fact, with atoms it's easy to reach full saturation; with molecules it is not so easy to get full saturation, owing particularly to population exchanges with adjacent rotational states, but saturation is feasible at $P \leq 1$ atm.

11.2.5 Typical Values for A and Q: Vibrational Transitions (IR)

1. $A_{21} \approx 10^1 - 10^2\,\mathrm{s}^{-1}$ ($30\,\mathrm{s}^{-1}$ for CO, $\Delta v = 1$) for strong IR bands
2. Q_{21} is either the vibration–translation (V–T) de-excitation rate or vibrational–vibrational (V–V) transfer rate (to another species). The dominant process is usually V–V; e.g. $Q_{\mathrm{V-V}} \approx 10^5\,\mathrm{s}^{-1}$ for CO with N_2 near STP, but higher at elevated temperatures.
3. Therefore, the fluorescent yield is typically in the range $10^{-3} - 10^{-5}$. This is sufficient for IR LIF to be a useful diagnostic, providing a means of imaging many species not accessible via electronic transitions.

11.3 Detection Limits (Pulsed Laser)

Since LIF is used to detect species, the minimum detectivity of this technique is an important characteristic.

11.3.1 Weak Excitation Limit

The fluorescence signal level in the weak excitation limit in terms of total photons is

$$S_F = \int S_F(t)dt \tag{11.26}$$

$$= n_1^0\, V \times \int I_\nu dt \times B_{12} \frac{A_{21}}{A_{21} + Q_{21}} \frac{\Omega}{4\pi} \tag{11.27}$$

where

$$n_1^0 = n\, X_i f_{v,J} \tag{11.28}$$

and

$$n = \text{total number density}$$

$$X_i = \text{mole fraction of species } i$$

$$f_{v,J} = \text{fraction of molecules initially in } v \text{ and } J$$

The integral term, $\int I_\nu dt$, is equal to the laser pulse energy, E_p, divided by the laser linewidth, $\Delta\nu_L$, and the cross-sectional area of the exciting beam, A_c.

$$\int I_\nu dt = \frac{E_p}{\Delta\nu_L \cdot A_c} \tag{11.29}$$

Thus, the signal level reduces to

$$\boxed{S_F \approx n_1^0\, L \frac{E_p}{\Delta\nu_L} B_{12} \frac{A}{Q} \frac{\Omega}{4\pi}} \tag{11.30}$$

where L is the length of the measurement zone ($L = V/A_c$). Here, we have assumed a common case where $A_{21} \ll Q_{21}$, and for simplicity, we have dropped the subscripts on Q_{21} and A_{21}.

Now, let's calculate the detection limit for representative weak excitation conditions using the new expression for the signal level. The following information is given:

Laser pulse energy	$E_P = 10^{-5}\,\text{J}\,(10\,\mu\text{J}) = 10^2$ ergs
Pulse length	$\tau_L = 10^{-8}$ s
Laser linewidth	$\Delta\nu_L = 1\ \text{cm}^{-1}$
Number density	$n = 5 \times 10^{18}$ molecules/cc (at 1 atm, 1620 K)
Boltzmann fraction	$f_{v,J} = 0.01$ (1 % of the species is in the absorbing state)
Measurement volume	$V = A_c \times L = (10^{-2}\,\text{cm}^2)(10^{-1}\,\text{cm}) = 0.001$ cc
Collection angle	$\Omega/4\pi = 10^{-3}$ (for optics with $f\# = 8$)
Einstein coefficients	$A/Q \approx 10^{-4}$, $B_{12} \approx 20 \cdot A_{21}\,[\text{cm}^2/\text{erg s}]$, $A_{21} = 10^6\,\text{s}^{-1}$

Using the typical values listed above and Eq. (11.30) gives

$$\begin{aligned} S_F &= \underbrace{(5\times10^{18}\cdot X_i\cdot10^{-2})}_{n_1^0}(10^{-1})\left(\frac{E_P\,[\text{ergs}]}{\Delta\nu_L\,[\text{s}^{-1}]}\right)\underbrace{(20\times10^6)}_{B_{12}}(10^{-4})(10^{-3}) \\ &= 10^{16}\cdot X_i\cdot\frac{E_P\,[\text{ergs}]}{1\ \text{cm}^{-1}\cdot3\times10^{10}\,\text{cm/s}} \\ &= 3\times10^{7}\cdot X_i \end{aligned}$$

where X_i is the mole fraction of the absorber i.

The total number of photoelectrons produced by this number of photons incident on a quantum detector is $S_F\eta$, where η is the quantum efficiency of the detector. The quantum efficiency is defined as

$$\eta = \frac{\#\ \text{of}\ e^-\ \text{produced}}{\#\ \text{of photons input}} \tag{11.31}$$

For ideal, shot-noise-limited detection, the signal-to-noise ratio (SNR) is simply the square root of the number of photoelectrons in the pulse

$$\text{SNR} = \sqrt{S_F\eta} \tag{11.32}$$

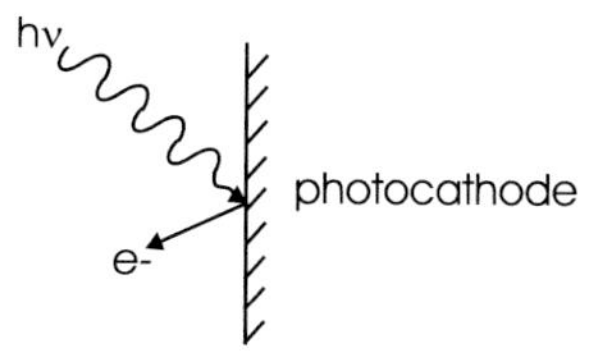

Thus, an estimate for the SNR for the prescribed conditions in the weak excitation limit (and assuming $\eta = 0.1$) is

$$\mathrm{SNR} = \sqrt{3 \times 10^6 \cdot X_i}$$

Therefore, the SNR for $X_i = 1\,\mathrm{ppm}$ ($X_i = 10^{-6}$) is 1.7. Or, $X_i = 0.3\,\mathrm{ppm}$ for a SNR = 1. This sensitivity is pretty impressive for a nonintrusive measurement made in a 1 mm^3 volume in 10 ns and with only 10 μJ of laser energy!

11.3.2 Saturation Limit

In the saturation limit, the signal level is

$$S_F = n_1^0 \, V \frac{g_2}{g_2 + g_1} A_{21} \frac{\Omega}{4\pi} \left(\int_0^{\tau_{laser}} dt = \tau_L \right) \;\; [\text{photons}] \qquad (11.33)$$

To find the detection limit for typical conditions, use the information and procedure from Sect. 11.3.1. The signal level in the saturated case is

$$\begin{aligned} S_F &= (5 \times 10^{18} \cdot X_i \cdot 10^{-2})(10^{-3}\,\mathrm{cc})(0.5 = g_2/(g_1 + g_2))(10^6)(10^{-3})(10^{-8}\,\mathrm{s}) \\ &= 2.5 \times 10^8 \cdot X_i \end{aligned}$$

Then, with $\eta = 0.1$,

$$\mathrm{SNR} = \sqrt{2.5 \times 10^7 \cdot X_i},$$

or SNR = 5 at $X_i = 1\,\mathrm{ppm}$. Thus, if we can saturate, the minimum detectivity is $X_{i,\min} = 0.04\,\mathrm{ppm}$ for a SNR = 1!

In molecules, the dominant collisional rate (in the rate equation analysis) tends to be rotational transfer, and because this rate is typically larger than Q_{elec}, saturation is relatively difficult to achieve except at reduced pressures.

11.4 Characteristic Times

With pulsed excitation, the question emerges of whether we can continue to use the same steady-state relations that were derived before. This section compares the time to reach steady-state with laser pulsewidth to show that these relations are generally acceptable (Fig. 11.8).

Recall that the fluorescence signal level at any time t is

$$(n_2)_{SS} = n_1 \frac{W_{12}}{W_{21} + A_{21} + Q_{21}} \tag{11.34}$$

$$= n_1^0 \frac{W_{12}}{W_{12} + W_{21} + A_{21} + Q_{21}} \tag{11.35}$$

$$(n_2)_{SS}^{weak} = n_1^0 \frac{W_{12}}{A_{21} + Q_{21}} \tag{11.36}$$

Are these results applicable for pulsed experiments? Yes, if there is time to reach steady-state, that is, if $\tau_{SS} < \tau_{laser}$. Consider idealized step changes in I_ν and define τ_{SS} (characteristic time to reach steady-state) as follows (Fig. 11.9):

$$\tau_{SS} = \frac{(n_2)_{SS}}{(\dot{n}_2)_{initial}} \tag{11.37}$$

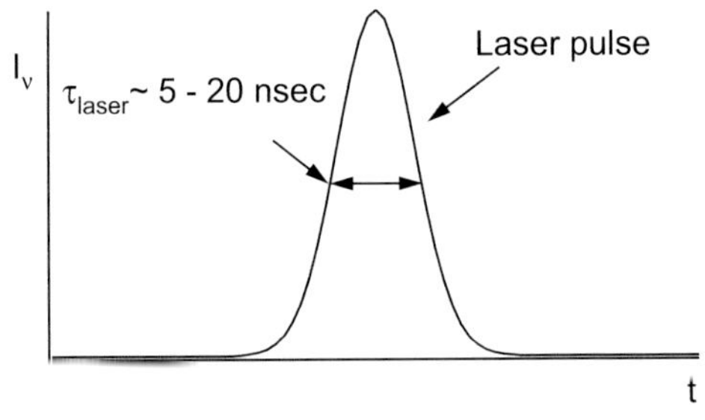

Fig. 11.8 Characteristic laser pulse times can be 5–20 ns

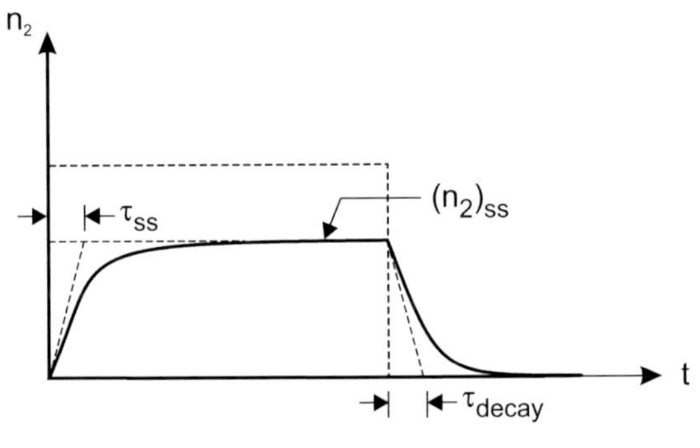

Fig. 11.9 Characteristic steady-state and decay times for the population in state 2

Then

$$\tau_{SS} = \frac{1}{W_{12} + W_{21} + A_{21} + Q_{21}} \tag{11.38}$$

and, similarly,

$$\tau_{decay} = \frac{1}{A_{21} + Q_{21}} \tag{11.39}$$

What, then, is $n_2(t)$?

Weak Excitation

For weak excitation, $W_{12} \ll Q_{21}$ and $A_{21} \ll Q_{21}$, so $\tau_{SS} \approx 1/Q_{21}$, e.g. $Q_{21} \approx 10^9 - 10^{10}\,\mathrm{s}^{-1}$ at STP, then $\tau_{SS} \approx 0.1 - 1$ ns, and the SS relation applies at virtually all values of time in the laser pulse.

Strong Excitation

For strong excitation, $W_{12} \gg Q_{21} \gg A_{21}$, so

$$\tau_{SS} \approx \frac{1}{W_{12} + W_{21}} \ll \frac{1}{Q_{21}}$$

so that for typical Q_{21},

$$\tau_{SS} \ll 10^{-9}\,\mathrm{s},$$

and the SS approximation is good on most time scales of interest for strong excitation (except, perhaps, for ultrafast lasers).

11.5 Modifications of the Two-Level Model

11.5.1 Hole-Burning Effects

Inhomogeneous (velocity) broadening can lead to "hole" burning (saturation due to depletion of a certain velocity class) with intense, spectrally narrow lasers.

11.5.2 Multi-Level Effects

Upper and lower levels of molecules are really manifold states (Fig. 11.10). Hence the LIF signal and the associated fluorescence quantum yield depend on collisional transfer rates among upper levels and the number of these monitored by the detection system. Two interesting limits are narrowband detection and broadband detection.

If emission occurs only to v'' and is collected *only* from v', J' (this is narrowband detection), then

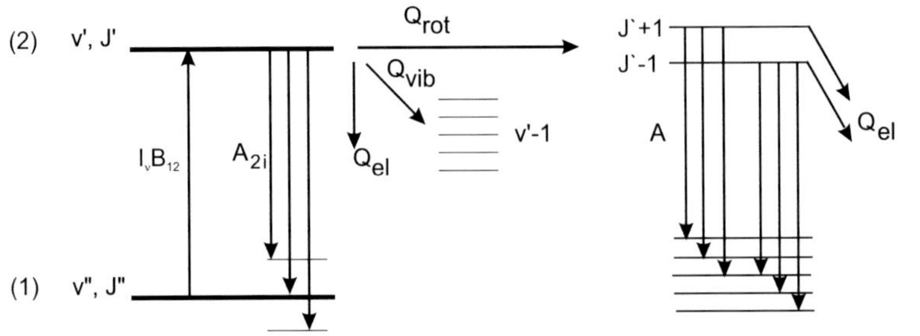

Fig. 11.10 Upper and lower manifold states

$$\dot{n}_2 = n_1 B_{12} I_\nu - n_2 (B_{21} I_\nu + A_{21} + Q_{\text{elec}} + Q_{\text{rot}} + Q_{\text{vib}}) \tag{11.40}$$

where A_{21} is the sum of A-coefficients (A_{2i}) for allowed transitions that are monitored by the detection system, shown here as the emission lines (e.g., P, Q and R). For weak excitation, i.e. $B_{21} I_\nu$ small relative to other decay terms,

$$(n_2)_{\text{SS}} = n_1 \frac{B_{12} I_\nu}{A_{21} + \Sigma Q}, \tag{11.41}$$

where ΣQ (the effective loss or "quenching" rate) includes Q_{rot}, Q_{vib}, and Q_{elec} (i.e., *all* processes *removing* molecules from the upper state observed by fluorescent emission). Hence, in the weak excitation limit,

$$S_F = n_1 \cdot V^o \cdot I_\nu B_{12} \cdot \frac{A_{21}}{A_{21} + \Sigma Q} \cdot \frac{\Omega}{4\pi} \tag{11.42}$$

What about broadband collection (e.g., from all J' with v'), again in the weak excitation limit? Then the "2" state is redefined to include all rotational levels, and the steady-state relation becomes

$$(n_2)_{\text{SS}} = n_1 \frac{B_{12} I_\nu}{A_{21} + \Sigma Q}, \tag{11.43}$$

where ΣQ now includes only Q_{vib} and Q_{elec} ! The fluorescence signal is

$$S_F = n_1 \cdot V^o \cdot I_\nu B_{12} \cdot \frac{A_{21}}{A_{21} + \Sigma Q} \cdot \frac{\Omega}{4\pi} \tag{11.44}$$

The important conclusion to be drawn is that collection bandwidth defines the upper state "2" being monitored by emission and the effective quenching rate. The LIF signal strength, and the fluorescence yield, are thus dependent on the collection bandwidth!

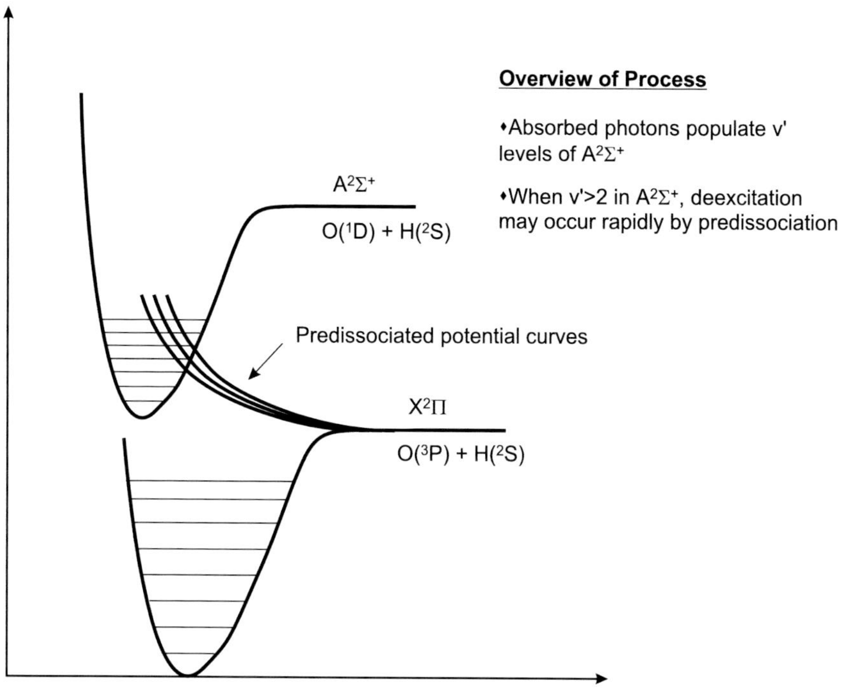

Fig. 11.11 Potential curves for ground, excited, and predissociative states of OH

(Note that A_{21}, i.e. the sum of A-coefficients from a given J', is effectively independent of J'; see Appendix F for details.)

11.5.3 Predissociation

For some excitation scenarios, new decay paths become available along predissociative potential curves (Fig. 11.11). Including a new decay path, Q_{pre}, the rate analysis yields

$$\dot{n}_2 = n_1 B_{12} I_\nu - n_2 (B_{21} I_\nu + A_{21} + Q_{\text{elec}} + Q_{\text{rot}} + Q_{\text{vib}} + Q_{\text{pre}}) \qquad (11.45)$$

Interestingly, Q_{pre} can be much larger than the other collision rates, i.e. $Q_{\text{pre}} \gg Q_{\text{rot}} > Q_{\text{vib}}$. For example, for O_2, $Q_{\text{pre}} \approx 10^{11} - 10^{12}\,\text{s}^{-1}$. In this case, we recover the two-level result, with $Q_{\text{eff}} = Q_{\text{pre}}$.

Advantage Q_{pre} is independent of composition, pressure, and temperature, so the fluorescence yield is more readily quantified.

Disadvantage A/Q_{pre} may be small! (i.e. low fluorescence yield)

11.6 Example: Acetone LIF

11.6.1 Background

In order to perform LIF in air, it is common to add a fluorescent tracer (neither O_2 nor N_2 are readily accessible). In the 1980s, researchers at the High Temperature Gasdynamics Laboratory (HTGL) at Stanford University used biacetyl ($(CH_3CO)_2$), a food additive with modest vapor pressure (40 Torr), but it had two problems:

1. smelly and hard to handle
2. highly quenched by O_2

In the early 1990s, Lozano's survey identified acetone (CH_3COCH_3), a member of the ketone family with desirable properties. Since then, acetone-based LIF has become a popular laser diagnostic in laboratories worldwide, with applications ranging from fundamental studies of fluid mechanics to internal combustion engines.

11.6.2 Acetone Photophysics

See Fig. 11.12.

- Vapor pressure is 180 Torr at room temperature
- cheap, non-toxic, easy to handle
- fluorescence lifetime is approximately 2 ns and is independent of O_2
- constant fluorescence yield of about 0.2 % at room temperature
- each state is a manifold of vibrational levels
- "intramolecular intersystem crossing," at rate Q_t, dominates, therefore the FY = $A_{10}/(A_{10} + Q_t) \approx A_{10}/Q_t \approx 0.2\,\%$!
- absorption and fluorescence are broadband

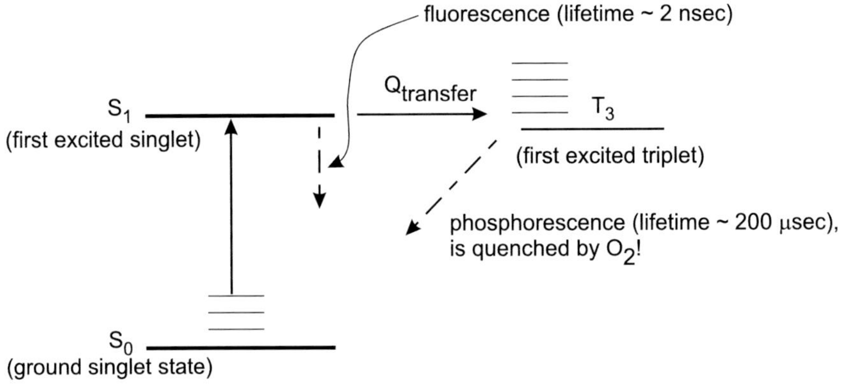

Fig. 11.12 Molecular levels for acetone

11.7 Applications of LIF

There are different strategies for measuring various properties with LIF.

11.7.1 Species Density

To measure species density (n_i [molecules/cc]) with LIF using linear excitation, we have (where Q_{21} and Q are used interchangeably)

$$S_F \text{ [photons/s]} = n_1^0 \cdot V \cdot I_\nu B_{12} \cdot \frac{A_{21}}{A_{21} + Q_{21}} \cdot \frac{\Omega}{4\pi} \tag{11.46}$$

$$S_F \propto n_i \frac{f_{v,J}(T)}{Q} \tag{11.47}$$

- assumed known: A_{21}, B_{12}, I_ν, V, and Ω are usually measured (i.e., by Rayleigh scattering)
- then $S_F \propto C \cdot n_i$, where $C \propto \frac{V\Omega f_{v,J}}{Q}$ is from a calibration point
- for T-varying systems, we can select v and J to give $f_{v,J}/Q$ nearly independent of T, in which case S_F is directly proportional to n_i, the desired quantity.

11.7.2 Species Mole Fraction

For measuring mole fraction X_i, use the following expression

$$S_F \propto n_i \frac{f_{v,J}(T)}{n\sigma\bar{c}} \tag{11.48}$$

$$= X_i \frac{f_{v,J}}{\sigma\sqrt{T}} \tag{11.49}$$

$$\approx C \cdot X_i \tag{11.50}$$

Note that $Q = n\sigma\bar{c}$, where $\bar{c}$ is the mean molecular speed, and C is obtained by calibration. This relation implies that selection of v and J to give $\frac{f_{v,J}}{\sigma\sqrt{T}} \neq f(T)$ leads to direct proportionality between S_F and X_i.

11.7.3 Temperature

There are two strategies for measuring temperature.

Strategy 1
The first strategy is a single-line method.

$$S_F \propto X_i \frac{f_{v,J}}{\sigma\sqrt{T}} \tag{11.51}$$

Therefore,

1. use a tracer with fixed X_i
2. select v and J for large T dependence of $\frac{f_{v,J}}{\sigma\sqrt{T}}$

Strategy 2

The second strategy is a two-line method (two excitation wavelengths) [1].

$$\frac{S_F(1)}{S_F(2)} = \frac{\left[n_i \frac{f_{v,J}(T)}{Q}\right]_1}{\left[n_i \frac{f_{v,J}(T)}{Q}\right]_2} \tag{11.52}$$

$$\approx \frac{f_{v,J}(T)_1}{f_{v,J}(T)_2} \tag{11.53}$$

$$= F(T) \tag{11.54}$$

Therefore, select λ_1 and λ_2 to probe states with strong T-dependence for $f_{v,J}(T)_1/f_{v,J}(T)_2$.

Example: OH

For OH, the $R_1(7)$ and $R_1(11)$ lines are located within 0.5 cm^{-1} of each other, so it is possible to scan over the entire pair with one laser sweep (Fig. 11.13). The laser at Stanford based on a fast-scanning ring dye laser can scan in approximately 200 μs.

$$\frac{S_F(\nu_{11}^0)}{S_F(\nu_7^0)} = \frac{n_{OH} f_b(N''=11)[I_\nu{}^{11} B_{12}{}^{11} \phi(\nu_{11}^0)(A/Q)]^{11}}{n_{OH} f_b(N''=7)[I_\nu{}^{7} B_{12}{}^{7} \phi(\nu_{7}^0)(A/Q)]^{7}} \tag{11.55}$$

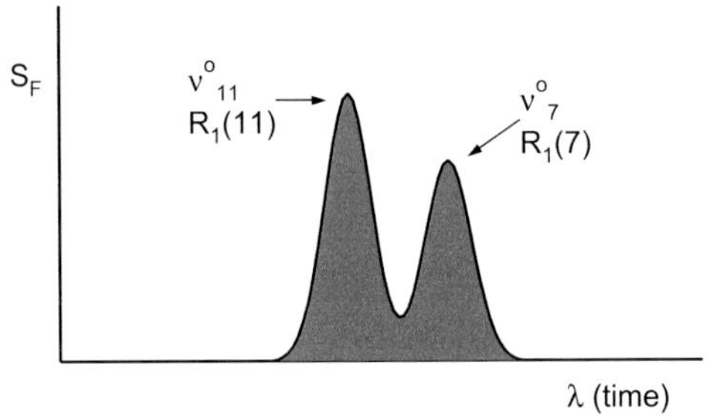

Fig. 11.13 OH line pair for temperature measurements

$$= c \cdot \frac{f_b(N'' = 11)[A/Q]^{11}}{f_b(N'' = 7)[A/Q]^7} \tag{11.56}$$

Therefore,

$$\frac{S_F(\nu_{11}^0)}{S_F(\nu_7^0)} \propto \frac{f_b(N'' = 11)}{f_b(N'' = 7)} \tag{11.57}$$

$$= \mathcal{F}(T) \tag{11.58}$$

With T known, the integrated area can be used to find n_{OH}, and the measured linewidth can give the pressure.

11.7.4 Velocimetry

Flow velocity information can be found from Doppler shift (Figs. 11.14 and 11.15).

$$\frac{\Delta\nu_{\text{Doppler shift}}}{\nu} = \frac{v\cos\theta}{c} \tag{11.59}$$

$$\Delta\nu_{\text{Doppler shift}} = \frac{v\cos\theta}{\lambda} \tag{11.60}$$

Therefore, we can infer $v\cos\theta$ from a measure of $\Delta\nu_{\text{Doppler}}$.

1. These measurements have been made in supersonic and subsonic flows
2. Pulsed (broad) laser sources have also been used for the measurements
3. With two laser components, the 2-D velocity can be resolved
4. The mass flux, $\dot{m} = \rho v$, is given by the product of ρ (obtained from the determination of total number density, n) and v (from the Doppler shift), assuming θ is known
5. The momentum flux, $T = \rho v^2$, is available from measurements of ρ and v

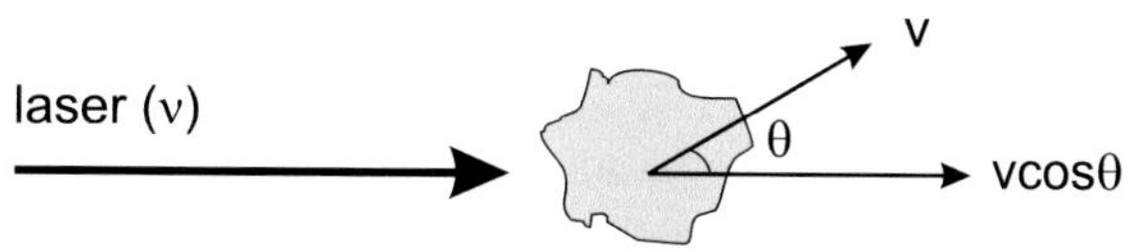

Fig. 11.14 The Doppler shift from a flowing gas can be used to measure the gas velocity along the direction of the laser beam

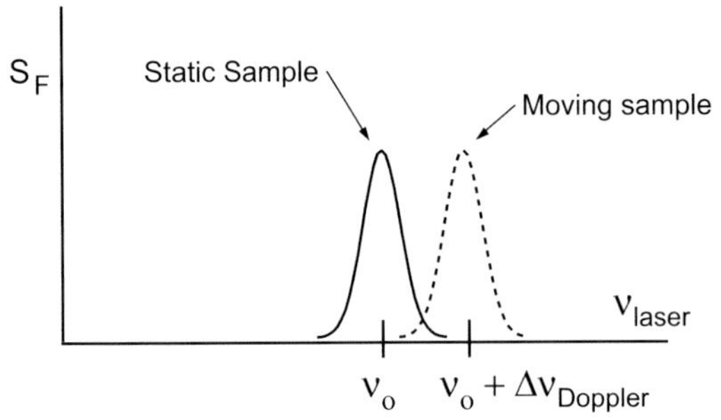

Fig. 11.15 The Doppler shift from a flowing gas can be used to measure a component of velocity

11.8 Exercises

1. What is the effective electronic temperature, T(K), for a 2-level atom with $E_2 > E_1$ in the limit of fully saturated LIF?
2. You are asked to study a molecule described by only 2 energy levels. A short duration, spectrally narrow, high-power laser pulse is used to saturate the absorption transition between levels 1 and 2. After the laser pulse is complete (i.e., $t > 0$), the gas begins to relax towards a new thermal equilibrium via the pathways shown below. The energy levels and pertinent constants are indicated.

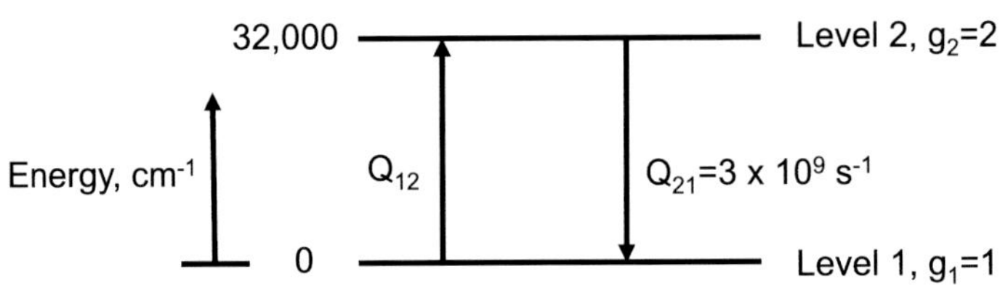

   (a) What are the fractional populations in each energy level (i.e., n_1/n and n_2/n) immediately after the laser pulse?
   (b) At equilibrium and 1000 K, what is Q_{12}?
   (c) Write a differential equation describing the fractional number density in level 2 (i.e., n_2/n) for $t > 0$.
   (d) 0.3 ns after the laser pulse, what is the fractional population in level 2? Hint: Neglect small terms.
3. A continuous-wave diode laser is used to excite a vibrational band of H_2O via laser absorption. Assuming weak-excitation and a two-level molecule and ignoring collisional excitation and stimulated emission:
   (a) Derive an equation for the fractional population in the excited vibrational state.

(b) Derive an expression for the LIF signal in photons/s cm^3 and the fluorescence quantum yield.

Note: Assume steady-state and write your equations in terms of n or n_i, Q, A, W_{12}, and Ω.

4. You are given a three-level model of a molecular system diagrammed below. Remember that $W_{i,j} = B_{i,j}I$, and for $i > j$, $T_{i,j} = A_{i,j} + Q_{i,j}$ and $T_{j,i} = Q_{j,i}$. Also for this problem $g_1 = g_2 = g_3$ and $N_1\ (t = 0) = N^0$.
   (a) Write the complete set of equations needed to characterize the rate of change of the population of each level.
   (b) Solve this system of equations to determine the steady-state population difference between levels 1 and 3 ($\Delta N_{13}^{SS} \equiv N_1^{SS} - N_3^{SS}$) in terms of N_0 and the transition rate coefficients. You may neglect all non-radiative excitation processes. In other words, $T_{13} = T_{12} = T_{23} = 0$.
   (c) Find the saturation intensity, $I_{13,\mathrm{sat}}$, for the $1 \rightarrow 3$ transition in terms of the stimulated absorption coefficient, B_{13} and the other rates, $T_{i,j}$. The saturation intensity is I such that $\Delta N_{13}^{SS} = \frac{1}{2}\Delta N_{13}^{t=0}$. Assume $\Delta N_{13}^{t=0} \approx N^{t=0} \approx N^0$.
   (d) Using the three-level rate equations, derive an expression for the rate of the fluorescence, S_{32} (number of photons/s emitted in all directions), at steady-state from level 3 to level 2. Assume $I \ll I_{13,\mathrm{sat}}$.

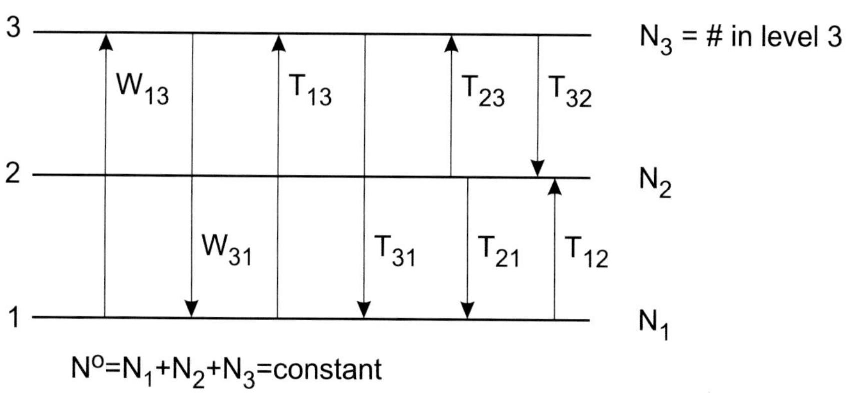

Reference

1. A.C. Eckbreth, *Laser Diagnostics for Combustion Temperature and Species* (Abacus Press, Kent, 1988)
2. F.P. Schafer, W. Schmidt, J. Volze, Organic Dye Solution Laser, *Applied Physics Letters* **9**, 306–309 (1966)

12 Diagnostic Techniques for Gaseous Flows

A variety of techniques exist that complement or enhance traditional absorption and fluorescence diagnostics. Here we introduce a number of these techniques.

12.1 Absorption Techniques

12.1.1 Frequency-Modulation Spectroscopy (FMS)

In FMS, the optical frequency of a laser is sinusoidally modulated, typically with an electro-optic modulator (EOM), at frequencies equal to or greater than the HWHM of the absorption transition lineshape. This typically requires modulation frequencies from 100 MHz to GHz rates for IR transitions at $\leq$ 1 atm (1 GHz = 0.033 cm^{-1}). The modulation effectively shifts absorption and dispersion information in frequency space to the harmonics of the modulation frequency which lie above lower frequency noise sources. As a result, FMS can provide highly sensitive measurements (down to absorbances of 10^{-8} [1]) of narrow absorption transitions.

Figure 12.1 illustrates a typical experimental setup for an FMS experiment. Frequency modulation is achieved by modulating the voltage applied to an EOM crystal which modulates its refractive index, and therefore, the phase or frequency of the laser beam passing through the EOM. The electric field of the laser beam exiting the EOM is given by:

$$E_o(t) = E_o \exp(i2\pi \nu_c t + i\beta \sin(2\pi f_m t)) \tag{12.1}$$

If the modulation index, β, is < 1, the electric field exiting the EOM predominantly consists of a carrier wave at ν_c and two out-of-phase sidebands located at $\nu_c \pm f_m$, known as a *triplet*. In the absence of wavelength-specific absorption, the triplet is *balanced* and the beat signals (of opposite sign) cancel each other out and go undetected by the photodetector. However, if the triplet is resonant with

R.K. Hanson et al., *Spectroscopy and Optical Diagnostics for Gases*,
DOI 10.1007/978-3-319-23252-2_12

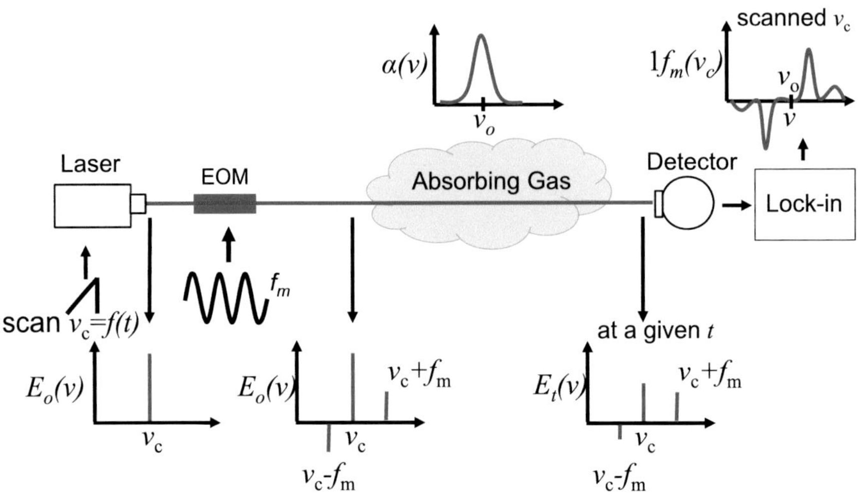

Fig. 12.1 Schematic for laser-absorption measurements using FMS

an absorption transition, the wavelength-specific absorption and dispersion (phase shift) cause unbalance within the triplet (since each component is attenuated and phase shifted a different amount) which introduces a beat signal at f_m in the detector. This beat signal can be extracted via homodyne detection (e.g., with a lockin-amplifier or double-balanced mixer) and compared with measured reference signals or simulated signals to infer gas properties. By scanning the carrier frequency of the laser across an absorption feature, FMS spectra can be measured. More information can be found in [2–6] and the references therein.

12.1.2 Wavelength-Modulation Spectroscopy (WMS)

In WMS, the laser's wavelength is sinusoidally modulated (typically via injection current modulation) about a given location on an absorption transition lineshape (Fig. 12.2). The wavelength modulation leads to intensity modulation, according to Beer's Law, which introduces frequency content centered at the harmonics of the modulation frequency in the detector signal. The harmonic signals can be extracted via lock-in filters during post-processing or data acquisition and can be compared with calibration-free WMS models [7–13] to infer gas conditions (T, P, χ, V) from the measured WMS harmonic signals. Like FMS, WMS enables improved measurement sensitivity via high-frequency modulation, down to absorbances of 10^{-5} [1], but is more versatile than FMS since lower modulation frequencies (f_m is typically 1 kHz to 1 MHz) and larger modulation depths O(0.1–1 cm^{-1}) can be used, the latter of which enables sensitive measurements at high pressures.

$$I_t(t) = I_o(t)\exp[-\alpha(\nu(t), T, P, \chi, L)] \tag{12.2}$$

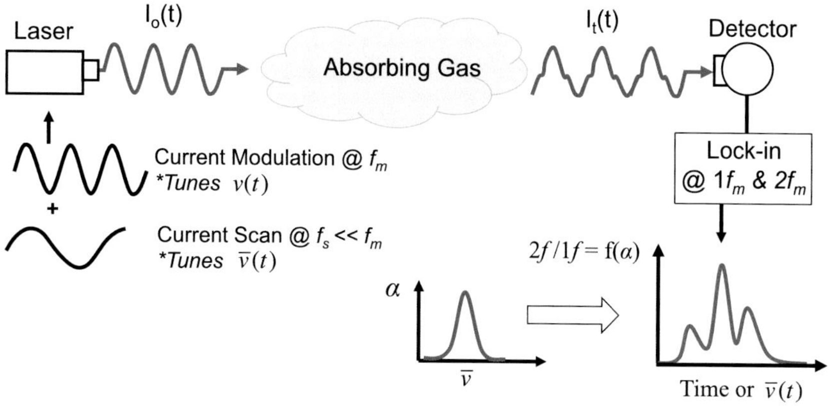

Fig. 12.2 Schematic for laser-absorption measurements using WMS

When using injection-current-tuned lasers (e.g., diode lasers, quantum cascade lasers) the current modulation introduces near-linear intensity modulation that can be exploited to normalize-out the dependence of WMS signals on the DC light intensity and electronic gain [10, 14]. This enables use of first-harmonic-normalized WMS signals that are insensitive to low-frequency (compared to f_m) variations in the light intensity impinging on the detector (e.g., resulting from beamsteering, window fouling, etc.) thereby improving measurement fidelity in harsh environments [10].

12.1.3 Cavity-Ringdown Spectroscopy (CRDS)

Cavity-ringdown spectroscopy enables increased sensitivity in laser absorption detection of a species by increasing the path length over which the light interacts with the absorbing species (Fig. 12.3).

In CRDS, a small fraction of pulsed light is coupled into a cavity through highly reflective mirrors ($R > 99.99\,\%$). This light reflects many times ($>$ 10,000 passes) so that the light circulating in this cavity has a long interaction path with any gases inside the cavity. The rate of decay of the transmitted intensity can be related to the reflectivity of the mirrors and any other loss processes (e.g., absorption, scattering) of light from the cavity. If a concentration of absorbing species is the dominant loss, the transmitted intensity can be expressed in the following form [15]:

$$I_t(t) = I_o \exp(-t/\tau) \tag{12.3}$$

$$= I_o \exp(-((1-R) + \alpha_{SP}(\nu))ct/L) \tag{12.4}$$

Cavity-ringdown spectroscopy has the potential to be 10,000 times or more sensitive than direct absorption [16]. Variations of pulsed CRDS (Fig. 12.3) include

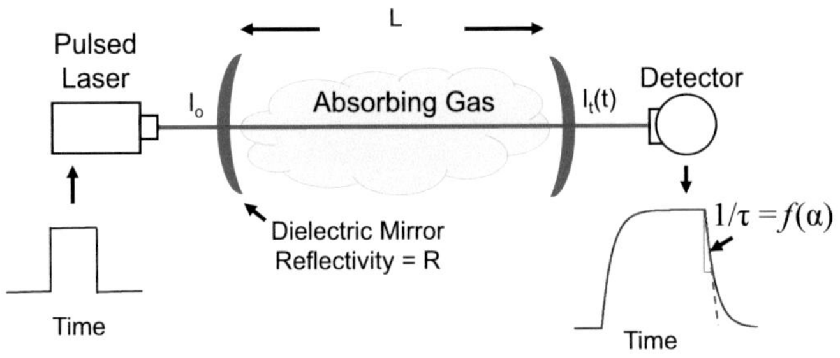

Fig. 12.3 Schematic for laser-absorption measurements using CRDS

CW (continuous wave), PS (phase shift) CRDS, and off-axis integrated-cavity output spectroscopy (OA-ICOS) [17, 18].

For optimum performance, the laser should be coupled into the TEM_{00} mode of the cavity. The spectral width of the laser is also important. If the laser bandwidth is wider than the cavity mode spacing or is not very narrow compared to the absorption linewidth, additional considerations are required [15, 17]. Additionally, a small amount of window fouling can create a significant interference when the beam is reflected from the mirror 10,000 times. Thus, maintaining clean mirrors is critical for sensitive measurements.

12.1.4 Off-Axis Integrated Cavity Output Spectroscopy

Similar to CRDS, OA-ICOS enables improved measurement sensitivity by increasing the effective path length over which light interacts with the absorbing species. However, in OA-ICOS cavity-mode-locking is not required since the reflections within the cavity are spatially separated (laser light travels around the perimeter of the mirrors in an elliptical pattern as it moves back and forth, see Fig. 12.4). At each interface, a small fraction of the light is transmitted through each mirror, and the spatially integrated signal is collected and focused onto a detector. The transmitted intensity exiting the mirror is given by a variant of Beer's Law, Eq. (12.5).

$$I_t(t) = I_o(t)\exp[-\alpha_{CEAS}(\nu)] = I_o(t)/(1 + G\alpha_{SP}(\nu)) \tag{12.5}$$

Here, $I_o(t)$ and $I_t(t)$ are the intensities exiting the cavity in the absence and presence of absorbers, respectively, α_{CEAS} is the effective cavity-enhanced absorbance, α_{SP} is the absorbance for a single pass (i.e., $S(T)P\chi_i\Phi L$), and G is the gain of the cavity.

By spatially separating the reflections, many roundtrips are required before the light rays overlap and retrace their original path through the cavity. The effective cavity length is now m times longer (up to O(1 km)) and the effective free-spectral range (FSR) of the cavity is now m times smaller (typically $<<$ HWHM of absorption transition at STP). These attributes enable scanned-wavelength measurements

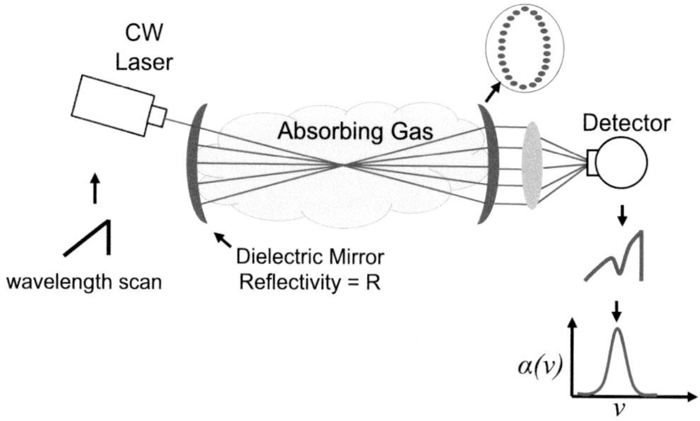

Fig. 12.4 Schematic for laser-absorption measurements using OA-ICOS

of absorption spectra and a cavity-enhanced path length that is much less sensitive to alignment compared to CRDS. More information regarding OA-ICOS can be found in [19–22] and the references therein.

12.2 Fluorescence Techniques

12.2.1 Planar Laser-Induced Fluorescence (PLIF)

One advantage of LIF is that it can easily be extended to two dimensions. By replacing the detector with a 2-D detector array (i.e., a digital camera), and expanding the excitation laser beam into a sheet, a two-dimensional image can be acquired (see Fig. 12.5). PLIF was first applied to combustion in 1982 [23,24]. Details of PLIF can be found in Kychakoff et al. [25] and the extension to high pressure was reported by Lee et al. [26]. PLIF has been used to measure concentration, temperature, pressure, and velocity [27, 28]. Pulsed lasers are normally employed to achieve single-shot imaging, and the short excitation pulse effectively freezes the flow field. While it is useful for measuring quantities in a plane, PLIF can suffer from lower SNR than 1-D LIF because the laser energy is spread out over a much larger area. For steady systems, PLIF measurements may use averaged measurements for improved SNR with either CW or pulsed laser excitation [27, 28].

12.2.2 Narrow Linewidth LIF

Stanford has pioneered the use of rapid-tuning, narrow-linewidth CW lasers for absorption and fluorescence spectroscopy (dye lasers, diode lasers (IR, NIR, VIS)). The LIF linear excitation equation for narrow-linewidth sources is

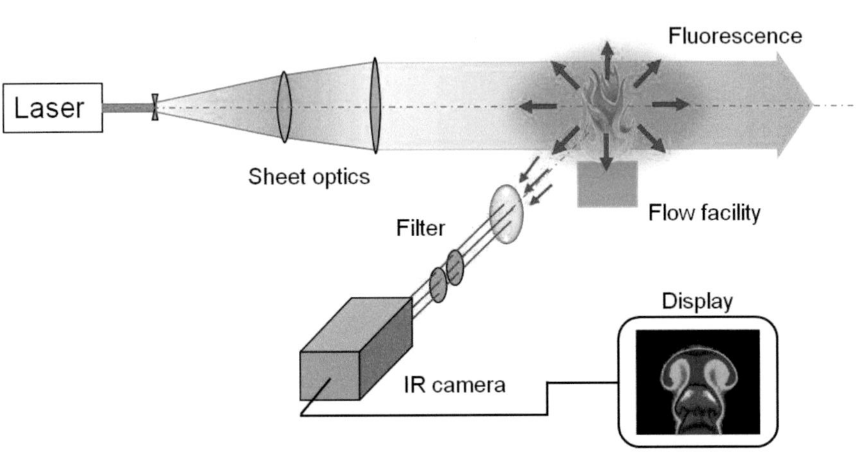

Fig. 12.5 Schematic of a PLIF experiment to measure species concentration

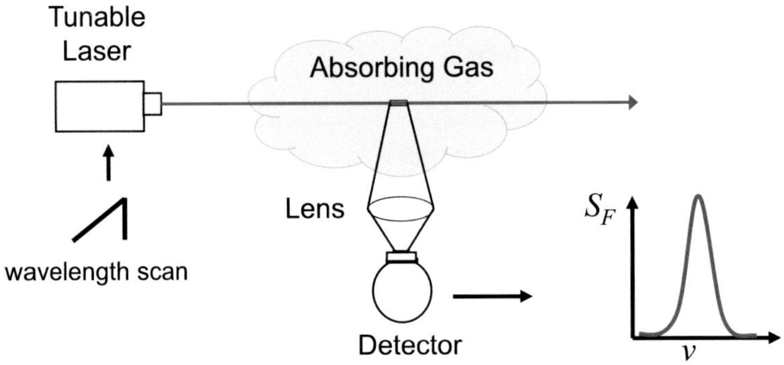

Fig. 12.6 A typical experimental schematic for narrow-linewidth LIF

$$S_F(\nu \rightarrow \nu + d\nu) = n_1^0 \cdot V \cdot \underbrace{I_\nu B_{12}[\phi(\nu)d\nu]}_{W_{12}=\text{prob/s of abs}} \cdot \frac{A}{A+Q} \cdot \frac{\Omega}{4\pi} \tag{12.6}$$

where $\phi(\nu)$ is the absorption lineshape function (see Chap. 8). Recall that $\int \phi(\nu)d\nu = 1$, therefore $\phi(\nu)d\nu$ = the fraction of total probability per second of absorption (I_ν B_{12}), for the frequency range $\nu \rightarrow \nu + d\nu$. What can we do with this? We can make spatially resolved measurements of the lineshape function and its integral (Fig. 12.6).

The shape of the LIF line provides opportunities for inferring pressure and temperature, while the integral (area) of the lineshape can provide a measure of $n_1 = n_i f_b(v'', J'')$.

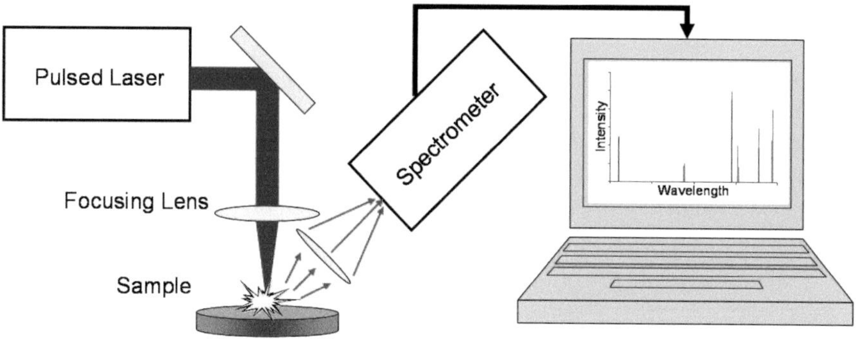

Fig. 12.7 Experimental schematic for LIBS analysis of a condensed-phase specimen

12.2.3 Laser-Induced Breakdown Spectroscopy (LIBS)

Laser-induced breakdown spectroscopy is a simple yet powerful analysis tool for particles and surfaces.

In LIBS, a high-energy laser pulse is focused onto a small sample of unknown composition, ablating a small amount of material from the sample and heating it to high temperatures (~10,000–20,000 K) [29]. At these high temperatures, the ablated material is broken down into the atoms and ions which constituted the original sample. These atoms emit light at very specific wavelengths, in proportion to the excited-state atomic concentration of the sample. Analysis of the emission spectrum can be used to identify the relative concentrations of the atoms present in the sample, which can then be matched to LIBS spectra of known specimens.

LIBS can be performed on stationary solid or liquid samples as shown in Fig. 12.7 or may be applied to flowing gaseous or particle-laden systems [30, 31]. Variations on standard LIBS include double-pulse and mixed-wavelength systems for enhanced identification [32, 33].

Advantages:

1. Requires little or no sample preparation
2. Can analyze very small sample sizes ($\sim 10^{-9}$ g)
3. Measurement time is typically short (~1 s)
4. Experimental setup is simple
5. Utilizes high-quality, inexpensive Vis/UV optical components

Disadvantages: LIBS is subject to variation in the pulse energy and can be subject to interference from plasma breakdown of the air surrounding the sample. Quantitative analysis can be complicated by electronic nonequilibrium and unknown, time-varying temperature.

12.3 Photothermal Techniques

Following laser absorption, the upper quantum energy states are depopulated via fluorescence or quenching. LIF is commonly limited by low fluorescence yields, an inefficiency caused by the more dominant quenching process. Conversely, photothermal techniques exploit rapid quenching by relating changes in the translational energy (i.e., temperature) of the gas to the absorbed photon energy. Two variants of photothermal diagnostic methods are presented below.

12.3.1 Photoacoustic Spectroscopy (PAS)

Recall the equation for laser-induced fluorescence signal for the simple two-level model:

$$S_F = (n_2)_{SS} \cdot A_{21} \cdot h\nu \ \ [\text{energy/vol/s}] \tag{12.7}$$

- Where did the other absorbed energy go? Heating.

$$S_{\text{heating}} = (n_2)_{SS}\, Q\, h\nu \tag{12.8}$$

$$= n_1^0 \cdot B_{12} I_\nu \cdot \frac{Q}{I_\nu B + A + Q} \cdot h\nu \tag{12.9}$$

$$\approx n_1^0 \cdot B_{12} I_\nu \cdot h\nu \ \ [\text{energy/vol/s}] \tag{12.10}$$

How can we measure the heating strength? Use a pulsed laser and measure the pressure pulse with a microphone! In a confined test cell, where density is constant, the absorbed energy per unit volume is related to the change in temperature and pressure by

$$E_{\text{abs}} = c_v \rho \Delta T \tag{12.11}$$

and

$$\Delta P = \rho R \Delta T \tag{12.12}$$

such that

$$E_{\text{abs}} = c_v (\Delta P / R) \tag{12.13}$$

A cylindrical gas cell may serve as a resonant cavity for sound waves, amplifying the signal when laser modulation is performed at acoustic frequencies. Due to signal dependencies on hardware components, PAS usually requires calibration for

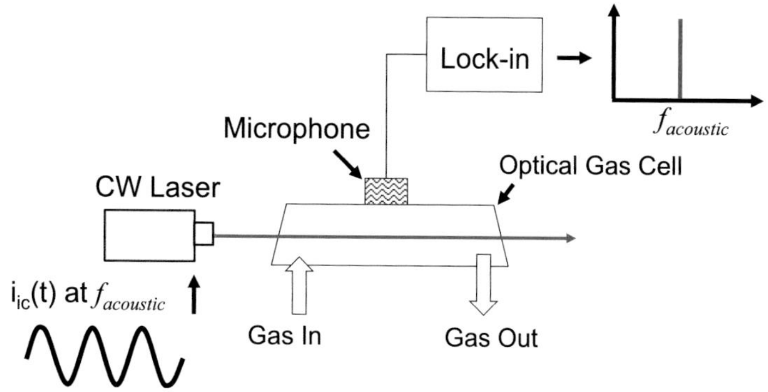

Fig. 12.8 Experimental configuration for optoacoustic measurements

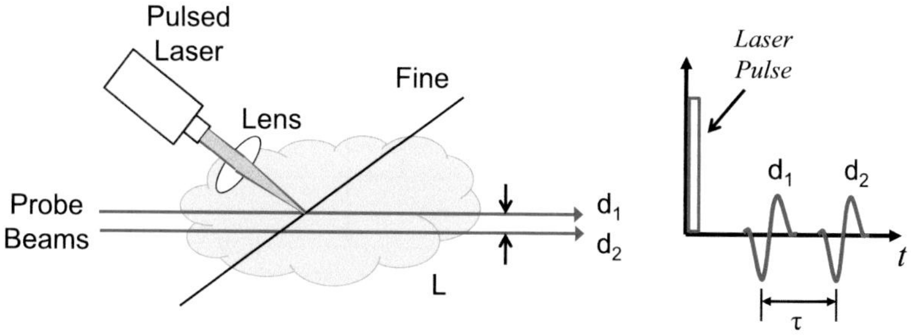

Fig. 12.9 Schematic and data traces for photothermal deflection

quantitative species concentration measurements. PAS is a common technique for species monitoring in relatively quiescent environments (Fig. 12.8).

12.3.2 Photothermal Deflection (PTD)

Temperature from the Speed of Sound

The speed of sound is proportional to $T^{1/2}$, and so a measurement of the speed of a weak pressure wave can be used to infer T. As one example, local heating on a wire from a pulsed laser leads to a pressure pulse that causes deflection of two HeNe probe beams (Fig. 12.9). The times at which the pressure pulse causes deflection at the two different HeNe probe beams is related to their separation distance, L, and the speed of sound in the flow.

$$a = \frac{L}{\tau} \tag{12.14}$$

$$= \sqrt{\frac{\bar{\gamma} R T}{\overline{M}}} \rightarrow T \tag{12.15}$$

An alternative is to use laser spark breakdown to produce a pressure wave.

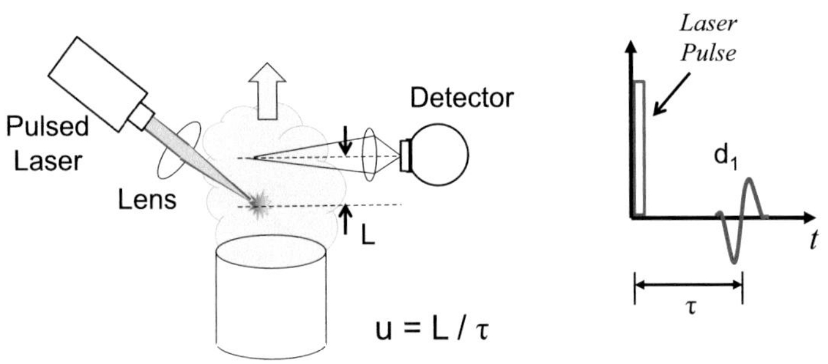

Fig. 12.10 Schematic velocity measurements using photothermal deflection

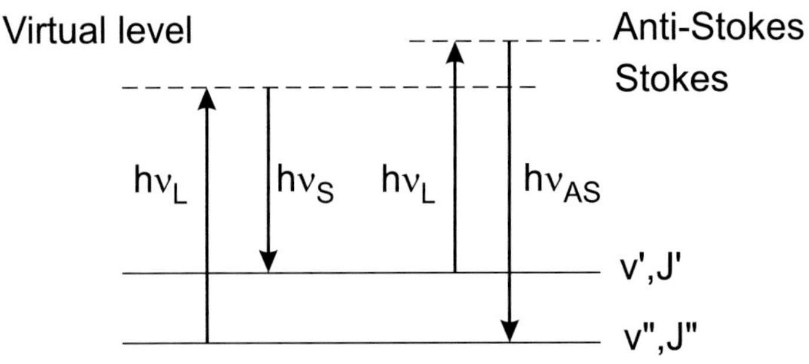

Fig. 12.11 Energy levels for spontaneous Raman scattering

Flow Velocity from Propagation of a Hot Spot

Flow velocity can be inferred from the convection of marked fluid, e.g. a thermal hot spot, as in Fig. 12.10. Alternatively, fluid may be marked by laser excitation to produce a new chemical species or excited state of a constituent species that can be tracked by a probe beam.

12.4 Scattering Techniques

Various forms of spectroscopic detection have been developed based on the Raman effect. Most notable are spontaneous Raman scattering (SRS) and coherent anti-Stokes Raman spectroscopy (CARS).

12.4.1 Spontaneous Raman Scattering

Rotational Raman:	($\Delta v = 0$); changes in J only ($\Delta J = \pm 2$)
Vibrational Raman:	changes in v and J (Fig. 12.11)
Set-up:	Similar to LIF (90° detection); CW and pulsed lasers

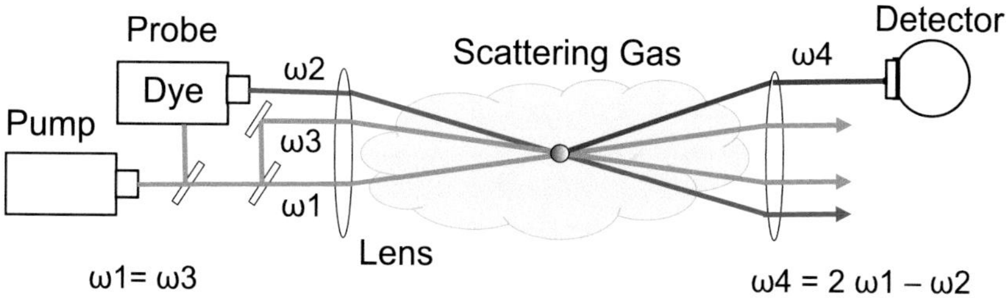

Fig. 12.12 Typical experimental setup for CARS (4-wave mixing)

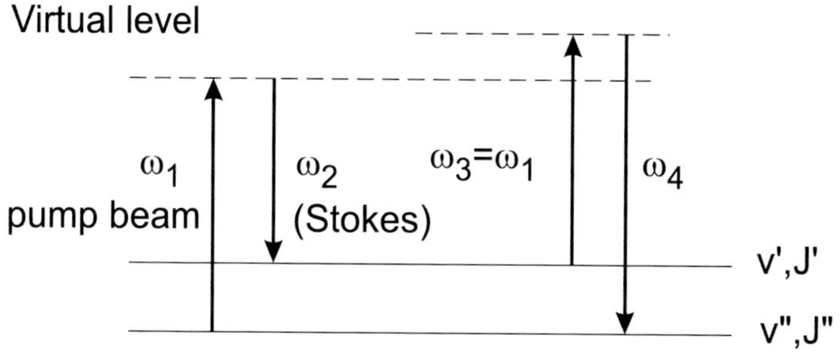

Fig. 12.13 Energy levels for CARS spectroscopy with 4-wave mixing

Uses: measure n_i and T (e.g., by use of filters and $\frac{I_{AS}}{I_S} \propto \exp(-\theta_v/T)$)

$$I(\nu_s) = \frac{cn_i(2J''+1)(v''+1)\exp(-E(v'',J'')/kT)}{Q_{r,v}}(\nu_L - \nu_R)^4 \qquad (12.16)$$

Note: ν^4-dependence favors short-wavelength lasers.

12.4.2 Coherent Anti-Stokes Raman Spectroscopy

Figure 12.12 depicts a typical experimental setup for a CARS experiment and Figs. 12.13 and 12.14 illustrate the energy levels and processes most relevant to CARS. Energy conservation requires that $\omega_4 = 2\omega_1 - \omega_2$, where ω is in frequency units. When $\omega_1 - \omega_2$ is resonant, there is strong, coherent generation of ω_4 via wave-mixing. In addition, conservation of momentum is in effect.

The spectrum and scanning ω_2 mode are depicted in the following figure (Figs. 12.15).

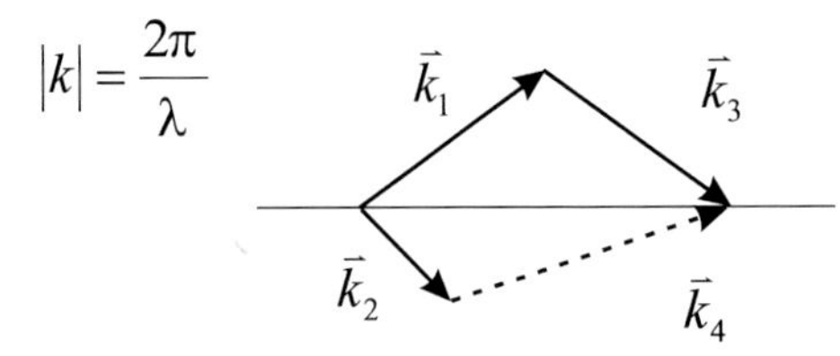

Fig. 12.14 Conservation of momentum for 4-wave mixing

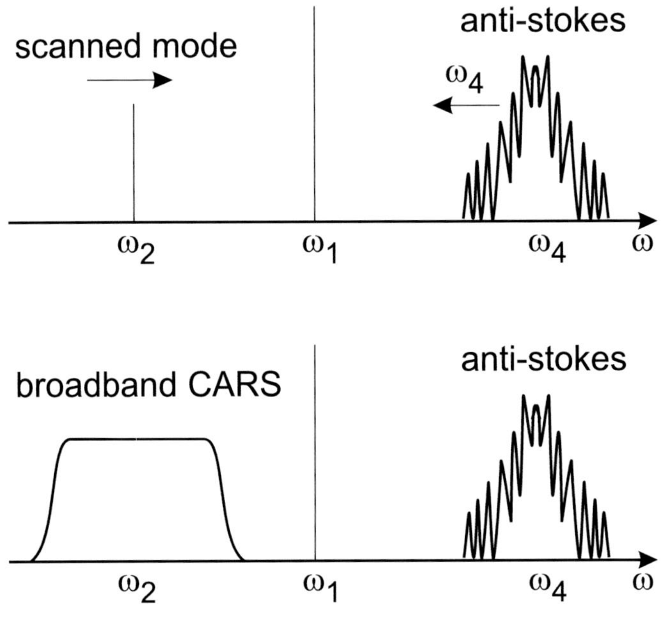

Fig. 12.15 CARS spectra for scanning and broadband sources

Advantages:	directed beam, efficient process
Uses:	thermometry/single shot in jet exhaust
Disadvantages:	complexity of theory/experiment/cost

Modern CARS techniques can provide high temporal and spatial resolution in complex flows [35, 36].

12.5 Laser Ionization Spectroscopy

Multiple analytical strategies have been developed based on the concept of laser-induced ionization (Fig. 12.16).

The lifetimes for laser ionization spectroscopy are as follows:

real levels	$\tau \approx 10^{-9}$
virtual levels	$\tau \approx 10^{-15}$

The experimental schematic for the measurements is shown below (Fig. 12.17).

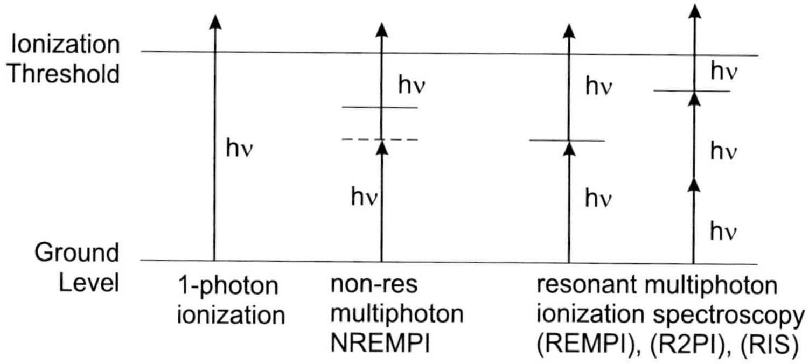

Fig. 12.16 Different energy levels for laser ionization spectroscopy [34]

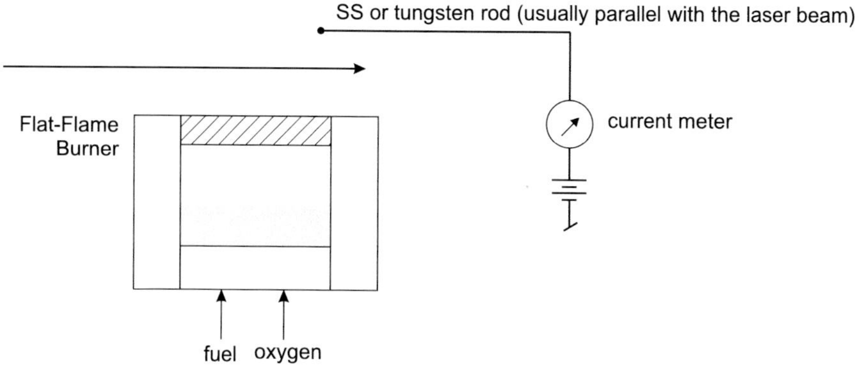

Fig. 12.17 Experimental schematic for laser ionization spectroscopy

Example applications for 3-photon processes:

H-atom detection	$(1S \to 2S)$ at approximately 121.6 nm
	$2 \times 243 \to 121.5$ nm
	$2S \to H^+ + e$ with third photon
O-atom detection	$2^3P \to 3^3P$
	$3^3P \to O^+ + e^-$ at 223 nm

References

1. J.A. Silver, Frequency-modulation spectroscopy for trace species detection: theory and comparison among experimental methods. Appl. Opt. **31**(6), 707–717 (1992)
2. G.C. Bjorklund, Frequency-modulation spectroscopy: a new method for measuring weak absorptions and dispersions. Opt. Lett. **5**(1), 15 (1980)
3. G.C. Bjorklund, M.D. Levenson, W. Lenth, C. Ortiz, Frequency modulation (FM) spectroscopy. Appl. Phys. B Photophys. Laser Chem. **32**(3), 145–152 (1983)

4. M. Gehrtz, W. Lenth, A.T. Young, H.S. Johnston, High-frequency-modulation spectroscopy with a lead-salt diode laser. Opt. Lett. **11**(3), 132 (1986)
5. G.E. Hall, S.W. North, Transient laser frequency modulation spectroscopy. Annu. Rev. Phys. Chem. **51**, 243–274 (2000)
6. G. Friedrichs, Sensitive absorption methods for quantitative gas phase kinetic measurements. Part 1: frequency modulation spectroscopy. Z. Phys. Chem. **222**(1), 1–30 (2008)
7. H. Li, G.B. Rieker, X. Liu, J.B. Jeffries, R.K. Hanson, Extension of wavelength-modulation spectroscopy to large modulation depth for diode laser absorption measurements in high-pressure gases. Appl. Opt. **45**(5), 1052–1061 (2006)
8. K. Duffin, A.J. McGettrick, W. Johnstone, G. Stewart, D.G. Moodie, Tunable diode-laser spectroscopy with wavelength modulation: a calibration-free approach to the recovery of absolute gas absorption line shapes. J. Lightwave Technol. **25**(10), 3114–3125 (2007)
9. A.J. McGettrick, K. Duffin, W. Johnstone, G. Stewart, D.G. Moodie, Tunable diode laser spectroscopy with wavelength modulation: a phasor decomposition method for calibration-free measurements of gas concentration and pressure. J. Lightwave Technol. **26**(4), 432–440 (2008)
10. G.B. Rieker, J.B. Jeffries, R.K. Hanson, Calibration-free wavelength-modulation spectroscopy for measurements of gas temperature and concentration in harsh environments. Appl. Opt. **48**(29), 5546–5560 (2009)
11. J.R.P. Bain, W. Johnstone, K. Ruxton, G. Stewart, M. Lengden, K. Duffin, Recovery of absolute gas absorption line shapes using tunable diode laser spectroscopy with wavelength modulation - part 2: experimental investigation. J. Lightwave Technol. **29**(7), 987–996 (2011)
12. K. Sun, X. Chao, R. Sur, C.S. Goldenstein, J.B. Jeffries, R.K. Hanson, Analysis of calibration-free wavelength-scanned modulation spectroscopy for practical gas sensing using tunable diode lasers. Meas. Sci. Technol. **24**(125203), 12 (2013)
13. C.S. Goldenstein, C.L. Strand, I.A. Schultz, K. Sun, J.B. Jeffries, R.K. Hanson, Fitting of calibration-free scanned-wavelength-modulation spectroscopy spectra for determination of gas properties and absorption lineshapes. Appl. Opt. **53**(3), 356–367 (2014)
14. D.T. Cassidy, J. Reid, Atmospheric pressure monitoring of trace gases using tunable diode lasers. Appl. Opt. **21**(7), 1185–1190 (1982)
15. A. McIlroy, J.B. Jeffries, Cavity ring-down spectroscopy for concentration measurements. In *Applied Combustion Diagnostics*, ed. by J.B. Jeffries, K. Kohse-Höinghaus (Taylor & Francis, New York, NY, 2002), pp. 98–127
16. P. Zalicki, Y. Ma, R.N. Zare, E.H. Wahl, J.R. Dadamio, T.G. Owano, C.H. Kruger, Methyl radical measurement by cavity ring-down spectroscopy. Chem. Phys. Lett. **234**(4/6), 269–274 (1995)
17. G. Berden, R. Peeters, G. Meijer, Cavity ring-down spectroscopy: experimental schemes and applications. Int. Rev. Phys. Chem. **19**(1), 565–607 (2000)
18. M. Gupta, T. Owano, D.S. Baer, A. O'keefe, Quantitative determination of the q(1) quadrupole hydrogen absorption in the near infrared via off-axis ICOS. Chem. Phys. Lett. **418**(1/3), 11–14 (2006)
19. A. O'Keefe, Integrated cavity output analysis of ultra-weak absorption. Chem. Phys. Lett. **293**(5–6), 331–336 (1998)
20. A.O. Keefe, J.J. Scherer, J.B. Paul, CW Integrated cavity output spectroscopy. Chem. Phys. Lett. **307**, 343–349 (1999)
21. J.B. Paul, J.J. Scherer, A.O. Keefe, L. Lapson, J.G. Anderson, C. Gmachl, F. Capasso, Bell Laboratories, Lucent Technologies, Mountain Ave, Murray Hill, Infrared cavity ringdown and integrated cavity output spectroscopy for trace species monitoring. **4577**, 1–11 (2002)
22. K. Sun, S. Wang, R. Sur, X. Chao, J.B. Jeffries, R.K. Hanson, Sensitive and rapid laser diagnostic for shock tube kinetics studies using cavity-enhanced absorption spectroscopy. Opt. Express **22**(8), 1–11 (2014)
23. G. Kychakoff, R.D. Howe, R.K. Hanson, J.C. Mcdaniel, Quantitative visualization of combustion species in a plane. Appl. Opt. **21**(18), 3225–3227 (1982)
24. M.J. Dyer, D.R. Crosley, Two-dimensional imaging of OH laser-induced fluorescence in a flame. Opt. Lett. **7**(8), 382–384 (1982)

25. G. Kychakoff, R.D. Howe, R.K. Hanson, Quantitative flow visualization technique for measurements in combustion gases. Appl. Opt. **23**(5), 704–712 (1984)
26. T. Lee, W.G. Bessler, C. Schulz, M. Patel, J.B. Jeffries, R.K. Hanson, UV planar laser induced fluorescence imaging of hot carbon dioxide in a high-pressure flame. Appl. Phys. B Lasers Opt. **79**(4), 427–430 (2004)
27. J.M. Seitzman, R.K. Hanson, P.A. Debarber, C.F. Hess, Application of quantitative two-line OH planar laser-induced fluorescence for temporally resolved planar thermometry in reacting flows. Appl. Opt. **33**(18), 4000–4012 (1994)
28. B. Hiller, R.K. Hanson, Simultaneous planar measurements of velocity and pressure fields in gas flows using laser-induced fluorescence. Appl. Opt. **27**(1), 33–48 (1988)
29. A.C. Samuels, F.C. Delucia, K.L. Mcnesby, A.W. Miziolek, Laser-induced breakdown spectroscopy of bacterial spores, molds, pollens, and protein: initial studies of discrimination potential. Appl. Opt. **42**(30), 6205–6209 (2003)
30. G.A. Lithgow, S.G. Buckley, Effects of focal volume and spatial inhomogeneity on uncertainty in single-aerosol laser-induced breakdown spectroscopy measurements. Appl. Phys. Lett. **87**(1), 11501–11501 (2005)
31. J.P. Singh, H.S. Zhang, F.Y. Yueh, K.P. Carney, Investigation of the effects of atmospheric conditions on the quantification of metal hydrides using laser-induced breakdown spectroscopy. Appl. Spectrosc. **50**(6), 764–773 (1996)
32. D.N. Stratis, K.L. Eland, S.M. Angel, Effect of pulse delay time on a pre-ablation dual-pulse libs plasma. Appl. Spectrosc. **55**(10), 1297–1303 (2001)
33. J. Scaffidi, J. Pender, W. Pearman, S.R. Goode, B.W. Colston, J.C. Carter, S.M. Angel, Dual-pulse laser-induced breakdown spectroscopy with combinations of femtosecond and nanosecond laser pulses. Appl. Opt. **42**(30), 6099–6106 (2003)
34. W. Demtröder, *Laser Spectroscopy: Basic Concepts and Instrumentation*, 2nd enl. edn. (Springer, New York, 1996)
35. S. Roy, J.R. Gord, A.K. Patnaik, Recent advances in coherent anti-Stokes Raman scattering spectroscopy: fundamental developments and applications in reacting flows. Prog. Energy Combust. Sci. **36**(2), 280–306 (2010)
36. C.N. Dennis, C.D. Slabaugh, I.G. Boxx, W. Meier, R.P. Lucht, Chirped probe pulse femtosecond coherent anti-Stokes Raman scattering thermometry at 5 kHz in a Gas Turbine Model Combustor. Proc. Combust. Inst. **35**, 3731–3738 (2015)

Spectroscopy Equipment 13

An important consideration for spectroscopic measurements is the availability and performance of light sources and detectors. Figure 13.1 shows the operating wavelength range for many common laser sources. This chapter is devoted to a discussion of laser light sources and detectors that operate between the UV and far-IR. While many of the sources and detectors find industrial uses, we concern ourselves here with the spectroscopic applications of these devices.

13.1 Sources

Laser light can be made at many wavelengths ranging from the UV to the far IR. Some lasers are quite large and expensive while others are small enough to fit in the palm of your hand and cost less than $100. Lasers can be pulsed or continuous wave (CW), with average powers ranging from microwatts to kilowatts or more. In this section we will discuss a few of the more popular sources and their uses.

13.1.1 The Helium–Neon Laser

The helium–neon (HeNe) laser is one of the simplest lasers and was one of the earliest lasers invented (the ruby laser was the first laser). The HeNe laser is so simple that there are books, web sites, and undergraduate classes that will teach motivated students to build their own [2]. The HeNe laser has a tube filled with a mixture of helium and neon ($\sim$0.1 Torr neon, 0.9 Torr helium) [3]. The helium is excited by an electrical discharge (which looks very much like a neon light). The excited helium atoms transfer energy to the neon atoms which produce the laser light. HeNe's are CW devices with typical powers ranging from 1 to 50 mW.

HeNe lasers are most commonly used to produce 632 nm light, but HeNe's are also available with output at 3.3903 μm, 1.15 μm, and several other visible

R.K. Hanson et al., *Spectroscopy and Optical Diagnostics for Gases*,
DOI 10.1007/978-3-319-23252-2_13

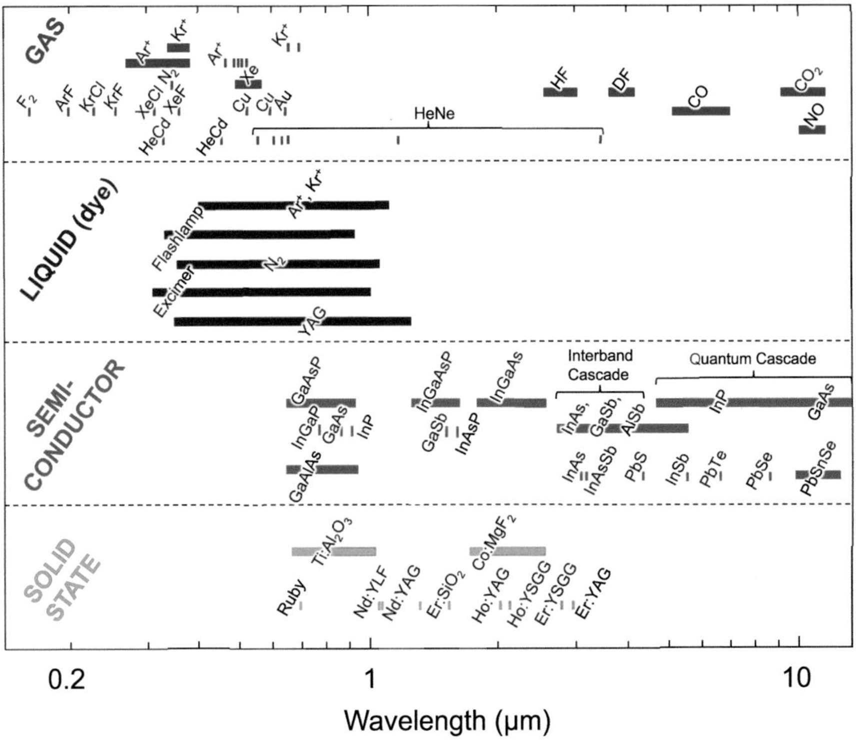

Fig. 13.1 Wavelengths of operation for many common lasers [1]

and IR wavelengths [4, 5]. Visible HeNe's are useful for optical alignment and optical scattering measurements. The 3.39 μm HeNe laser is useful in absorption spectroscopy for detection of hydrocarbons because the C–H stretch oscillates near 3.4 μm [6].

Argon-ion lasers are similar to HeNe lasers because they are gas lasers that are pumped by an electrical discharge. However, in an argon-ion laser, the argon is ionized in the plasma and ions act as the gain medium. Argon-ion lasers are typically CW devices, but can provide significantly more power than the HeNe laser. Argon-ion lasers can lase at many discrete wavelengths between 275 and 550 nm [7].

Argon-ion lasers can be used for direct absorption measurements at their fundamental wavelength [8] or at wavelengths of higher order harmonics [9]. They can also be used to pump dye lasers for absorption or fluorescence measurements [9].

13.1.2 The Nd:YAG Laser

The Nd:YAG laser (i.e., neodymium-doped yttrium aluminum garnet: $Nd{:}Y_3Al_5O_{12}$) is a common laser for high-power applications. The YAG crystal is doped with triply ionized neodymium, which replaces another element of roughly the same size, typically yttrium. To emit coherent radiation, the neodymium must first be pumped from the equilibrium state into a higher energy level. For these solid state lasers, this is often done using flashlamps, but diode lasers are also used. Nd:YAG lasers can be CW or pulsed. Higher peak powers are available when operated in pulsed mode, which is important for frequency conversion (i.e., frequency doubling, difference frequency generation, and sum frequency generation) and fluorescence.

Nd:YAG lasers are useful for their high power output. Nonlinear processes are often used to convert the IR light (usually at 1064 nm) to mid-IR light (2–4 μm) using an optical parametric oscillator (OPO) [10], or to visible and UV light (532, 355, 266 nm) through harmonic generation [11, 12]. Nd:YAG lasers can also be used to pump dye lasers [13]. Because of the high achievable powers and wide range of wavelengths that can be accessed using the Nd:YAG laser, they are a common component in many fluorescence and absorption experiments, both in the UV [12] and IR [10].

13.1.3 The Excimer Laser

Excimer lasers (short for *Exci*ted Di*mer*) are pulsed UV lasers with a gain medium that is a short-lived molecule made up of one rare-gas atom and one halogen (e.g., ArF, KrCl, KrF, XeCl, XeF) [4]. Excimer lasers are used for LIF because they have high output powers in the UV [12]. A UV laser photolysis technique has also been demonstrated using the excimer laser as the photolyzing source [14].

13.1.4 The CO_2 Laser

In the CO_2 laser, the molecular vibrations are pumped by a plasma discharge. CO_2 lasers can be either pulsed or CW. CO_2 lasers can operate at discrete wavelengths between 9.3 and 11.4 μm, however the 10.6 μm transitions are popular choices [5]. High-power CO_2 lasers are useful in manufacturing for cutting, welding, and other processing technologies [3]. CO_2 lasers are also utilized in spectroscopy for fluorescence and absorption measurements [15].

13.1.5 Semiconductor Lasers

Semiconductor lasers (e.g., diode lasers, quantum cascade lasers, external cavity diode lasers, distributed feedback (DFB) lasers) have become increasingly important in spectroscopy and remote sensing. Semiconductor lasers are available in wavelengths from the near UV (375 nm) to the far-IR ($\sim$11 μm) and produce powers from $\sim$1 to 500 mW. Near-IR diode lasers are well developed and commercially available due to significant investment from the telecom industry. These lasers are compact and rugged and can often be purchased in a prepackaged, fiber-coupled unit. Many DFB lasers can be rapidly tuned over several wavenumbers by changing the injection current or laser temperature. Some external cavity diode lasers can be tuned more than 100 cm^{-1}.

Due to their lower power outputs, tunable diode lasers are used primarily in absorption spectroscopy. Wavelength-tunability enables the user to interrogate an individual absorption feature of an atom (e.g., potassium or cesium) or small molecule (e.g., CO, CO_2, H_2O, or CH_4). Absorption measurements at multiple wavelengths have been used for simultaneous measurements of temperature and multiple species [16] or pressure, temperature, and velocity [17].

13.1.6 Dye Lasers

Dye lasers use an organic dye as the gain medium. Because the dye is a liquid, the absorption and fluorescence spectra are broadband, but the laser cavity is designed to emit "monochromatic" laser light somewhere within this emission band. Dye lasers are tunable over this emission band; each specific dye provides a range of wavelengths (10 to 70 nm), mostly in the visible and UV. Dye lasers are pumped by other UV lasers (such as an argon-ion laser) or by flashlamps.

Laser dyes can be found that emit between 300 and 1200 nm [18]. Harmonic generation of these tunable lasers facilitates spectroscopic work in the UV [9, 13].

13.1.7 Non-Laser Sources

While lasers act as valuable spectroscopic tools, and cover much of the electromagnetic spectrum from the UV to the far-IR, some wavelengths cannot be accessed using laser sources. Broadband lamps are useful tools for accessing wavelengths that cannot be easily obtained with lasers. Lamps offer the additional benefit of being simpler and less expensive than lasers, however, the lower spectral intensities reduce the SNR for spectrally resolved measurements.

13.2 Detectors

In an optical experiment, the detection system is an important consideration, second only to the optical source. Thus, we will devote some time to describing the different types of detectors and their characteristics. Before examining detectors in detail, some basic detector vocabulary is reviewed. Next, three common types of detectors are described. Finally, detector wavelength range, time response, and signal-to-noise ratio are compared for several common detectors. While this is only an introductory treatment, more information can be found in [19–21].

Shot noise refers to noise that is a result of random fluctuations in the time of arrival of the photons or electrons.

Dark current is the current produced by the detector with no signal from the source. Sources of dark current include background radiation from the environment and random thermal excitation of electrons within the detector itself.

Johnson noise is the noise generated by the equilibrium fluctuations of the electric current inside an electrical conductor, which happens without any applied voltage, due to the random thermal motion of the charge carriers (i.e., the electrons). In other words, Johnson noise is caused by random electron motion in the circuit.

Generation-recombination noise occurs in photoconductors because the photoexcited carrier has a characteristic lifetime, τ. The lifetime of each carrier is not exact, but described statistically. Thus, there is uncertainty in the precise time that the carrier recombines and this uncertainty results in random noise.

Background limited infrared performance (BLIP) occurs when the detector noise is limited by the background shot noise and not by intrinsic detector noise. Infrared detectors are often cryogenically cooled to approach this ideal condition.

Bandgap energy is the energy difference between the top of the valence band and the bottom of the conduction band in a semiconductor. Photons with energy that is lower than this bandgap energy are not detected by the semiconductor.

The detectors we will consider here are referred to as quantum detectors. Quantum detectors respond to individual photons and offer a fast response time. Thermal detectors are not discussed here, but more information can be found in [19]. It is sufficient to say that thermal detectors offer broad wavelength sensitivity, but at the cost of reduced time response.

13.2.1 The Photomultiplier

The photomultiplier is a popular optical detector used to measure radiation in the UV, visible, and near-IR wavelength regions. The photomultiplier consists of a photocathode, a series of electrodes, called dynodes, and an anode. When a photon strikes the photocathode, there is some probability that an electron will be emitted and accelerated towards the first dynode. Each dynode is held at a higher electrical potential than the previous so that it can attract the electrons. When an electron strikes the dynode, it sheds multiple electrons which accelerate to the next dynode. In this way, the signal is amplified at each dynode. Primary sources of noise in a photomultiplier are shot noise (generated by both the signal and the dark current) and Johnson noise.

13.2.2 Photoconductive Detectors

A photoconductive (PC) detector is a semiconductor-based detector with an electrical resistance that is sensitive to the light incident on it. A voltage placed across the detection element is used to measure the resistance. Photoconductive detectors are available in the near-IR to the far-IR (wavelengths from 1 to 50 μm).

Photoconductive detectors (and other semiconductor-based detectors) have a bandgap energy associated with them. Photons with energy that is smaller than this bandgap energy are not observed by the detector. For photons with energy that is greater than the bandgap energy, the current generated is proportional to the number of photons striking the active area. Thus, for fixed power, the detector sensitivity increases linearly with wavelength until the bandgap energy is reached, then the sensitivity drops off rapidly.

When a photon is absorbed by the material, an electron is excited from the valence band to an acceptor atom. There is a minimum photon energy required to excite the electron, placing a lower limit on the photon energy (or an upper limit on the detectable wavelength). The electron hole is acted on by the electric field (the applied voltage) and drifts along the field direction, resulting in a current. The contribution of a particular hole to the total current ends when an electron recombines with the hole. This process is referred to as electron-hole recombination. Photoconductive detectors are often cryogenically cooled so that thermal excitation of the carriers does not dominate over the photoexcitation process. There are two primary sources of noise in photoconductors. The first noise source is shot noise which can originate from the dark current or from the measured signal. The second source of noise is generation-recombination noise, which is a result of the finite lifetime of the photoexcited carrier.

13.2.3 Photodiode Detectors

A photodiode is a semiconductor that generates a voltage (or current) when light is incident on it and is often called a photovoltaic (PV) detector. Like photoconductors, photodiodes have a minimum photon energy associated with the energy bandgap of the semiconductor. Noise in a photodiode is dominated by the Johnson noise and therefore is not usually shot-noise limited.

A variation of the standard photodiode is an avalanche photodiode. In an avalanche photodiode, an electric voltage is placed across the diode. When a carrier is produced, it is accelerated by the electric field to energies great enough to knock new electrons out of the valence band. Thus, the signal is amplified resulting in more sensitive detection. The shot noise of the avalanche photodiode is also amplified in this system, but because a standard photodiode is Johnson noise limited, the SNR increases with increasing amplification until the shot noise dominates.

13.2.4 Selecting a Detector

There are many criteria that might be used to choose the correct detector for a specific application including wavelength, time response, noise characteristics, simplicity, and cost. Different types of detectors are sensitive to different wavelengths of light. Figure 13.2 shows the wavelength range for many common detectors.

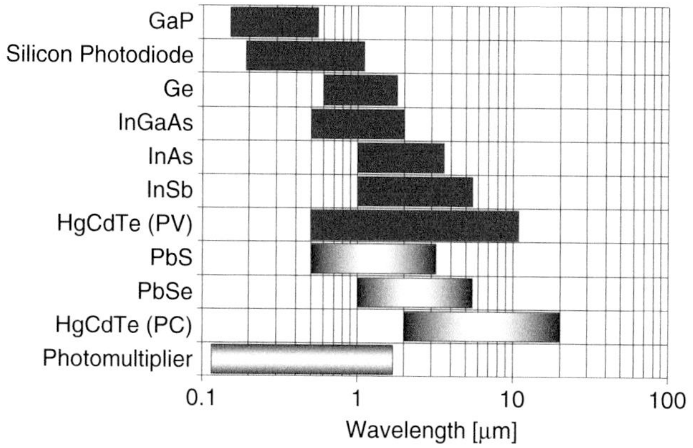

Fig. 13.2 Wavelength range for common detectors. *Black boxes* indicate photovoltaic detectors (photodiodes). *Gradient from left to right* indicates a photoconductor and *gradient from top to bottom* indicates a photomultiplier

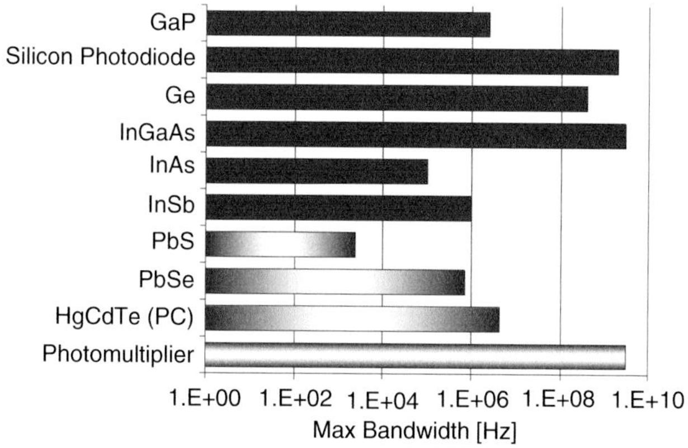

Fig. 13.3 Typical bandwidth for common detectors. *Black boxes* indicate photovoltaic detectors (photodiodes). *Gradient from left to right* indicates a photoconductor and *gradient from top to bottom* indicates a photomultiplier

By choosing a specific wavelength or set of wavelengths, the choice of detectors will be some subset of the detectors indicated in the figure.

As shown in Fig. 13.2, there are often several detectors available for a particular wavelength, so more selection criteria can be used to reduce this list. Detectors have a frequency bandwidth which is important for time-resolved measurements. Bandwidth can be dependent on the detector area and temperature as well as the preamplifier gain, and of course, the detector material. By increasing the detector area or the preamplifier gain, the frequency bandwidth will generally be reduced. Typical commercially available bandwidth for some common detector types is plotted in Fig. 13.3.

Detector noise can also be an important issue, especially when measuring weak signals (such as with fluorescence). Detector noise is characterized by the detectivity (D^*):

$$D^* = \frac{\sqrt{A_{\text{Detector}} \Delta f}}{\text{NEP}} \tag{13.1}$$

where A_{Detector} is the detector area, Δf is the bandwidth, and NEP, the noise equivalent power, is the amount of optical power required to equal the magnitude of the detector noise. The signal-to-noise ratio for a measurement dominated by detector noise can be calculated using this equation:

$$\text{SNR} = \frac{P_{\text{incident}}}{\text{NEP}} = \frac{P_{\text{incident}} D^*}{\sqrt{A_{\text{Detector}} \Delta f}} \tag{13.2}$$

where P_{incident} is the incident optical power. Thus a high D^* is required for sensitive optical measurements.

Cost and complexity should also be considered when choosing detectors. Most near-IR, visible, and UV photodiodes are available in compact packages at a relatively low price while photomultipliers typically are bulkier and more fragile and expensive. Many IR detectors are mounted in large dewars and require cryogenic cooling which can be inconvenient and makes the detector package more expensive and fragile. However, some IR detectors operate at room temperature or can be thermo-electrically cooled, making these detectors more compact, rugged, and portable.

Spatially uniform responsivity is an important factor that should not be ignored. IR detectors (particularly HgCdTe and room temperature and TE cooled PV detectors) can have a spatially nonuniform response across the active detector surface. This nonuniform response can manifest itself as noise in a poorly controlled experiment. Oftentimes smaller detectors are more uniform than large detectors and cooled detectors are more uniform than uncooled detectors. In the end, there is often a tradeoff between cost, sensitivity, bandwidth, noise, size, and complexity.

References

1. R.W. Waynant, M.N. Ediger, *Electro-Optics Handbook*, 2nd edn. (McGraw-Hill, New York, 2000)
2. G. Mccomb, *Lasers, Ray Guns, and Light Cannons: Projects From the Wizard's Workbench* (McGraw-Hill, New York, 1997)
3. J.T. Luxon, D.E. Parker, *Industrial Lasers and Their Applications*, 2nd edn. (Prentice-Hall, Englewood Cliffs, NJ, 1992)
4. J. Hecht, *Understanding Lasers: An Entry Level Guide*, 2nd edn. (IEEE Press, Piscataway, NJ, 1994)
5. K.J. Kuhn, *Laser Engineering* (Prentice-Hall, Upper Saddle River, NJ, 1998)
6. A.E. Klingbeil, J.B. Jeffries, R.K. Hanson, Temperature- and pressure-dependent absorption cross sections of gaseous hydrocarbons at 3.39 μm. Meas. Sci. Technol. **17**, 1950–1957 (2006)

7. W.B. Bridges, A.N. Chester, Visible and uv laser oscillation at 118 wave lengths in ionized neon, argon, krypton, xenon, oxygen and other gases. Appl. Opt. **4**(5), 573–580 (1965)
8. M. Rohrig, E.L. Petersen, D.F. Davidson, R.K. Hanson, Shock tube study of the pyrolysis of NO_2. Int. J. Chem. Kinet. **29**(7), 483–493 (1997)
9. M.A. Oehlschlaeger, D.F. Davidson, J.B. Jeffries, R.K. Hanson, Ultraviolet absorption cross-sections of hot carbon dioxide. Chem. Phys. Lett. **399**(4/6), 490–495 (2004)
10. B.J. Kirby, R.K. Hanson, Linear excitation schemes for ir planar-induced fluorescence imaging of CO and CO_2. Appl. Opt. **41**(6), 1190–1201 (2002)
11. J. Egermann, T. Seeger, A. Leipertz, Application of 266-nm and 355-nm nd:yag laser radiation for the investigation of fuel-rich sooting hydrocarbon flames by raman scattering. Appl. Opt. **43**(29), 5564–5574 (2004)
12. J.D. Koch, R.K. Hanson, Temperature and excitation wavelength dependencies of 3-pentanone absorption and fluorescence for PLIF applications. Appl. Phys. B Lasers Opt. **76**(3), 319–324 (2003)
13. T. Lee, J.B. Jeffries, R.K. Hanson, Experimental evaluation of strategies for quantitative laser-induced-fluorescence of nitric oxide in high-pressure flames (1–60 bar). Proc. Combust. Inst. **31**, 757–764 (2007)
14. D.F. Davidson, R.K. Hanson, High temperature reaction rate coefficients derived from n-atom ARAS measurements and excimer photolysis of NO. Int. J. Chem. Kinet. **22**(8), 843–861 (1990)
15. B.J. Kirby, R.K. Hanson, CO_2 imaging with saturated planar laser-induced vibrational fluorescence. Appl. Opt. **40**(33), 6136–6144 (2001)
16. D.S. Baer, R.K. Hanson, M.E. Newfield, N.K.J.M. Gopaul, Multiplexed diode-laser sensor system for simultaneous H_2O, O_2, and temperature measurements. Opt. Lett. **19**(22), 1900–1902 (1994)
17. L.C. Philippe, R.K. Hanson, Laser diode wavelength-modulation spectroscopy for simultaneous measurement of temperature, pressure, and velocity in shock-heated oxygen flows. Appl. Opt. **32**(30), 6090–6103 (1993)
18. W. Demtröder, *Laser Spectroscopy: Basic Concepts and Instrumentation*, 2nd enl. edn. (Springer, New York, 1996)
19. J.D. Vincent, *Fundamentals of Infrared Detector Operation and Testing* (Wiley, New York, 1989)
20. A Yariv, *Optical Electronics in Modern Communications*, 5th edn. (Oxford University Press, Oxford, 1996)
21. E.L. Dereniak, G.D. Boreman, *Infrared Detectors and Systems* (Wiley, New York, 1996)

Case Studies: Molecular Spectroscopy 14

Here we present spectroscopic case studies of specific molecules, spanning the ultraviolet to long-wave infrared. These case studies are intended to reinforce the concepts of line positions, line strengths, and line shapes, while also exposing the reader to some of the many species-specific details that contribute to the uniqueness of observed spectra. A detailed study of the OH molecule is followed by more brief reviews of O_2, H_2O, and hydrocarbons.

14.1 Ultraviolet OH Spectroscopy: The $A^2\Sigma^+ - X^2\Pi$ System

Contributing Author: Jerry M. Seitzman

As a practical example of material previously covered, we will examine the ultraviolet (UV) absorption spectrum of the hydroxyl molecule. In order to understand the electronic absorption spectrum of a molecule, there are at least three things one must consider: (1) the energy levels of the molecule (for line positions and Boltzmann populations), (2) the specific radiative transitions which are allowed (determined from selection rules) and, finally, (3) the proper calculation of an absolute spectral absorption coefficient.

14.1.1 OH Energy Levels

In this first section, we briefly examine the structure of the two OH electronic levels involved in the OH spectrum. Based on the structure of each, we then consider the proper term energies to apply. For a more complete discussion of this material, see Chap. V, Sect. 2 of Herzberg (Spectra of Diatomic Molecules).

R.K. Hanson et al., *Spectroscopy and Optical Diagnostics for Gases*,
DOI 10.1007/978-3-319-23252-2_14

Term Energies

Recall, the term energy $E(n, v, J)$ of a specific energy level with electronic quantum number n, vibrational quantum number v, and angular momentum quantum number J is given in general by:

$$E(n, v, J) = T_e(n) + G(v) + F(J), \tag{14.1}$$

where T_e is the electronic term energy, $G(v)$ is the vibrational term energy, and $F(J)$ is the angular momentum term energy (also commonly referred to as the rotational term energy even though it comprises more motions than just nuclear rotations). In writing this equation, we have applied the Born–Oppenheimer approximation, which assumes the different motions associated with each energy mode are separable.

Generally, explicit equations for each term energy mode can be written. These equations have as variables the appropriate quantum number and other molecular parameters. For example, recall that the vibrational term energies are often modelled by an equation of the form

$$G(v) = \omega_e(v + 1/2) - \omega_e x_e(v + 1/2)^2 \tag{14.2}$$

Because of the complexity of modelling electronic term energies, the values for T_e are usually obtained from appropriate tables.

In the remainder of this section, we focus on proper modelling of the appropriate expressions for $F(J)$. As discussed previously, this depends on how we choose to model the coupling between the various forms of angular momentum (e.g., nuclear rotation, electron spin, and electron orbital angular momentum). Generally, a model based on one of Hund's Cases is chosen. For most diatomics of interest here, either Hund's Case (a), Hund's Case (b) or an intermediate model representing a transition from Hund's (a) to Hund's (b) is appropriate (at least for values of J up to some cutoff).

Spin-Splitting

According to standard spectroscopic notation, the $A^2\Sigma^+ - X^2\Pi$ system of OH involves a transition between the ground electronic configuration, $X^2\Pi$ (where X denotes the ground level), and the $A^2\Sigma^+$ level, the lowest energy excited electronic configuration with the same spin multiplicity as the ground level. Likewise, a $B - X$ transition would usually indicate a transition between the ground and the second excited electronic energy level with the same spin multiplicity.

In this case for OH, the two electronic levels are doublets, i.e. $2S + 1 = 2$ or $S = \frac{1}{2}$ (which results from the configurations having an unpaired electron in a suborbital). Therefore, each electronic configuration actually consists of two spin sublevels, one having an electron with spin "up" and one with the electron spin "down." Instead of writing a different electronic term energy T_e for the two spin sublevels, it is more convenient to denote two *spin-split* levels with the same

J but different spin with the angular momentum term energies F_1 and F_2. Recall, the quantum number N (when defined) is given by

$$N = \begin{cases} J - 1/2 \text{ for } F_1 \\ J + 1/2 \text{ for } F_2 \end{cases} \quad (14.3)$$

The Excited $A^2\Sigma^+$ Level

The designation Σ implies that for this configuration, Λ, the component of the electronic orbital angular momentum (**L**) along the internuclear axis, is zero. Since Λ is zero, there is no induced magnetic field to couple the electron spin vector **S** to the internuclear axis and, therefore, Hund's case (b) is the proper coupling model. In that model, **L** couples (aligns) directly with the *B*-axis of nuclear rotation and we get the following quantum number rules

$$N = \Lambda, \Lambda + 1, \Lambda + 2, \Lambda + 3 \ldots$$

$$\text{or in this case} \quad N = 0, 1, 2, 3, \ldots$$

and then coupling to *S*, we get the two spin-split levels,

$$J = N \pm S \text{ or } J = N \pm 1/2$$

The rotational term energies for such a $^2\Sigma$ state are well modelled by:

$$F_1(N) = B_v N(N+1) - D_v[N(N+1)]^2 + \gamma_v N$$

$$F_2(N) = B_v N(N+1) - D_v[N(N+1)]^2 - \gamma_v(N+1) \quad (14.4)$$

$$(\gamma_v \approx 0.1 \text{ cm}^{-1} \text{ for OH } A^2\Sigma^+)$$

The energies of two states having the same *N* but different *J* are nearly identical, with a spin-splitting (constant in this model) of $\gamma_v(2N+1)$. (Recall the $_v$ subscripts denote molecular "constants" which can be a function of vibrational level.) See Table 14.1 for values of γ_v.

Figure 14.1 shows the first few energy levels of the $A^2\Sigma$ configuration. Note, only one state exists with a value of $N = 0$.

As a final note, the symbol + in the $A^2\Sigma^+$ name denotes that even numbered *N* levels have a positive parity (that is, they have wave functions which are symmetric). For a Σ^- state, the even *N* levels have − parity, and the odd levels are now +.

The Ground $X^2\Pi$ Level

Hund's ***a***	$\Lambda \neq 0, S \neq 0$
Hund's ***b***	$\Lambda = 0, S \neq 0$

Here, the symbol Π designates that the ground electronic configuration has a value of $\Lambda = 1$. Because **S** can now couple to the internuclear axis Hund's Case (a) may be appropriate, although this is not a strict rule, especially for hydrides and

Table 14.1 OH term energy constants (in cm^{-1})

	T	ω_e	$\omega_e x_e$
	32,682.0	3184.28	97.84
$A^2\Sigma^+$	B_v	D_v	γ_v
$v = 3$	14.422	0.206E−2	0.0980
$v = 2$	15.287	0.208E−2	0.0997
$v = 1$	16.129	0.203E−2	0.1056
$v = 0$	16.961	0.204E−2	0.1122
	T_e	ω_e	$\omega_e x_e$
	0.0	3735.21	82.21
$X^2\Pi$	B_v	D_v	Y_v
$v = 3$	16.414	0.182E−2	−8.568
$v = 2$	17.108	0.182E−2	−8.214
$v = 1$	17.807	0.182E−2	−7.876
$v = 0$	18.515	0.187E−2	−7.547

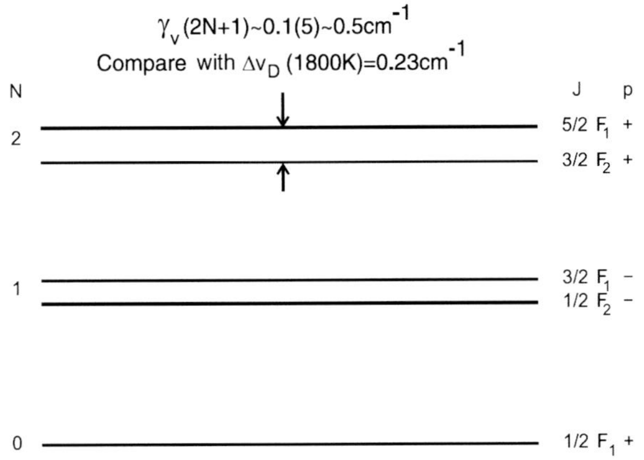

Fig. 14.1 Energy level progression for $A^2\Sigma^+$; Hund's Case (b). The symbol "p" denotes parity (+ or −)

light molecules (see Herzberg, p. 221). In fact, for OH, a transition occurs from Case (a) to Case (b) as the nuclear rotation increases (i.e., the coupling between **L** and the nuclear rotation (N) becomes stronger, and the coupling between **S** and the internuclear axis becomes weaker). Because the ground level has some characteristics of Case (a) and Case (b), we use a mix of quantum numbers to describe it. With the definitions for Hund's (b) already given above, now recall in Hund's Case (a), the coupling between **L** and **S** along the internuclear axis forms a new quantum number.

$$\Omega = |\Lambda + \Sigma|, \ |\Lambda + \Sigma - 1|.$$

With $\Lambda = 1$ and $\Sigma = 1/2$, the two spin-split levels have different quantum number values

$$\Omega = 1/2, 3/2.$$

and the total angular quantum numbers for each is given by

$$J = \Omega, \Omega + 1, \Omega + 2, \Omega + 3, \ldots$$

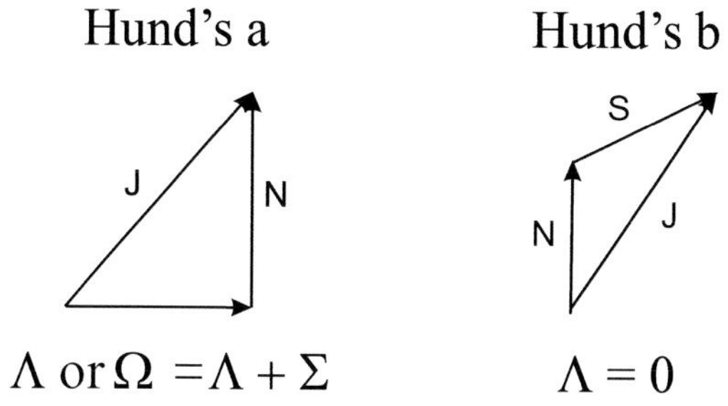

The rotational term energies for a $^2\Pi$ state in the intermediate case are relatively well modelled by:

$$\begin{aligned} F_1(N) &= B_v\{(N+1)^2 - \Lambda^2 - \frac{1}{2}[4(N+1)^2 \\ &\quad + Y_v(Y_v - 4)\Lambda^2]^{1/2}\} - D_v[N(N+1)]^2 \\ F_2(N) &= B_v\{N^2 - \Lambda^2 + \frac{1}{2}[4N^2 + Y_v(Y_v - 4)\Lambda^2]^{1/2}\} - D_v[N(N+1)]^2 \end{aligned} \quad (14.5)$$

where $Y_v \equiv A/B_v$ (< 0 for OH) and A (spin-orbit coupling constant) is like the parameter A (from the moment of inertia) in the symmetric top model; see Eq. (10.13). At large N, the spin-splitting $F_2(N) - F_1(N)$ approaches zero, indicating the transition to the Hund's (b) case. Note also that the approach to zero is from the positive side, i.e. $F_1(N) \leq F_2(N)$ according to the model represented by (14.5).

By substituting the definitions of (14.3) into (14.5) and neglecting the centrifugal correction terms (D_v),

$$\begin{aligned} F_1(J) &= B_v\{(J+1/2)^2 - \Lambda^2 - \tfrac{1}{2}[4(J+1/2)^2 + Y_v(Y_v-4)\Lambda^2]^{1/2}\} \\ F_2(J) &= B_v\{(J+1/2)^2 - \Lambda^2 + \tfrac{1}{2}[4(J+1/2)^2 + Y_v(Y_v-4)\Lambda^2]^{1/2}\} \end{aligned} \quad (14.6)$$

it is also evident that $F_1(J) < F_2(J)$ for all J.

So far we have identified two spin levels with different quantum numbers Ω, and two term energies F_1 and F_2. Which term energy expression belongs to which Ω? Recall the Hund's case (a) expressions which showed two levels with the same J but different Ω were separated by a term $\sim A\Omega^2$, with the $\Omega = 1/2$ state lower in energy for $A > 0$. Similarly here, for $Y > 2$ or so-called *regular* states, a given J level in $\Omega = 1/2$ should have a lower energy than the corresponding level in $\Omega = 3/2$; in this case, the F_1 expression corresponds to the $\Omega = 1/2$ levels. Likewise, for $Y < 2$ or *inverted* states, as is the case here for OH, F_1 is associated with the $\Omega = 3/2$ levels. One might wonder what is the importance of the Ω designation since we have already defined the values of N and J for a given F_i. Recall that the lowest allowed value of J in a given Ω level is $J = \Omega$. Therefore, the regular or inverted designation determines the identity of the lowest (energy) lying J levels.

Because $\Lambda \neq 0$, the Π configuration also is lambda-doubled. Each specific (J, N) configuration actually consists of two lambda-doubled sublevels, designated c and d (recently, the slightly different designations e and f have begun to be used). The lambda-doubling corrections to (14.6) can be adequately modelled by

$$F_{ic} = F_i(J) + \delta_c J(J+1)$$
$$F_{id} = F_i(J) + \delta_d J(J+1) \qquad (14.7)$$

where δ_c and δ_d are the lambda-doubling constants and $F_{id}(J) < F_{ic}(J)$. The lambda-doubling is usually quite small ($[F_{ic}(J) - F_{id}(J)] \sim 0.04\,\mathrm{cm}^{-1}$ for the X state of OH). Finally, each lambda-doubled sublevel for a given N and J has a different parity, i.e. one is $+$ and the other is $-$.

Figure 14.2 shows the first few energy levels of the inverted OH $X^2\Pi$ configuration ($Y_0 = -7.547$). Figure 14.3 shows what the assignment of levels would be if the ground configuration was regular. Note, the smallest value of N changes and only one J is assigned to $N = 0$ and to $N = 1$.

14.1.2 Allowed Radiative Transitions

In this section, we examine the general selection rules to determine which rotational levels can be radiatively coupled, or in other words, what rotational branches will occur in the OH spectrum. The standard notation for rotational branches is also reviewed.

General Selection Rules

Here we summarize the general selection rules for single photon dipole transitions that are derived from quantum mechanics:

(1) parity must change $+ \to -$ or $- \to +$;
(2) angular momentum can only change by $\Delta J = 0, \pm 1$; and
(3) no $Q(J = 0)$ transition, $J = 0 \to J = 0$ not allowed.

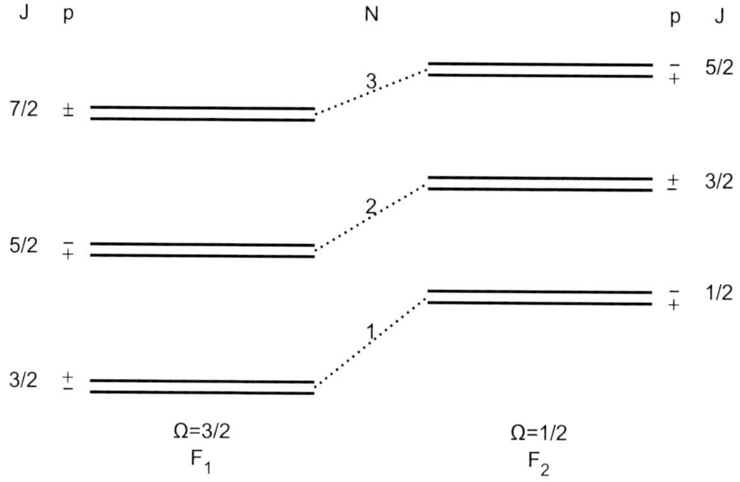

Fig. 14.2 Energy level progression for OH $X^2\Pi$ (inverted)

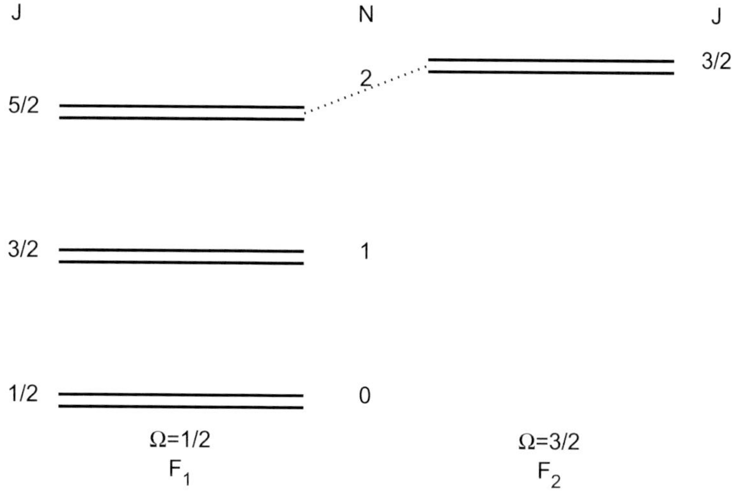

Fig. 14.3 Energy level progression for a regular $X^2\Pi$ configuration

Transition Notation

In general, a specific radiative absorption transition for a diatomic molecule is denoted by a symbol of the form, $A^2\Sigma^+(v') \leftarrow X^2\Pi(v'')\ {}^{Y}X_{\alpha\beta}(N'' \text{ or } J'')$. The electronic and vibrational systems involved in the transition are given first, while the last term denotes the specific (N, J) states being considered. In the ${}^{Y}X_{\alpha\beta}(N'' \text{ or } J'')$ notation,

(1) Y represents a symbol for ΔN ($O, P, Q, R,$ or S for $\Delta N = -2$ to 2);
(2) X represents ΔJ ($P, Q,$ or R for $\Delta J = -1$ to 1);

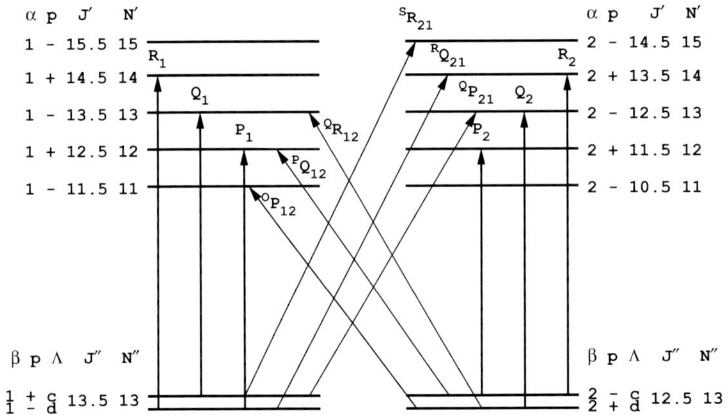

Fig. 14.4 Allowed rotational transitions from $N'' = 13$ in the $A^2\Sigma^+ \leftarrow X^2\Pi$ system

(3) α gives i for F_i' (1 or 2 for spin doublets); and
(4) β gives i for F_i''.

For example, the symbol ${}^PQ_{12}(13)$, in a spin-doublet transition, represents a transition from $N'' = 13$, $J'' = 12.5$ (in F_2'') to $N' = 12$, $J' = 12.5$ (in F_1'). For transitions between two lambda-doubled electronic configurations, the identity of the lambda-doubled states is given by symbol of the form ${}^PQ_{12cd}(13)$ (here the upper level is the c state and the lower one is the d state).

One should note the following shorthand forms that can be confusing:

(1) when $\Delta N = \Delta J$, the Y symbol is suppressed;
(2) when $\alpha = \beta$, the β symbol is suppressed; and
(3) often authors will always suppress either the Y or X symbol.

Finally, though N'' or J'' can be used in the transition notation, N'' is found more frequently in the recent literature.

Figure 14.4 illustrates the application of the selection rules and transition notation to the OH system. There, all possible transitions from all states with $N'' = 13$ are shown.

Shown in the figure are the four states with $N'' = 13$ and all excited states to which they can be radiatively excited (by single photon transitions). In total, 12 bands are possible with 3 bands originating from each *lambda-doubled, spin-split X state*. Two kinds of rotational branches are evident: (1) the *main* branches for which α and β are equal and (2) the *cross-branches*, with α not equal to β. As N increases, the cross-branches weaken, and the 6 main branches contain the only significant transition strengths. Note for transitions between two lambda-doubled levels (e.g., a $\Pi - \Pi$), there would be twice as many transitions, 24, with 6 connected to any given lambda-doubled, spin-split state.

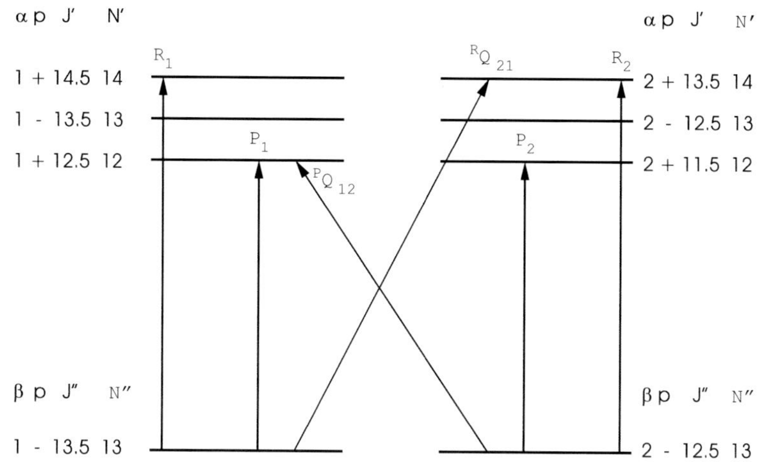

Fig. 14.5 Allowed rotational transitions from $N'' = 13$ in the $A^2\Sigma^+ \leftarrow X^2\Sigma^+$ system

As another example of the general selection rules, consider the $^2\Sigma^+ -^2 \Sigma^+$ system shown in Fig. 14.5. Note the effect of the parity selection rule in reducing the number of allowed main branches to 4.

14.1.3 Absorption Coefficient

In order to perform quantitative spectroscopy (e.g., absorption or laser-induced fluorescence), one often needs to calculate an absolute absorption coefficient. Recall, the spectral absorption coefficient k_ν [cm^{-1}] at frequency ν [s^{-1}] for a single, isolated radiative transition from an initial molecular state 1 to a final state 2 is given by

$$k_\nu = \frac{\pi e^2}{m_e c}(N_1 f_{12} - N_2 f_{21})\phi(\nu - \nu_0) \tag{14.8}$$

where N_1 and N_2 are the number density [cm^{-3}] of molecules in the *specific* states involved, f_{12} and f_{21} are the oscillator strengths for stimulated absorption and stimulated emission, respectively, and $\phi(\nu - \nu_0)$ [s] is the lineshape of the absorption profile. (Note, the value of $\pi e^2/m_e c$ is written in cgs units here, and is approximately 2.651×10^{-2} cm^2/s. If wavenumber units are preferred for ν, replace ϕ[s] using the equality $\phi[\text{s}] = (1/c)\phi[\text{cm}^{-1}]$.) Though it is common to work with and find tabulations of oscillator strengths, Einstein coefficients (A_{21}, B_{21}, B_{12}) are frequently used, and helpful conversion formulas are

$$\frac{B_{21}}{A_{21}} = \frac{c^2}{8\pi h\nu_{21}^3} \tag{14.9}$$

$$f_{21} = \frac{m_e c^3}{8\pi^2 \nu_{21}^2 e^2} A_{21} \tag{14.10a}$$

$$f_{21} = \frac{\epsilon_o m_e c^3}{2\pi\, \nu_{21}^2 e^2} A_{21} \tag{14.10b}$$

where e is in cgs units (an esu, Eq. (14.10a)) or in SI units (Coulomb, Eq. (14.10b)), where we have multiplied by $4\pi\epsilon_o$ to convert units.

In the case of thermal equilibrium at temperature T, the number densities are related by a Boltzmann distribution, and the expression for k_ν reduces to

$$k_\nu = \frac{\pi e^2}{m_e c} N_a \frac{N_1}{N_a} f_{12} \left(1 - e^{\frac{-h\nu_0}{kT}}\right) \phi(\nu - \nu_0) \quad (14.11)$$

where ν_0 is the transition center frequency (s^{-1}) given by $(E_2 - E_1)c$, N_a is the total number density of absorber molecules, and N_1/N_a is the fraction of absorber molecules contained in the specific transition level 1. For visible or ultraviolet transitions, $E_2 - E_1 \geq 14{,}000\,\text{cm}^{-1}$ (or in equivalent temperature, 20,160 K). Thus, the exponential term can be neglected in most temperature environments.

Oscillator Strengths

In order to calculate the absorption coefficient, it is useful to consider discrete energy states and again assume the molecular motions are separable into distinct modes. For a transition between two specific states (n'', v'', Σ'', J'', Λ'') and (n', v', Σ', J', Λ') where n characterizes the electronic configuration, v represents the vibrational level, Σ identifies the spin multiplet, J is the angular momentum, and Λ discriminates between lambda-doubled levels, the absorption oscillator strength is given by

$$f_{(n'',v'',\Sigma'',J'',\Lambda''),(n',v',\Sigma',J',\Lambda')} = f_{n''n'} q_{v''v'} \frac{S_{J''J'}}{2J''+1} \quad (14.12a)$$

$$\text{or} \quad f_{J''J'} = f_{n''n'} q_{v''v'} \frac{S_{J''J'}}{2J''+1} \quad (14.12b)$$

where we have used the common shorthand notation $f_{J''J'}$ to describe the oscillator strength for a *specific* rotational (absorption) transition. In (14.12a) and (14.12b), $f_{n''n'}$ is the electronic part of the oscillator strength (proportional to the electronic dipole moment $|R_e|^2$), $q_{v''v'}$ is the Franck–Condon (vibrational overlap) factor, and $S_{J''J'}$ is the Hönl–London (rotational strength) factor. The electronic and vibrational parts of the oscillator strength are often combined to form the so-called band oscillator strength

$$f_{v''v'} = f_{n''n'} q_{v''v'}, \quad (14.13)$$

or

$$f_{J''J'} = f_{v''v'} \frac{S_{J''J'}}{2J''+1}.[1] \quad (14.14)$$

Equivalently, the band oscillator strength is simply $\sum_{J'} f_{J''J'}$.

[1] Because of the normalization chosen for the Hönl–London factors (see Eq. (14.17) below) and because of a common, but different, way of defining $f_{v'v''}$, transitions from an excited Π configuration and a lower energy Σ level may follow a different equation, e.g. $f_{J''J'} = \frac{1}{2} f_{v''v'} \frac{S_{J''J'}}{2J'+1}$.

For some molecules, like OH, this separation of the motions of the molecule into discrete modes is deficient. For OH, the band oscillator strength is, in fact, a function of J'' and J'. Empirical corrections, denoted $T_{J''J'}$, are used as follows.

$$f_{J''J'} = f_{v''v'} \frac{S_{J''J'} T_{J''J'}}{2J'' + 1} \tag{14.15}$$

Since the $q_{v''v'}$ and $S_{J''J'}$ terms represent subdivisions of the total electronic oscillator strength, they must conform to certain sum rules. These rules are

$$\sum_{v'} q_{v''v'} = 1 \tag{14.16}$$

and for transitions *from all states* with a specific (n'', v'', J'') *to all* J' levels with a specific (n', v')

$$\sum_{J'} S_{J''J'} = (2J'' + 1)(2S + 1)\delta \tag{14.17}$$

where δ accounts for lambda-doubling; it is unity for $\Sigma - \Sigma$ transitions and two otherwise. (Note that $2S + 1$ is equal in the upper and lower states, for the spin-allowed transitions of interest here.) Using these sum rules and the law of detailed balance (for transitions between discrete states with degeneracies g_i, $f_{12}g_1 = f_{21}g_2$), it can be shown that

$$\sum_{v''} q_{v''v'} = 1 \tag{14.18}$$

and

$$\sum_{J''} S_{J'J''} = (2J' + 1)(2S + 1)\delta \tag{14.19}$$

where

$$S_{J''J'} = S_{J'J''} \tag{14.20}$$

and

$$f_{v''v'} = f_{v'v''}. \tag{14.21}$$

Note that the sum of $S_{J'J''}$ terms from a single upper state (i.e., specific J', N') will now equal $2(2J' + 1)$, but as there are 2 states with the same J', the overall sum of $S_{J'J''}$ will obey Eq. (14.19) as required, i.e. $4(2J' + 1)$.

Table 14.2 Band oscillator strengths for the OH $A^2\Sigma^+ - X^2\Pi$ System

(v',v'')[a]	$f_{v'v''}$
(0,0)	0.00096
(1,0)	0.00028

Table 14.2 lists band oscillator strengths for two OH bands and Table 14.3 lists $S_{J''J'}/(2J'' + 1)$ values for $J'' = 0.5$, 1.5, 2.5, 3.5, and 9.5. [a]Note, the usual convention for a band is to specify the vibrational quantum numbers in the order (v', v''), so that (1,0) corresponds to $v' = 1, v'' = 0$.

The normalization is that described by (14.17). Note for the $J = 0.5$, only the F_2 levels exist, so that the sum over all J' is only 2. Also note that while the cross-branch transitions are important at low J, they decrease in strength at higher rotational levels. The ${}^{O}P_{12}$ and ${}^{S}R_{21}$ cross-branches disappear most rapidly, since $\Delta N = \pm 2$ for them.

Population Fractions

In calculating the absorption coefficient for transitions from a *specific* energy level, we must be careful to consider the number of molecules in that level. For example, the $Q_1(13)$ transition of OH (see Fig. 14.4) involves molecules in one state only, the $F_1(N = 13, J = 13.5, \Lambda = c)$ state. We do not want to include the population of the $\Lambda = d$ level or the F_2 levels.

In thermal equilibrium, the fractional population of all molecules N_a which are in a specific energy level N_1 is given by the Boltzmann distribution

$$\frac{N_1}{N_a} = \frac{g_i e^{\frac{-hcE_i}{kT}}}{Q} \tag{14.22}$$

where the energies are in (cm^{-1}) units, and the partition function Q is found by summing over all possible energy levels,

$$Q = \sum_i g_i e^{\frac{-hcE_i}{kT}} \tag{14.23}$$

For example, the population fraction $N_{(n,v,\Sigma,J,\Lambda)''}/N_a$ of a specific energy level $(n'', v'', \Sigma'', J'', \Lambda'')$ would be

$$\frac{N_{(n,v,\Sigma,J,\Lambda)''}}{N_a} = \frac{(2J''+1)e^{\frac{-hcE(n'',v'',\Sigma'',J'',\Lambda'')}{kT}}}{\sum_{(n,v,\Sigma,J,\Lambda)''}(2J''+1)e^{\frac{-hcE(n'',v'',\Sigma'',J'',\Lambda'')}{kT}}} \tag{14.24}$$

Table 14.3 Hönl–London factors for selected OH transitions

Transition	$SJ''J'/(2J''+1)$	$\Sigma F_1(J)$	$\Sigma F_2(J)$	$\Sigma[F_1(J)+F_2(J)]$
$Q_{12}(0.5)$	0.667	0	2	2
$Q_2(0.5)$	0.667			
$R_{12}(0.5)$	0.333			
$R_2(0.5)$	0.333			
$P_1(1.5)$	0.588	2	2	4
$P_{12}(1.5)$	0.078			
$P_{21}(1.5)$	0.392			
$P_2(1.5)$	0.275			
$Q_1(1.5)$	0.562			
$Q_{12}(1.5)$	0.372			
$Q_{21}(1.5)$	0.246			
$Q_2(1.5)$	0.687			
$R_1(1.5)$	0.165			
$R_{12}(1.5)$	0.235			
$R_{21}(1.5)$	0.047			
$R_2(1.5)$	0.353			
$P_1(2.5)$	0.530	2	2	4
$P_{12}(2.5)$	0.070			
$P_{21}(2.5)$	0.242			
$P_2(2.5)$	0.358			
$Q_1(2.5)$	0.708			
$Q_{12}(2.5)$	0.263			
$Q_{21}(2.5)$	0.214			
$Q_2(2.5)$	0.757			
$R_1(2.5)$	0.256			
$R_{12}(2.5)$	0.173			
$R_{21}(2.5)$	0.050			
$R_2(2.5)$	0.379			
$P_1(3.5)$	0.515	2	2	4
$P_{12}(3.5)$	0.056			
$P_{21}(3.5)$	0.167			
$P_2(3.5)$	0.405			
$Q_1(3.5)$	0.790			
$Q_{12}(3.5)$	0.195			
$Q_{21}(3.5)$	0.170			
$Q_2(3.5)$	0.814			
R_1 (3.5)	0.316			
$R_{12}(3.5)$	0.131			
$R_{21}(3.5)$	0.044			
$R_2(3.5)$	0.402			
$P_1(9.5)$	0.511	2	2	4
$P_{12}(9.5)$	0.016			
$P_{21}(9.5)$	0.038			

(continued)

Table 14.3 (continued)

Transition	$SJ''J'/(2J''+1)$	$\Sigma F_1(J)$	$\Sigma F_2(J)$	$\Sigma[F_1(J)+F_2(J)]$
$P_2(9.5)$	0.488			
$Q_1(9.5)$	0.947			
$Q_{12}(9.5)$	0.050			
$Q_{21}(9.5)$	0.048			
$Q_2(9.5)$	0.950			
$R_1(9.5)$	0.441			
$R_{12}(9.5)$	0.035			
$R_{21}(9.5)$	0.014			
$R_2(9.5)$	0.462			

While (14.24) is a rigorous definition of the population fraction, it requires calculating the energy of every possible state (at least until the sum converges for some given T). While computers can do this rapidly, it is still often convenient to divide the energy into separable modes and produce a population fraction for each using simplified partition functions. In that case,

$$\frac{N_{(n,v,\Sigma,J,\Lambda)''}}{N_a} = \frac{N_{n''\Sigma''\Lambda''}}{N_a} \times \frac{N_{v''}}{N_{n''\Sigma''\Lambda''}} \times \frac{N_{J''}}{N_{v''}} \tag{14.25}$$

where we have effectively defined each spin and lambda level of a particular configuration to be a different electronic level. Next, we would write a Boltzmann fraction (14.22) for each. Since we usually combine the energies for the different spins and lambdas into the angular momentum term energy $F(J)$, it is more reasonable to write

$$\frac{N_{(n,v,\Sigma,J,\Lambda)''}}{N_a} = \frac{e^{\frac{-hcT_e(n'')}{kT}}}{Q_e} \times \frac{e^{\frac{-hcG(v'')}{kT}}}{Q_v} \times \frac{(2J''+1)e^{\frac{-hcF(J'')}{kT}}}{Q_r} \tag{14.26}$$

where Q_e, Q_v, and Q_r are the electronic, vibrational, and rotational partition functions and the *only degeneracy* to be included in the numerator is $(2J''+1)$.

At temperatures $\gg B_v(hc/k)$, the rigid rotor result

$$Q_r = \frac{T}{B_v \frac{hc}{k}} \tag{14.27}$$

is often a reasonable approximation to the complete sum. (It is useful to remember $hc/k = 1.44\,\mathrm{K/cm^{-1}}$.) While Q_v can also be represented simply for a harmonic oscillator,

$$Q_v = \frac{e^{\frac{-hc\omega_e}{2kT}}}{1 - e^{\frac{-hc\omega_e}{kT}}} \tag{14.28}$$

accurate values of Q_v for realistic anharmonic oscillators at elevated temperatures may require performing the sum (14.23), from $v = 0$ to a high enough v that the sum converges.

Finally, we consider Q_e. For many molecules, the excited electronic levels, i.e. levels other than the ground X configuration, have sufficiently high energies that they do not contribute to the sum except at extremely high temperatures (they are negligibly populated at lower temperatures). When only the ground electronic configuration is populated, we can simply model Q_e as the "degeneracy" of the ground electronic configuration

$$Q_e = g_e = (2S+1)(2-\delta_{0,\Lambda}) \tag{14.29}$$

where $\delta_{0,\Lambda}$ is 1 if $\Lambda = 0$ (Σ states), otherwise it is 0. We have to remember, however, that the splitting between different spin levels F_i can actually be appreciable (perhaps 100–200 cm^{-1} or, equivalently, a few hundred Kelvin). At low temperatures, where this splitting is important, we can approximate the effect with

$$Q_e = (2-\delta_{0,\Lambda}) \sum_{1\ldots(2S+1)} \exp \frac{-hc[F_i(J_m) - F_1(J_m)]}{kT} \tag{14.30}$$

where J_m is the minimum J value for the F_1 level. This works well for a Hund's case (a) state, where the spin-splitting is a constant value. Unfortunately for many molecules, this is not the case. From the Eq. (14.5) for states intermediate between Hund's cases (a) and (b), we have already seen that the splitting varies with J.

Combined with the inaccuracy of the rigid rotor approximation (14.27) at low temperatures, the full summation (14.24) should be performed when the temperature is low and when high degrees of accuracy are required. It should also be noted that comparisons between *different transitions at the same temperature* can be carried out with ease and accuracy, since the ratio of the absorption coefficients is independent of the partition functions.

Example Calculation

As an example, consider the spectral absorption coefficient of the $(0,0)Q_1(9)$ line in the OH $A^2\Sigma^+ - X^2\Pi$ system, at line center. The transition occurs at ~309.6 nm or roughly 32,300 cm^{-1}. For the calculation, we will use a temperature of 2000 K, a collision fullwidth of 0.05 cm^{-1}, and we will express the answer as a function of OH partial pressure. Again from (14.11) the spectral absorption coefficient is given by

$$k_\nu\ [\text{cm}^{-1}] = 2.651 \times 10^{-2} \frac{\text{cm}^2}{\text{s}} \frac{P_a}{kT} \frac{N_{(n,v,\Sigma,J,\Lambda)''}}{N_a} f_{J''J'} \phi(\nu_o)$$

where we have written the number density of absorber molecules in terms of their partial pressure ($N_a = P_a/kT$). So, we need to calculate the absorption oscillator strength, the population fraction, and the lineshape factor at linecenter.

The oscillator strength from Tables 14.2 and 14.3 is

$$f_{Q_1(9)} = f_{v''v'} \frac{S_{J''J'}}{2J''+1} = 0.00096 \times 0.947 = 9.09 \times 10^{-4}$$

The population fraction in the absorbing state ($v'' = 0,\ F_{1c}(9.5)$), according to (14.26) and with the term energies from (14.2), (14.5) and the values in Table 14.1, is

$$\begin{aligned}
\frac{N_{f_{1c}(9.5)}}{N_a} &= \frac{e^{\frac{-hcT_e(0)}{kT}}}{Q_e} & \frac{e^{\frac{-hcG(0)}{kT}}}{Q_v} & \frac{(2J''+1)e^{\frac{-hcF_1(9.5)}{kT}}}{Q_r} \\
&= \frac{e^0}{4} & \frac{e^{\frac{-2660\,K}{T}}}{0.287} & \frac{20e^{\frac{-2313\,K}{T}}}{\frac{T}{26.66\,K}} \\
&= \frac{1}{4} & \frac{0.264}{0.287} & \frac{6.29}{75.0} \\
&= 0.25 & 0.920 & 0.0839 \\
&= 0.0193
\end{aligned}$$

where Q_v is from performing the sum, which converges by about $v = 3$, and Q_e is the electronic "degeneracy" (14.29). It is interesting to note that the harmonic oscillator approximation gives $Q_v = 0.280$, a 2.4 % difference from the converged sum. Also, the final result, 1.93 % of the OH molecules in ($v = 0$) $F_{1c}(9.5)$ of the X state, is very close to the value of 1.91 % obtained from performing the complete sum over states (14.24).

The final component in the absorption coefficient calculation is the lineshape factor. At 2000 K, the OH Doppler width is 0.25 cm^{-1}. With the collision width given earlier, 0.05 cm^{-1}, this gives a Voigt a parameter of 0.17. At line center ($x = 0$), the Voigt profile ϕ is 3.13 cm or 1.04×10^{-10} s (ignoring the collision broadening would have given us $\phi = 3.75$ cm). So we have

$$k_{\nu 0} = \left(2.651 \times 10^{-7}\ \frac{\text{cm}^2}{\text{s}}\right)(P_a\ [\text{atm}])\left(3.66 \times 10^{18}\ \frac{\text{cm}^{-3}}{\text{atm}}\right)$$
$$\times (1.93\,\%)(9.09 \times 10^{-4})(1.04 \times 10^{-10}\text{s}) = 177 \frac{\text{cm}^{-1}}{\text{atm}}\ (P_a\ [\text{atm}])$$

which may be rounded to $k_{\nu 0} = (180\ \text{cm}^{-1}/\text{atm})(P_a\ [\text{atm}])$ for present purposes. Recalling Beer's law,

$$I_\nu = I_\nu^0 \exp(-k_\nu L)$$

we calculate that a narrowband laser centered on the $Q_1(9)$ transition would see approximately 59 % absorption in 5 cm of gas containing 1000 ppm of OH at 2000 K, 1 atm (total static pressure).

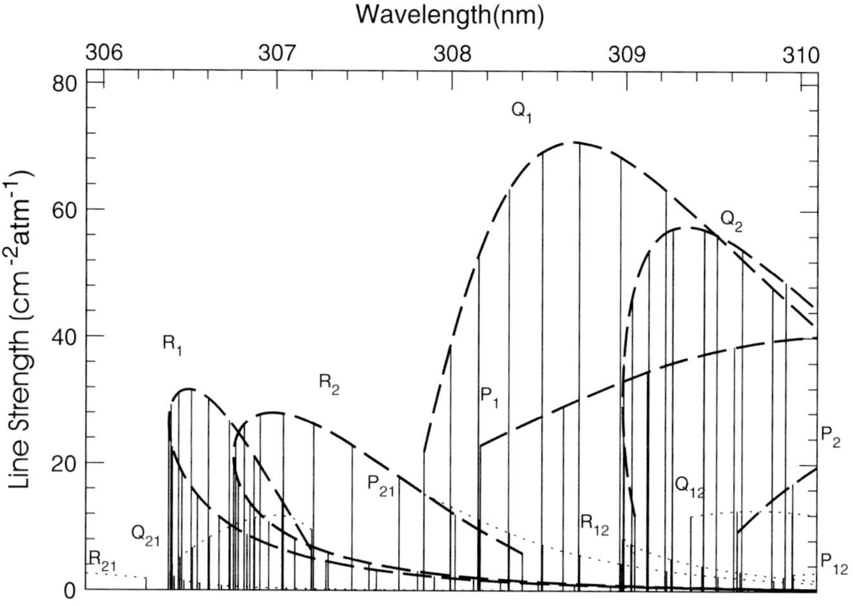

Fig. 14.6 Plot of OH line strengths for a selected region of the $A^2\Sigma^+ \leftarrow X^2\Pi(0,0)$ band at 2000 K

As a more comprehensive illustration of the information presented here, Fig. 14.6 shows 2000 K line strengths, $k_\nu/\phi(\nu)$ (also known as integrated line intensity), for lines in the (0,0) band between 305.9 and 310.1 nm. Unlike the example calculation shown above, the units of the line strength are presented in wavenumbers (e.g., $\phi(\nu)$ has units of [cm] instead of [s]). To convert into the units given earlier, simply divide by c [cm/s].

In the figure, vertical lines are plotted for each transition, with the height representing the line strength. Lines belonging to a specific branch are connected (at the top) with a dashed or dotted curve. The main branches are those connected by thicker dashed lines, while the cross-branch lines are identified by thin dotted lines. Each branch is labelled, using the standard notation where the ΔN symbol has been dropped.

The bandheads associated with the R branches are clearly seen. (Remember, the bandheads can be predicted by comparing the B_v values of the upper and lower states; if $B_{v'} < B_{v''}$, we get a bandhead in the R branches and if $B_{v'} > B_{v''}$, we get a bandhead in the P branches.) The Q_{21} branch also has a bandhead; recall it is properly denoted ${}^RQ_{21}$. The reverse holds for the ${}^QR_{12}$ branch.

The following sections provide more brief case studies of spectra in the visible, near-infrared, and mid-infrared.

14.2 Visible/Near-Infrared O_2 Spectroscopy

Quantitative detection of molecular oxygen is of interest for monitoring biological and chemical processes, and for planetary sciences. Due to oxygen's symmetry, absorption and emission interactions are generally constrained to electronic or rovibronic transitions. Although electronic spectra are predominant in the ultraviolet, some transitions may be observed in the visible or near-infrared wavelength domain, enabling utilization of more convenient and less expensive light sources, such as tunable diode lasers, to conduct spectroscopic measurements.

The observation of the solar spectrum around 760 nm in telescopes has shown the fine structure of a weak, spin-forbidden, atmospheric oxygen absorption band. The analysis of the line positions and their absolute wavelength measured precisely by interferometry has shown that this band corresponds to transitions between the $X^3\Sigma_g^-(v'' = 0)$ electronic ground state of O_2 and the $b^1\Sigma_g^+(v' = 0)$ excited state. Potential curves of the electronic states of the O_2 are shown in Fig. 14.7.

The rotational levels of the O_2 electronic states are designated by the quantum numbers N and J, corresponding, respectively, to the rotational angular momentum of the nuclei and the total angular momentum (with$''$ for the ground state and$'$ for the excited state). The ground electronic state, which is a triplet state with a spin of unity, is split into levels corresponding to $J'' = N''$, $J'' = N'' - 1$, and

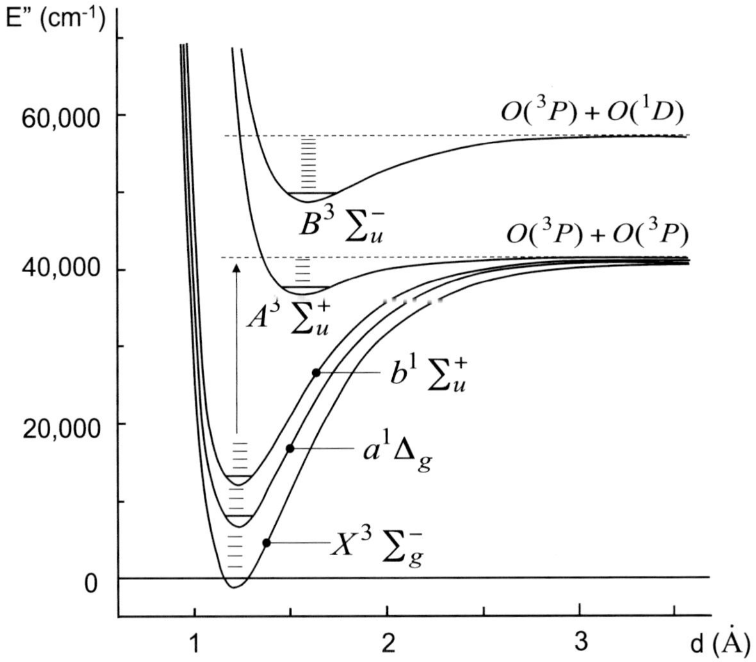

Fig. 14.7 Potential energy diagram for diatomic molecular oxygen [1]

Table 14.4 Spectroscopic constants for O_2

State	$T_e[\text{cm}^{-1}]$	$\omega_e[\text{cm}^{-1}]$	$\omega_e X_e[\text{cm}^{-1}]$
$X^3\sum_g^-(v''=0)$	0	1580.36	12.07
$b^1\sum_g^+(v'=0)$	13,195.22	1432.69	13.95

$J'' = N'' + 1$, with only odd values of N'' allowed. The excited state (zero spin) is composed of singlet states with $J' = N'$, where only even values of N' are allowed. The selection rules allow four types of transitions, denoted $\Delta N \Delta J(N'', J'')$, with $\Delta N \Delta J = PP, PQ, RR, RQ$.

> Example: $RQ(7, 7)$ is a transition from $N'' = 7, J'' = 7$ state to the $N' = 8$, $J' = 7$.

For the $X^3\sum_g^-(v'' = 0)$ electronic ground state of O_2 and the $b^1\sum_g^+(v' = 0)$ excited state, Herzberg gives the constants listed in Table 14.4:

The band center for this transition can be calculated from the energy difference between $X^3\sum_g^-(v'' = 0)$ and $b^1\sum_g^+(v' = 0)$. Use the following equation, we can find the band center:

$$\nu_0 = T_e + \frac{\omega_e'}{2} - \frac{\omega_e X_e'}{4} - \frac{\omega_e''}{2} + \frac{\omega_e X_e''}{4}$$

The atmospheric A-band band center is calculated to be 13,120.915 cm^{-1}.

The set of the PP and PQ lines form the P branch, at frequencies smaller than 13,120 cm^{-1}; the set of the RR and RQ lines form the R branch, located at frequencies higher than 13,120 cm^{-1}. At 13,165 cm^{-1}, the R lines gather to form a bandhead (see Fig. 14.8).

14.3 Near-Infrared H_2O Spectroscopy: Lineshapes

Accurate lineshape models are needed to simulate absorption and emission spectra and, therefore, are often required for quantitative absorption- or emission-based measurements of gas conditions. Complicating matters are the large number of variables that influence the transition lineshape including temperature, pressure, collision partner (i.e., composition), and even transition states, to name only a few. The Voigt profile often provides satisfactory accuracy (typically within 1–2 % of peak absorbances) despite only accounting for Doppler and collisional broadening and assuming them to be independent processes. However, H_2O and other molecules with large rotational-energy-level spacing (e.g., HF, HBr, HCl) often require the use of more advanced lineshape models that address the assumptions of the Voigt profile and account for additional collision physics. This section will briefly present a few cases where Rautian (RP) [2], Galatry (GP) [3], and quadratic speed-dependent Voigt (qSDVP) [4] profiles were used to more accurately model the lineshapes of

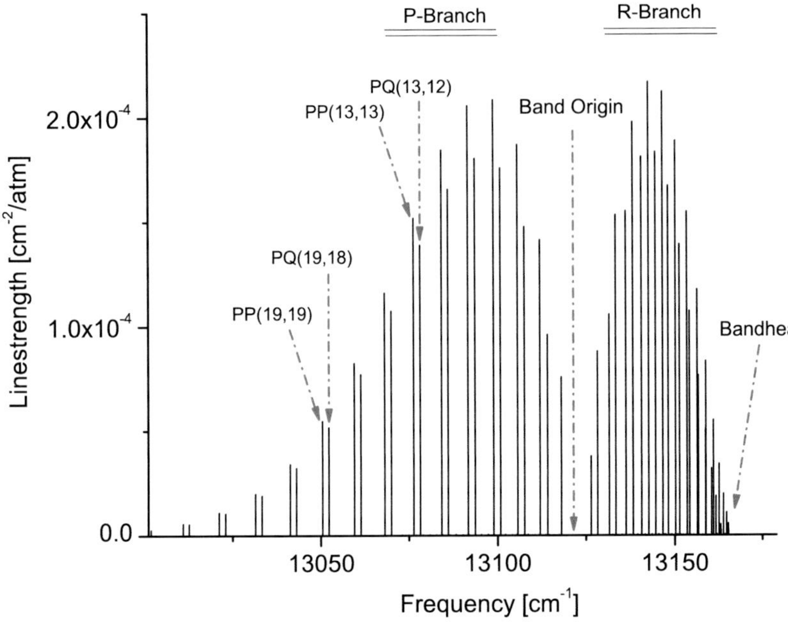

Fig. 14.8 The *A*-Band of O_2

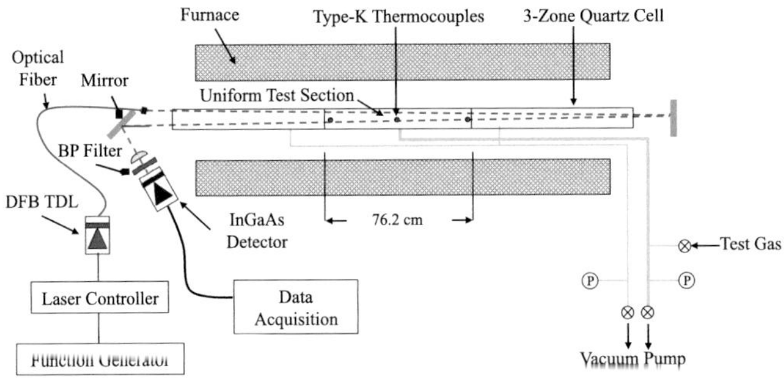

Fig. 14.9 Experimental setup used to measure H_2O lineshapes

high-*J* H_2O absorption transitions and discuss the influence of temperature upon H_2O lineshapes. More details regarding this work can be found in [5] and similar lineshape studies can be found in [6–10].

14.3.1 Experimental Setup

Figure 14.9 shows a schematic of the experimental setup used. Tunable diode lasers (TDLs) with a linewidth <5 MHz were scanned across 7 high-*J* rovibrational H_2O transitions near 1.4 μm at temperatures and pressures up to 1300 K and 800 Torr, respectively. Measurements were acquired in pure H_2O and H_2O-N_2 mixtures in

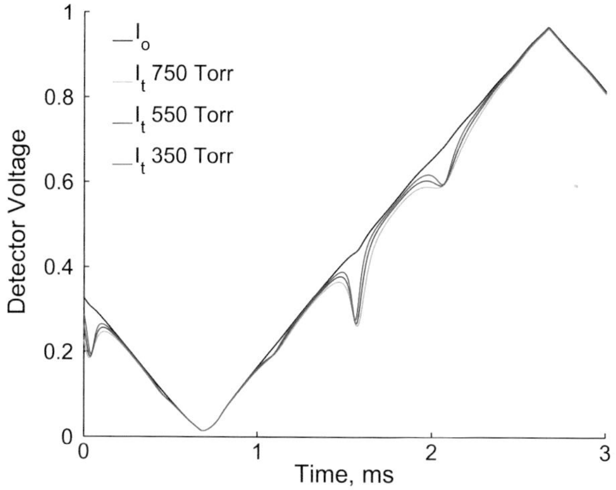

Fig. 14.10 Measured I_o and I_t for single scans across H_2O transitions near 1345 nm at various pressures. $I_t = I_o$ at non-resonant wavelengths

a heated static cell. Each TDL was injection-current tuned approximately 1 cm^{-1} with a 100 Hz triangle wave generated by an analogue function generator. The laser light was collimated and pitched through 3-zone quartz cell with a 76.2 cm test section using a double-pass arrangement. The quartz cell was located within a 3-zone furnace to provide a near-uniform test section at up to 1300 K. The transmitted laser light was measured using an InGaAs photodetector with a 3 mm diameter active area and 3 MHz bandwidth.

A fiber-coupled fused-silica etalon with a free-spectral range (FSR) of 0.02 cm^{-1} was used to characterize the wavelength tuning of each laser and convert the data from the time domain to the optical frequency domain. The measured spectral absorbance, $\alpha(\nu)$, was then calculated using Beer's Law after the background emission was subtracted from $I_o(t)$ and $I_t(t)$. Figure 14.10 shows measured I_o and I_t for a single scan across transitions near 1345 nm at a few different pressures.

14.3.2 Experimental Results

Influence of Lineshape Model

Figure 14.11 shows measured absorbance spectra and best-fit lineshape residuals for H_2O transitions near 7413.02 cm^{-1} with $J'' = 8$ and 6919.95 cm^{-1} with $J'' = 14$. Upon immediate comparison it is clear that transition near 7413.02 cm^{-1} has a much larger collisional-broadening coefficient and the best-fit Voigt profile is more accurate compared to that of the transition near 6919.95 cm^{-1}. This is because the larger collisional broadening of the transition near 7413.02 cm^{-1} mitigates the influence of Dicke narrowing and the speed-dependence of collisional broadening. For the transition near 6919.95 cm^{-1} at the conditions studied, the best-fit Voigt profile recovers the measured absorbance spectra within only 6 % of the peak absorbances, and exhibits a systematic gull-wing shaped residual with the largest

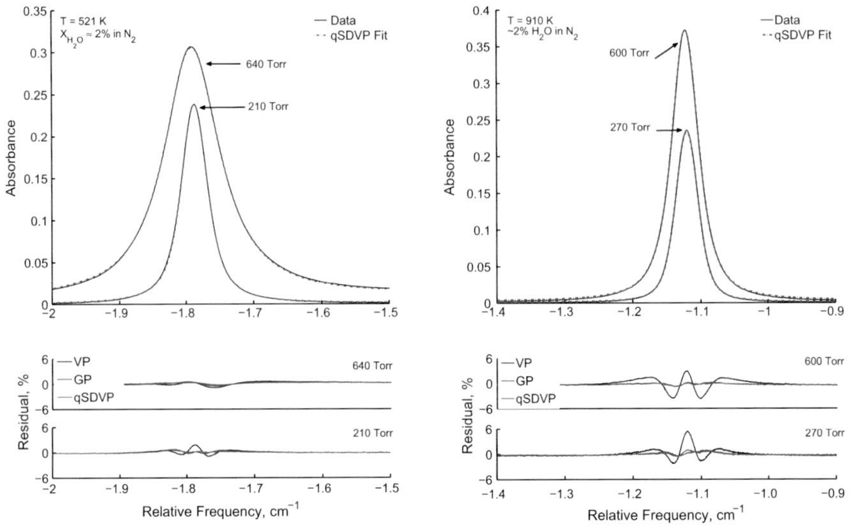

Fig. 14.11 Measured absorbance spectra and best-fit lineshape residuals for H_2O transitions near 7413 cm^{-1} with $J'' = 8$ (*left*) and 6919.9 cm^{-1} with $J'' = 14$ (*right*). The Voigt profile is significantly less accurate for the transition near 6919.9 cm^{-1}

errors near ν_o and in the *near-wings* of the transition. In comparison, the more advanced Rautian, Galatry, and speed-dependent Voigt profiles provide significantly smaller residuals and recover the measured spectra within 0.5–1 % of the peak absorbance. This suggests that Dicke narrowing and/or speed-dependent broadening are primarily responsible for the non-Voigt shape. This trend was consistent for all transitions studied and the accuracy of the Voigt profile steadily declined with increasing J''; however, this is not always expected as collisional broadening depends on more than J'' [11, 12].

The best-fit lineshape parameters can vary significantly with the lineshape model used. Figure 14.12 shows the best-fit lineshape parameters as a function of temperature for the transitions near 7413.02 cm^{-1} (left) and 7471.6 cm^{-1} with $J'' = 16$ (right) perturbed by N_2. The best-fit N_2-broadening coefficients are nearly equal for the Voigt, Galatry, Rautian, and speed-dependent Voigt profile for the transition near 7413.02 cm^{-1}, but vary greatly for the transition near 7471.6 cm^{-1}. This is because non-Voigt effects are relatively insignificant for the transition near 7413.02 cm^{-1} (in which case all lineshape models studied here reduce to the VP), but play a significant role in the lineshape of the transition near 7471.6 cm^{-1} due to its larger energy-level spacing and smaller collisional broadening.

Influence of Temperature on Collision Broadening

Temperature can influence collision broadening by modifying: (1) the relative velocity distributions and the probability of collision-induced state changes, (2) the trajectories of the perturber relative to the absorber, and (3) the population

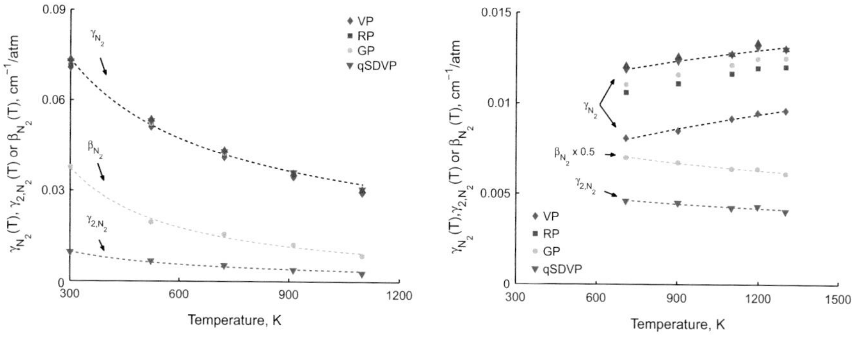

Fig. 14.12 Measured γ_{N_2}, γ_{2,N_2}, and β_{N_2} for the H_2O transitions near 7413.02 cm^{-1} with $J'' = 8$ (*left*) and 7471.6 cm^{-1} with $J'' = 16$ (*right*). *Dotted lines* correspond to best-fit power-law model

distribution of the perturber across its rotational and vibrational states [13]. Depending on the species and energy levels involved, these processes compete with one another and can lead to collisional-broadening coefficients with a complex temperature dependence.

If the spacing between rotational states is small, collisions easily shuffle molecules between rotational states and the collisional-broadening coefficient decreases with increasing temperature, typically with n near 0.75, primarily due to the reduced collision frequency. If the rotational-energy spacing is large, only strong collisions can shuffle molecules between rotational states and, therefore, collisional-broadening coefficients may increase with increasing temperature due to the corresponding increase in thermal energy and, potentially, increased collisional resonance (i.e. matching rotational energy-level spacing between absorber and collision partner).

Figure 14.13 shows the best-fit γ_{N_2} as a function of temperature for all transitions studied using the speed-dependent Voigt profile. In general, as J'' is increased, γ_{N_2} and n decrease. Around $J'' = 10$, n becomes negative (i.e., γ_{N_2} increases with increasing temperature) indicating the importance of increased thermal energy and collisional resonance on N_2-broadening of these high-J H_2O transitions. A detailed discussion regarding this phenomenon can be found in [13].

14.4 Mid-Infrared Spectroscopy of Hydrocarbons and Other Organic Compounds

Detection of volatile organic compounds (VOCs) is of interest for energy, health, and environmental monitoring. Time-resolved sensing of fuels and fuel intermediates, largely hydrocarbons, are of particular importance to understanding the chemical kinetics of fuels. Here we review vibrational spectroscopy (infrared) of hydrocarbons and other organic species.

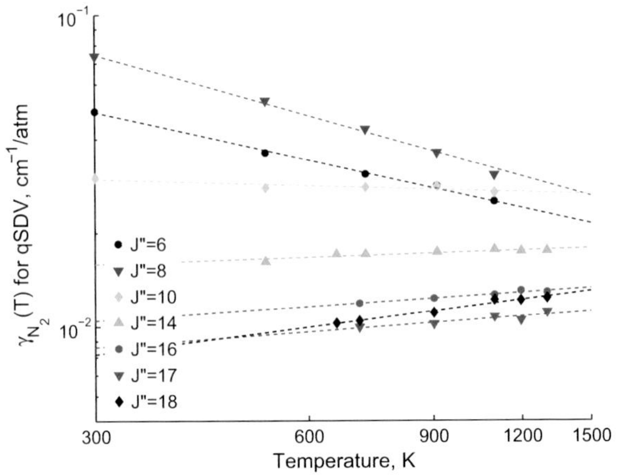

Fig. 14.13 Measured $\gamma_{N_2,\text{qSDVP}}$ and corresponding best-fit power-law model for all transitions studied. Error bars are too small to be seen

Table 14.5 Typical characteristic frequencies for hydrocarbon vibrations

Vibration	Frequency range [cm^{-1}]	Functional group
–C≡C–H: C–H bend	600–700	Alkynes
C–H rock	720–740	Alkanes
C–H out-of-plane	675–900	Aromatics
=C–H bend	650–1000	Alkenes
C–O stretch	1000–1320	Alcohols, esters
C–N stretch	1250–1340	Amines
C–H rock	1300–1370	Alkanes
C–H bend	1450–1470	Alkanes
C–C stretch	1400–1600	Aromatics
–C=C– stretch	1640–1700	Alkenes
C=O stretch	1665–1760	Aldehydes, ketones
C–H stretch	2690–3330	All hydrocarbons
O–H stretch	2500–3650	Alcohols

As detailed in previous chapters, the characteristic frequencies of absorption or emission for a given molecule are unique to the chemical bond structure of the molecule. Because many hydrocarbons have similar bonds or groups of bonds, these species have similar vibrational modes and correspondingly similar characteristic infrared spectra. Table 14.5 shows a list of characteristic vibrational modes and frequencies along with the associated functional groups.

The infrared absorption spectra of methane and octane [14], both saturated hydrocarbons (alkanes), are shown in Fig. 14.14. Some characteristic vibrational bands are labeled. Comparison of the two spectra highlights a few important fea-

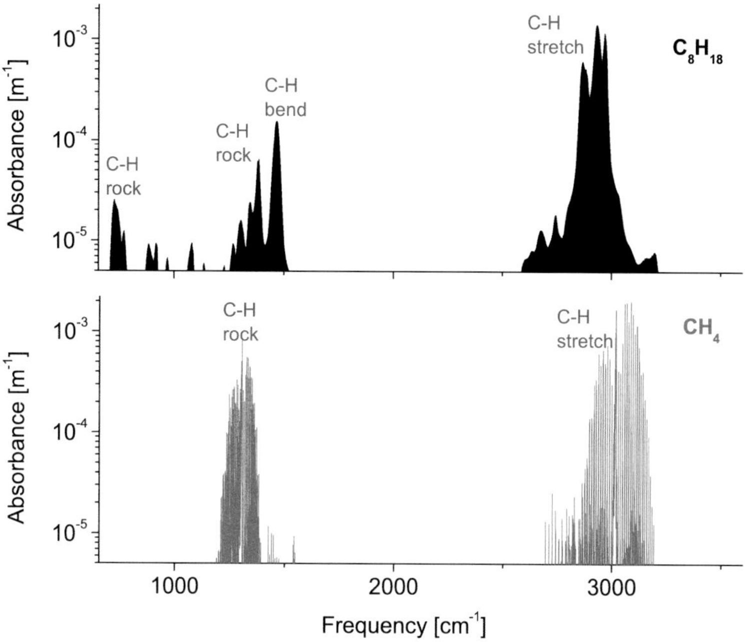

Fig. 14.14 FTIR measurements of octane and methane at 1 atm and 25 °C

tures. The C–H stretch vibration near 3000 cm^{-1} provides the strongest resonance for both species, and this is common to all alkanes. Discrete rotational lines are not observed for octane, whereas they can be observed for methane. This results directly from the much larger moment of inertia for octane due to its larger size and correspondingly smaller line spacing between transitions. Similarly, for many organic compounds of moderate size ($N > 6$ atoms), discrete rovibrational transitions are blended, and only the vibrational bands, which include an aggregate of blended lines, are observed. Since vibrations like the C–H stretch near 3000 cm^{-1} can be common to many organic species (e.g., alkanes, alkynes, aromatics, alkenes, aldehydes), the observed spectra of this band in a complex gas mixture offers poor differentiation for large molecules of interest for gas detection. Conversely, the spectral window between 700 and 1300 cm^{-1} is only active for the larger octane molecule, a result of interacting vibrational modes which can be difficult to interpret but are unique to a given organic compound. This spectral window is commonly referred to as the fingerprint region for large hydrocarbons and provides good differentiation at the expense of weaker absorption.

Figure 14.15 provides two examples of the distinct spectral features observed when hydrocarbons are oxygenated (e.g., alcohols, ethers, ketones) [14]. The inclusion of an oxygen atom yields different chemical bond structure and unique vibrational spectra, as shown here for ethanol and acetone.

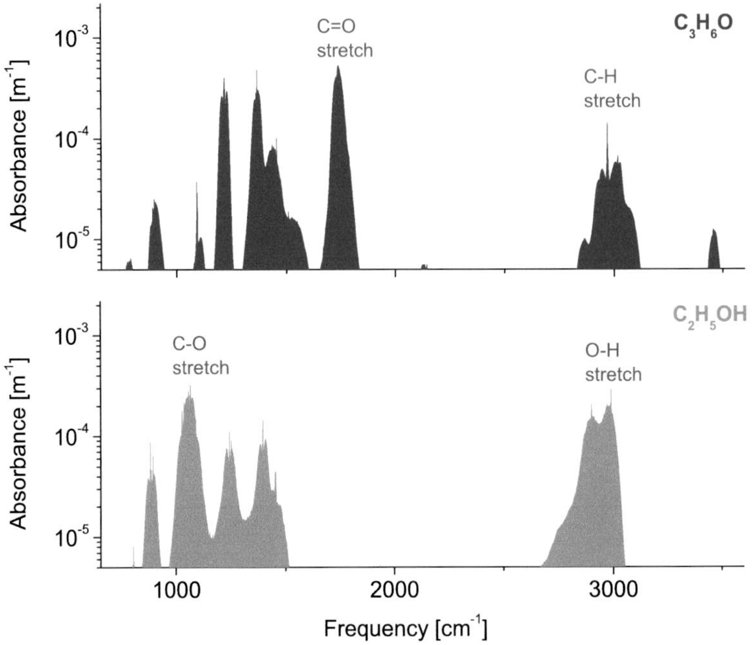

Fig. 14.15 FTIR measurements of acetone and ethanol at 1 atm and 25 °C

References

1. G. Herzberg, in *Molecular Spectra and Molecular Structure, Volume 1. Spectra of Diatomic Molecules*, 2 edn. (Krieger Publishing Company, Malabar, 1950)
2. S.G. Rautian, J.I. Sobel'man, The effect of collisions on the Doppler broadening of spectral lines. Sov. Phys. Usp. **9**, 701–716 (1967)
3. L. Galatry, Simultaneous effect of Doppler and foreign gas broadening on spectral lines. Phys. Rev. **122**, 1218–1223 (1961)
4. F. Rohart, H. Mäder, H.-W. Nicolaisen, Speed dependence of rotational relaxation induced by foreign gas collisions: studies on CH_3F by millimeter wave coherent transients. J. Chem. Phys. **101**(8), 6475 (1994)
5. C.S. Goldenstein, R.K. Hanson, Diode-laser measurements of linestrength and temperature-dependent lineshape parameters for H_2O transitions near 1.4 μm using Voigt, Rautian, Galatry, and speed-dependent Voigt profiles. J. Quant. Spectrosc. Radiat. Transf. **152**, 127–139 (2015)
6. R.S. Eng, Collisional narrowing of infrared water-vapor transitions. Appl. Phys. Lett. **21**(7), 303–305 (1972)
7. C. Claveau, A. Henry, D. Hurtmans, A. Valentin, Narrowing and broadening parameters of H_2O lines perturbed by He , Ne , Ar , Kr and nitrogen in the spectral range 1850–2140 cm. J. Quant. Spectrosc. Radiat. Transf. **68**, 273–298 (2001)
8. C.D. Boone, K.A. Walker, P.F. Bernath, Speed-dependent Voigt profile for water vapor in infrared remote sensing applications. J. Quant. Spectrosc. Radiat. Transf. **105**(3), 525–532 (2007)
9. N.H. Ngo, N. Ibrahim, X. Landsheere, H. Tran, P. Chelin, M. Schwell, J.-M. Hartmann, Intensities and shapes of H_2O lines in the near-infrared by tunable diode laser spectroscopy. J. Quant. Spectrosc. Radiat. Transf. **113**(11), 870–877 (2012)

10. C.S. Goldenstein, J.B. Jeffries, R.K. Hanson, Diode laser measurements of linestrength and temperature-dependent lineshape parameters of H_2O-, CO_2-, and N_2-perturbed H2O transitions near 2474 and 2482 nm. J. Quant. Spectrosc. Radiat. Transf. **130**, 100–111 (2013)
11. R.R. Gamache, A.L. Laraia, N_2-, O_2-, and air-broadened half-widths, their temperature dependence, and line shifts for the rotation band of $H_2{}^{16}O$. J. Mol. Spectrosc. **257**(2), 116–127 (2009)
12. R.R. Gamache, J.-M. Hartmann, Collisional parameters of H_2O lines: effects of vibration. J. Quant. Spectrosc. Radiat. Transf. **83**, 119–147 (2004)
13. J.M. Hartmann, J. Taine, J. Bonamy, B. Labani, D. Robert, Collisional broadening of rotation-vibration lines for asymmetric-top molecules. II. H_2O diode laser measurements in the 400-900 K range; calculations in the 300-2000 K range. J. Chem. Phys. **86**(1), 144 (1987)
14. S.W. Sharpe, T.J. Johnson, R.L. Sams, P.M. Chu, G.C. Rhoderick, P.A. Johnson, Gas-phase databases for quantitative infrared spectroscopy. Appl. Spectrosc. **58**(12), 1452–1461 (2004)

Appendices

Glossary

A

α_ν	Spectral absorptivity [no units] at frequency ν or over the frequency range $\nu \rightarrow \nu + d\nu$
ε_ν	Spectral emissivity [no units] at frequency ν or over the frequency range $\nu \rightarrow \nu + d\nu$
k_ν	Spectral absorption coefficient [cm^{-1}] at frequency ν or over the frequency range $\nu \rightarrow \nu + d\nu$
I_ν	Spectral intensity with units of power per unit area per unit spectral interval $\left[\frac{\text{W/cm}^2}{\text{cm}^{-1}}\right]$ in the frequency range $\nu \rightarrow \nu + d\nu$
I	Total radiant intensity [W/cm^2]
I_ν^0	Incident radiation intensity
$I(\nu)$	Radiant intensity at frequency ν [W/cm^2]
ν	Frequency [cm^{-1} or s^{-1}]
$\overline{\nu}$	Frequency [cm^{-1}]
P	Power [W]
T_ν	Spectral transmissivity [no units] at frequency ν or over the frequency range $\nu \rightarrow \nu + d\nu$

R.K. Hanson et al., *Spectroscopy and Optical Diagnostics for Gases*,
DOI 10.1007/978-3-319-23252-2

Voigt Tables

B

These tables contain the values of the Voigt function, $V(a, w)$, for a variety of values of "a" and "w". Each page list values of "a" across the top and values of "w" down the left column.[1]

$$V(a,w) = \frac{a}{\pi}\int_{-\infty}^{+\infty} \frac{\exp(-y^2)dy}{a^2 + (w-y)^2} \tag{B.1}$$

[1]Tabulated by Liebeskind, 1/30/92.

R.K. Hanson et al., *Spectroscopy and Optical Diagnostics for Gases*,
DOI 10.1007/978-3-319-23252-2

$w \backslash a$	0.10	0.20	0.30	0.40	0.50	0.60	0.70	0.80	0.90	1.00
0.00	0.8965	0.8090	0.7346	0.6708	0.6157	0.5678	0.5259	0.4891	0.4565	0.4276
0.10	0.8885	0.8026	0.7293	0.6665	0.6121	0.5648	0.5234	0.4870	0.4547	0.4260
0.20	0.8650	0.7835	0.7138	0.6537	0.6015	0.5560	0.5160	0.4807	0.4494	0.4215
0.30	0.8272	0.7529	0.6887	0.6330	0.5843	0.5416	0.5039	0.4705	0.4407	0.4140
0.40	0.7773	0.7121	0.6552	0.6053	0.5613	0.5222	0.4876	0.4566	0.4288	0.4038
0.50	0.7176	0.6632	0.6149	0.5717	0.5332	0.4986	0.4675	0.4395	0.4142	0.3912
0.60	0.6511	0.6083	0.5692	0.5336	0.5011	0.4715	0.4444	0.4198	0.3972	0.3766
0.70	0.5807	0.5497	0.5202	0.4923	0.4661	0.4417	0.4190	0.3979	0.3783	0.3602
0.80	0.5093	0.4897	0.4695	0.4492	0.4294	0.4103	0.3919	0.3745	0.3580	0.3425
0.90	0.4394	0.4303	0.4187	0.4058	0.3920	0.3780	0.3640	0.3502	0.3368	0.3239
1.00	0.3732	0.3732	0.3694	0.3630	0.3549	0.3456	0.3357	0.3254	0.3151	0.3047
1.20	0.2574	0.2709	0.2792	0.2834	0.2846	0.2835	0.2807	0.2767	0.2718	0.2662
1.40	0.1684	0.1892	0.2047	0.2157	0.2233	0.2280	0.2306	0.2314	0.2308	0.2292
1.60	0.1058	0.1289	0.1473	0.1617	0.1728	0.1812	0.1872	0.1914	0.1940	0.1954
1.80	0.0651	0.0871	0.1055	0.1208	0.1333	0.1434	0.1514	0.1576	0.1623	0.1657
2.00	0.0402	0.0595	0.0764	0.0909	0.1034	0.1138	0.1226	0.1298	0.1356	0.1402
2.20	0.0257	0.0419	0.0566	0.0697	0.0812	0.0912	0.0999	0.1074	0.1137	0.1189
2.40	0.0174	0.0308	0.0432	0.0546	0.0649	0.0741	0.0823	0.0896	0.0959	0.1013
2.60	0.0126	0.0237	0.0341	0.0438	0.0529	0.0612	0.0687	0.0755	0.0815	0.0869
2.80	0.0098	0.0189	0.0277	0.0361	0.0439	0.0513	0.0580	0.0643	0.0699	0.0750
3.00	0.0079	0.0156	0.0231	0.0303	0.0371	0.0436	0.0497	0.0553	0.0605	0.0653
3.20	0.0067	0.0132	0.0196	0.0259	0.0318	0.0376	0.0430	0.0481	0.0529	0.0573
3.40	0.0057	0.0114	0.0170	0.0224	0.0277	0.0327	0.0376	0.0422	0.0465	0.0506
3.60	0.0050	0.0100	0.0148	0.0196	0.0243	0.0288	0.0332	0.0373	0.0413	0.0450
3.80	0.0044	0.0088	0.0131	0.0274	0.0215	0.0256	0.0295	0.0333	0.0369	0.0403
4.00	0.0039	0.0078	0.0117	0.0155	0.0192	0.0229	0.0264	0.0298	0.0331	0.0363
4.20	0.0035	0.0070	0.0105	0.0139	0.0173	0.0206	0.0238	0.0269	0.0299	0.0328
4.40	0.0032	0.0063	0.0095	0.0126	0.0156	0.0186	0.0216	0.0244	0.0272	0.0299
4.60	0.0029	0.0058	0.0086	0.0114	0.0242	0.0169	0.0196	0.0222	0.0248	0.0273
4.80	0.0026	0.0052	0.0078	0.0104	0.0130	0.0155	0.0179	0.0204	0.0227	0.0250
5.00	0.0024	0.0048	0.0072	0.0096	0.0119	0.0142	0.0165	0.0187	0.0209	0.0230
5.50	0.0020	0.0039	0.0059	0.0078	0.0097	0.0116	0.0135	0.0154	0.0172	0.0190
6.00	0.0016	0.0033	0.0049	0.0065	0.0081	0.0097	0.0113	0.0128	0.0144	0.0159
6.50	0.0014	0.0028	0.0041	0.0055	0.0069	0.0082	0.0096	0.0109	0.0122	0.0135
7.00	0.0012	0.0024	0.0036	0.0047	0.0059	0.0071	0.0082	0.0094	0.0105	0.0116
7.50	0.0010	0.0021	0.0031	0.0041	0.0051	0.0061	0.0072	0.0081	0.0091	0.0101
8.00	0.0009	0.0018	0.0027	0.0036	0.0045	0.0054	0.0063	0.0071	0.0080	0.0089
8.50	0.0008	0.0016	0.0024	0.0032	0.0040	0.0048	0.0055	0.0063	0.0071	0.0079
9.00	0.0007	0.0014	0.0021	0.0028	0.0035	0.0042	0.0049	0.0056	0.0063	0.0070
9.50	0.0006	0.0013	0.0019	0.0025	0.0032	0.0038	0.0044	0.0050	0.0057	0.0063
10.00	0.0006	0.0011	0.0017	6.0023	0.0029	0.0034	0.0040	0.0046	0.0051	0.0057

$w \backslash a$	1.10	1.20	1.30	1.40	1.50	1.60	1.70	1.80	1.90	2.00
0.00	0.4017	0.3785	0.3576	0.3387	0.3216	0.3060	0.2917	0.2786	0.2665	0.2554
0.10	0.4004	0.3774	0.3566	0.3379	0.3208	0.3053	0.2911	0.2780	0.2660	0.2550
0.20	0.3965	0.3740	0.3537	0.3353	0.3186	0.3033	0.2893	0.2765	0.2646	0.2537
0.30	0.3900	0.3684	0.3488	0.3311	0.3148	0.3000	0.2864	0.2739	0.2624	0.2517
0.40	0.3813	0.3608	0.3422	0.3252	0.3097	0.2955	0.2824	0.2703	0.2592	0.2488
0.50	0.3704	0.3513	0.3339	0.3180	0.3034	0.2899	0.2774	0.2659	0.2552	0.2453
0.60	0.3576	0.3402	0.3242	0.3095	0.2958	0.2832	0.2715	0.2606	0.2505	0.2410
0.70	0.3434	0.3278	0.3133	0.2998	0.2873	0.2756	0.2647	0.2546	0.2451	0.2362
0.80	0.3279	0.3142	0.3013	0.2892	0.2779	0.2672	0.2572	0.2479	0.2390	0.2307
0.90	0.3115	0.2997	0.2885	0.2779	0.2678	0.2582	0.2492	0.2406	0.2325	0.2248
1.00	0.2946	0.2847	0.2752	0.2660	0.2571	0.2487	0.2406	0.2329	0.2255	0.2185
1.20	0.2602	0.2540	0.2476	0.2412	0.2349	0.2286	0.2224	0.2164	0.2106	0.2049
1.40	0.2268	0.2237	0.2202	0.2163	0.2123	0.2080	0.2037	0.1993	0.1949	0.1906
1.60	0.1957	0.1952	0.1941	0.1923	0.1902	0.1878	0.1851	0.1822	0.1792	0.1761
1.80	0.1680	0.1694	0.1700	0.1700	0.1695	0.1685	0.1672	0.1656	0.1637	0.1617
2.00	0.1438	0.1465	0.1485	0.1497	0.1504	0.1506	0.1504	0.1499	0.1490	0.1480
2.20	0.1233	0.1268	0.1296	0.1317	0.1333	0.1344	0.1350	0.1353	0.1353	0.1350
2.40	0.1060	0.1099	0.1132	0.1159	0.1181	0.1198	0.1211	0.1220	0.1226	0.1229
2.60	0.0916	0.0957	0.0992	0.1023	0.1048	0.1069	0.1086	0.1100	0.1111	0.1118
2.80	0.0796	0.0837	0.0873	0.0905	0.0932	0.0956	0.0976	0.0993	0.1007	0.1018
3.00	0.0697	0.0736	0.0772	0.0804	0.0832	0.0857	0.0879	0.0897	0.0914	0.0927
3.20	0.0614	0.0652	0.0686	0.0717	0.0745	0.0771	0.0793	0.0813	0.0830	0.0846
3.40	0.0544	0.0580	0.0612	0.0643	0.0670	0.0695	0.0718	0.0738	0.0756	0.0773
3.60	0.0486	0.0519	0.0550	0.0578	0.0605	0.0629	0.0652	0.0672	0.0691	0.0709
3.80	0.0436	0.0467	0.0496	0.0523	0.0548	0.0572	0.0594	0.0614	0.0632	0.0649
4.00	0.0393	0.0422	0.0449	0.0475	0.0499	0.0521	0.0542	0.0562	0.0580	0.0597
4.20	0.0356	0.0383	0.0408	0.0432	0.0455	0.0477	0.0497	0.0516	0.0534	0.0550
4.40	0.0324	0.0349	0.0373	0.0396	0.0417	0.0438	0.0457	0.0475	0.0492	0.0508
4.60	0.0297	0.0320	0.0342	0.0363	0.0383	0.0403	0.0421	0.0439	0.0455	0.0471
4.80	0.0272	0.0294	0.0314	0.0334	6.0354	0.0372	0.0389	0.0406	0.0422	0.0437
5.00	0.0251	0.0271	0.0290	0.0309	0.0327	0.0344	0.0361	0.0377	0.0392	0.0406
5.50	0.0207	0.0224	0.0240	0.0256	0.0272	0.0287	0.0302	0.0316	0.0329	0.0342
6.00	0.0174	0.0188	0.0202	0.0216	0.0230	0.0243	0.0256	0.0268	0.0280	0.0292
6.50	0.0148	0.0160	0.0173	0.0185	0.0196	0.0208	0.0219	0.0230	0.0241	0.0252
7.00	0.0127	0.0138	0.0149	0.0160	0.0170	0.0180	0.0190	0.0200	0.0209	0.0219
7.50	0.0111	0.0120	0.0130	0.0139	0.0148	0.0157	0.0166	0.0175	0.0183	0.0192
8.00	0.0097	0.0106	0.0114	0.0122	0.0131	0.0139	0.0147	0.0154	0.0162	0.0169
8.50	0.0086	0.0094	0.0101	0.0109	0.0116	0.0123	0.0130	0.0137	0.0144	0.0151
9.00	0.0077	0.0084	0.0090	0.6097	0.0103	0.0110	0.0116	0.0123	0.0129	0.0135
9.50	0.0069	060075	0.0081	0.0087	0.0093	0,0099	0.0105	0.0110	0.0116	0.0122
10.00	0.0062	0.0068	0.0073	0.0079	0.0084	0.0089	0.0095	0.0100	0.0105	0.0110

$w \backslash a$	2.20	2.40	2.60	2.80	3.00	3.20	3.40	3.60	3.80	4.00
0.00	0.2356	0.2185	0.2036	0.1905	0.1790	0.1687	0.1595	0.1513	0.1438	0.1370
0.10	0.2353	0.2182	0.2034	0.1904	0.1788	0.1686	0.1594	0.1512	0.1437	0.1369
0.20	0.2343	0.2174	0.2027	0.1898	0.1784	0.1682	0.1591	0.1509	0.1434	0.1367
0.30	0.2326	0.2160	0.2016	0.1889	0.1776	0.1675	0.1585	0.1504	0.1430	0.1363
0.40	0.2303	0.2142	0.2000	0.1876	0.1765	0.1666	0.1577	0.1497	0.1424	0.1358
0.50	0.2274	0.2118	0.1981	0.1859	0.1751	0.1654	0.1567	0.1488	0.1417	0.1352
0.60	0.2240	0.2090	0.1957	0.1839	0.1734	0.1640	0.1555	0.1478	0.1408	0.1344
0.70	0.2200	0.2057	0.1930	0.1817	0.1715	0.1623	0.1541	0.1466	0.,1397	0.1334
0.80	0.2155	0.2020	0.1899	0.1791	0.1693	0.1605	0.1525	0.1452	0.1385	0.1324
0.90	0.2107	0.1980	0.1866	0.1762	0.1669	0.1584	0.1507	0.1436	0.1371	0.1312
1.00	0.2055	0.1936	0.1829	0.1731	0.1643	0.1561	0.1487	0.1419	0.1357	0.1299
1.20	0.1942	0.1842	0.1749	0.1663	0.1584	0.1511	0.1444	0.1381	0.1323	0.1270
1.40	0.1821	0.1740	0.1662	0.1589	0.1520	0.1456	0.1395	0.1339	0.1286	0.1237
1.60	0.1697	0.1633	0.1571	0.1510	0.1451	0.1396	0.1343	0.1293	0.1245	0.1201
1.80	0.1573	0.1526	0.1477	0.1428	0.1380	0.1333	0.1288	0.1244	0.1202	0.1162
2.00	0.1452	0.1420	0.1384	0.1346	0.1308	0.1269	0.1231	0.1193	0.1157	0.1121
2.20	0.1337	0.1317	0.1293	0.1265	0.1235	0.1204	0.1173	0.1141	0.1110	0.1080
2.40	0.1228	0.1219	0.1204	0.1185	0.1164	0.1140	0.1115	0.1089	0.1063	0.1037
2.60	0.1126	0.1126	0.1120	0.1109	0.1094	0.1077	0.1058	0.1037	0.1016	0.0994
2.80	0.1033	0.1040	0.1041	0.1036	0.1028	0.1016	0.1002	0.0986	0.0969	0.0951
3.00	0.0947	0.0960	0.0966	0.0967	0.0964	0.0957	0.0948	0.0937	0.0924	0.0909
3.20	0.0870	0.0887	0.0897	0.0903	0.0904	0.0901	0.0896	0.0888	0.0879	0.0868
3.40	0.0799	0.0819	0.0833	0.0842	0.0847	0.0848	0.0846	0.0842	0.0836	0.0828
3.60	0.0736	0.0758	0.0774	0.0786	0.0794	0.0798	0.0799	0.0798	0.0794	0.0789
3.80	0.0678	0.0702	0.0720	0.0734	0.0744	0.0751	0.0754	0.0756	0.0754	0.0752
4.00	0.0626	0.0651	0.0670	0.0686	0.0698	0.0707	0.0712	0.0715	0.0717	0.0716
4.20	0.0580	0.0604	0.0625	0.0642	0.0655	0.0665	0.0673	0.0678	0.0680	0.0681
4.40	0.0537	0.0562	0.0583	0.0601	0.0615	0.0627	0.0635	0.0642	0.0646	0.0649
4.60	0.0499	0.0524	0.0545	0.0563	0.0578	0.0591	0.0600	0.0608	0.0614	0.0618
4.80	0.0464	0.0489	0.0510	0.0528	0.0544	0.0557	0.0568	0.0576	0.0583	0.0588
5.00	0.0433	0.0457	0.0478	0.0496	0.0512	0.0526	0.0537	0.0547	0.0554	0.0560
5.50	0.0367	0.0389	0.0409	0.0427	0.0443	0.0457	0.0469	0.0480	0.0489	0.0496
6.00	0.0314	0.0334	0.0353	0.0370	0.0386	0.0400	0.0412	0.0423	0.0433	0.0441
6.50	0.0271	0.0290	0.0307	0.0323	0.0338	0.0351	0.0364	0.0375	0.0385	0.0394
7.00	0.0236	0.0253	0.0269	0.0284	0.0298	0.0311	0.0323	0.0334	0.0344	0.0353
7.50	0.0208	0.0223	0.0238	0.0251	0.0264	0.0277	0.0288	0.0298	0.0308	0.0317
8.00	0.0184	0.0198	0.0211	0.0224	0.0236	0.0247	0.0258	0.0268	0.0277	0.0286
8.50	0.0164	0.0177	0.0189	0.0201	0.0212	0.0222	0.0232	0.0242	0.0251	0.0259
9.00	0.0147	0.0159	0.0170	0.0181	0.0191	0.0201	0.0210	0.0219	0.0227	0.0236
9.50	0.0132	0.0143	0.0153	0.0163	0.0173	0.0182	0.0191	0.0199	0.0207	0.0215
10.00	0.0120	0.0130	0.0139	0.0148	0.0157	0.0166	0.0174	0.0182	0.0189	0.0197

w\a	4.50	5.00	5.50	6.00	6.50	7.00	7.50	8.00	9.00	10.00
0.00	0.1225	0.1107	0.1010	0.0928	0.0858	0.0798	0.0746	0.0700	0.0623	0.0561
0.10	0.1224	0.1107	0.1009	0.0928	0.0858	0.0798	0.0746	0.0700	0.0623	0.0561
0.20	0.1223	0.1105	0.1008	0.0927	0.0857	0.0797	0.0745	0.0699	0.0623	0.0561
0.30	0.1220	0.1103	0.1007	0.0926	0.0856	0.0797	0.0745	0.0699	0.0622	0.0561
0.40	0.1216	0.1101	0.1005	0.0924	0.0855	0.0796	0.0744	0.0698	0.0622	0.0561
0.50	0.1211	0.1097	0.1002	0.0922	0.0853	0.0794	0.0743	0.0697	0.0621	0.0560
0.60	0.1206	0.1093	0.0999	0.0919	0.0851	0.0792	0.0741	0.0696	0.0620	0.0559
0.70	0.1199	0.1088	0.0995	0.0916	0.0849	0.0790	0.0740	0.0695	0.0619	0.0559
0.80	0.1191	0.1082	0.0990	0.0913	0.0846	0.0788	0.0738	0.0693	0.0618	0.0558
0.90	0.1182	0.1075	0.0985	0.0909	0.0843	0.0786	0.0736	0.0691	0.0617	0.0557
1.00	0.1173	0.1068	0.0980	0.0904	0.0839	0.0783	0.0733	0.0689	0.0616	0.0556
1.20	0.1151	0.1052	0.0967	0.0894	0.0831	0.0776	0.0728	0.0685	0.0613	0.0554
1.40	0.1127	0.1033	0.0952	0.0883	0.0822	0.0769	0.0722	0.0680	0.0609	0.0551
1.60	0.1099	0.1012	0.0936	0.0870	0.0811	0.0760	0.0715	0.0674	0.0605	0.0548
1.80	0.1070	0.0989	0.0918	0.0855	0.0800	0.0751	0.0707	0.0667	0.0600	0.0544
2.00	0.1039	0.0965	0.0899	0.0840	0.0787	0.0740	0.0698	0.0660	0.0595	0.0540
2.20	0.1007	0.0939	0.0879	0.0823	0.0774	0.0729	0.0689	0.0652	0.0589	0.0536
2.40	0.0973	0.0913	0.0857	0.0806	0.0760	0.0717	0.0679	0.0644	0.0583	0.0532
2.60	0.0939	0.0886	0.0835	0.0788	0.0745	0.0705	0.0668	0.0635	0.0576	0.0527
2.80	0.0905	0.0858	0.0812	0.0769	0.0729	0.0692	0.0658	0.0626	0.0569	0.0521
3.00	0.0871	0.0830	0.0789	0.0750	0.0713	0.0679	0.0646	0.0616	0.0562	0.0516
3.20	0.0837	0.0802	0.0766	0.0731	0.0697	0.0665	0.0635	0.0606	0.0555	0.0510
3.40	0.0803	0.0774	0.0743	0.0711	0.0680	0.0651	0.0622	0.0596	0.0547	0.0504
3.60	0.0770	0.0746	0.0719	0.0692	0.0664	0.0636	0.0610	0.0585	0.0539	0.0498
3.80	0.0738	0.0719	0.0696	0.0672	0.0647	0.0622	0.0598	0.0574	0.0531	0.0492
4.00	0.0707	0.0692	0.0673	0.0652	0.0630	0.0607	0.0585	0.0563	0.0523	0.0485
4.20	0.0677	0.0666	0.0651	0.0633	0.0613	0.0593	0.0572	0.0552	0.0514	0.0479
4.40	0.0648	0.0641	0.0629	0.0613	0.0596	0.0578	0.0560	0.0541	0.0505	0.0472
4.60	0.0621	0.0616	0.0607	0.0595	0.0580	0.0564	0.0547	0.0530	0.0497	0.0465
4.80	0.0594	0.0593	0.0586	0.0576	0.0563	0.0549	0.0534	0.0519	0.0488	0.0458
5.00	0.0568	0.0570	0.0566	0.0558	0.0547	0.0535	0.0521	0.0507	0.0479	0.0451
5.50	0.0510	0.0516	0.0517	0.0514	0.0508	0.0500	0.0490	0.0480	0.0457	0.0433
6.00	0.0458	0.0468	0.0472	0.0473	0.0471	0.0466	0.0460	0.0452	0.0434	0.0415
6.50	0.0412	0.0424	0.0432	0.0436	0.0437	0.0435	0.0431	0.0426	0.0413	0.0397
7.00	0.0371	0.0385	0.0395	0.0401	0.0404	0.0405	0.0404	0.0401	0.0391	0.0379
7.50	0.0336	0.0351	0.0362	0.0370	0.0375	0.0377	0.0378	0.0377	0.0371	0.0362
8.00	0.0305	0.0320	0.0332	0.0341	0.0347	0.0351	0.0354	0.0354	0.0351	0.0345
8.50	0.0278	0.0293	0.0305	0.0315	0.0322	0.0328	0.0331	0.0333	0.0332	0.0328
9.00	0.0254	0.0269	0.0281	0.0292	0.0300	0.0306	0.0310	0.0313	0.0314	0.0312
9.50	0.0232	0.0247	0.0260	0.0270	0.0279	0.0235	0.0290	0.0294	0.0297	0.0297
10.00	0.0213	0.0228	0.0240	0.0251	0.0259	0.0267	0.0272	0.0276	0.0281	0.0283

C Matlab Voigt Fitting Program

The following algorithm was first published by Humlíček (*J. Quant. Spectrosc. Radiat. Transfer*, Vol. 27, No. 4, pp. 437–444, 1982).

```
function [W] = Voigt(X,Y)
%VOIGT  Normalized Voigt profile
%
%   [W]=Voigt(X,Y)
%
%  Uses Humlicek's algorithm for calculating the
    Voigt profile
%
%  X = position/frequency
%  Y= Voigt ''a" parameter (ratio of Lorentz to
    Doppler widths)
%  W = Voigt value
%  area = sqrt(pi)
%  width (FWHM) = (Y+sqrt(Y*Y+4*ln(2)))(approximation)
%  linecenter is at X=0
%  amplitude = Voigt(0,Y)
%
%   To use with curve-fitting or for simulating
    absorption spectra,
%       use this function the following way:
%
%   Lineshape = amp* Voigt((2*sqrt(log(2))/WG)
    *(x-x0),a)
%   amp = 2*(sqrt(ln(2))*S*P*xj/(sqrt(pi)*WG)
%   log(2) = ln(2)
%   WG = Doppler FWHM
```


R.K. Hanson et al., *Spectroscopy and Optical Diagnostics for Gases*,
DOI 10.1007/978-3-319-23252-2

```
%   x = frequency position at which to calculate the
    Voigt function
%   x0 = linecenter
%   a = Voigt ''a" parameter
%
%   For calculating the amplitude, S is the
    integrated linestrength
%       [cm^(-2)/atm], P is the pressure [atm],
        and xj is the
%       molefraction of species of interest

T = complex(Y,-X); S = abs(X)+Y;
if S >= 15                                  %Region I
   W= T*0.5641896/(0.5+T*T);
else
   if S >= 5.5                              %Region II
      U= T*T;
      W= T*(1.410474+U*0.5641896)/(0.75+U*(3+U));
   else
      if Y >= (0.195*abs(X)-0.176)  %Region III
         Wnum= (16.4955+T*(20.20933+T*(11.96482+T*...
            (3.778987+T*0.5642236))));
         Wden= (16.4955+T*(38.82363+T*(39.27121+T*...
            (21.69274+T*(6.699398+T)))));
         W=Wnum/Wden;
      else                                  %Region IV
         U= T*T;
         Wnum=T*(36183.31-U*(3321.9905-U*
            (1540.787-U*...(  219.0313-U*(35.76683-U*
            (1.320522-U*0.56419))))));
         Wden =(32066.6-U*(24322.84-U*(9022.228-U*
            (2186.181...-U*(364.2191-U*
            (61.57037-U*(1.841439-U)))))));
         W=Wnum/Wden;
         W= complex(exp(real(U))*cos(imag(U)),0)-W;
      end
   end
end W = real(W);
```

HITRAN Database

D

The HITRAN database (updated $\approx$ every 4 years) is a popular compilation of spectroscopic data. This database provides line positions and strengths, collisional-broadening coefficients, and other spectroscopic parameters based on molecular models and experimental data. Using a database such as HITRAN gives useful predictive information about the spectral features for many different molecules. The molecules for which the current edition of HITRAN (HITRAN2012) has line-by-line parameters are the following:

H_2O	CO_2	O_3	N_2O	CO	CH_4	O_2
NO	SO_2	NO_2	NH_3	HNO_3	OH	HF
HCl	HBr	HI	ClO	OCS	H_2CO	HOCl
N_2	HCN	CH_3Cl	H_2O_2	C_2H_2	C_2H_6	PH_3
COF_2	SF_6	H_2S	HCOOH	HO_2	O	$ClONO_2$
NO^+	HOBr	C_2H_4	CH_3OH	CH_3Br	CH_3CN	CF_4
C_4H_2	HC_3N	H_2	CS	SO_3		

HITRAN is organized by Rothman et al. [1] and can be visualized using free online simulation tools such as *HITRAN on the Web* (hitran.iao.ru) or *SpectraPlot* (SpectraPlot.com).

Though HITRAN is not entirely comprehensive, it does have information for many rovibrational bands. For molecules that have been well investigated, such as CO, CO_2 and H_2O, the database is quite thorough. For high-temperature studies, one can also use the HITEMP database [2] which uses the same format as HITRAN, but includes more high-energy transitions for CO, CO_2, H_2O, NO, and OH. A small portion of the HITRAN2012 database is shown in Tables D.1 and D.2.

Most databases list pertinent spectroscopic information with respect to a reference temperature. For the HITRAN and HITEMP databases, the reference temperature is $T_0 = 296$ K. The linestrengths can be scaled for different

R.K. Hanson et al., *Spectroscopy and Optical Diagnostics for Gases*,
DOI 10.1007/978-3-319-23252-2

Table D.1 Sample HITRAN2012 output for CO_2 for v_o from 2311.105 to 2311.12 cm^{-1}

Molec	Isotope	ν_0	$S(T_o)$	A	γ_{air}	γ_{self}	E''	n	δ	v'	v''	Q'	Q''
2	1	2311.105404	4.749e−19	2.059e+02	0.06780	0.079	704.3005	0.75	−0.002938	0 0 0 11	0 0 0 01		P 42e
2	3	2311.105622	6.385e−25	1.910e+02	0.07480	0.102	2002.1799	0.69	−0.002557	1 1 1 12	1 1 1 02		R 16f
2	1	2311.109065	6.926e−26	1.859e+02	0.06880	0.090	3864.8738	0.75	−0.002897	2 1 1 11	2 1 1 01		R 30f
2	3	2311.111845	3.576e−25	1.929e+02	0.07060	0.096	2192.6426	0.73	−0.002774	0 3 3 11	0 3 3 01		R 23e
2	3	2311.111846	3.576e−25	1.929e+02	0.07060	0.096	2192.6426	0.73	−0.002774	0 3 3 11	0 3 3 01		R 23f
2	1	2311.112179	1.865e−27	8.017e−02	0.06620	0.071	3134.5498	0.71	−0.004059	2 2 2 01	1 1 1 02		R 55e
2	1	2311.113564	8.251e−28	6.538e−02	0.06570	0.069	3271.0081	0.70	−0.004137	2 2 2 01	1 1 1 02		R 58f
2	3	2311.114731	1.309e−29	1.792e−03	0.06790	0.079	2023.9087	0.75	−0.005979	1 0 0 11	0 2 2 01		P 43e
2	1	2311.117019	2.306e−25	3.729e−01	0.06860	0.088	2345.9209	0.76	−0.002897	1 1 1 12	1 1 1 02		Q 32f
2	1	2311.118396	1.158e−27	1.811e+02	0.06770	0.079	4767.5566	0.75	−0.002951	3 0 0 12	3 0 0 02		R 42e
2	1	2311.121386	1.886e−23	1.964e+01	0.07840	0.105	2074.6543	0.69	−0.002392	0 3 3 11	0 3 3 01		Q 13e

Uncertainty and reference codes are not shown

Table D.2 Key to symbols in Table D.1

Molec:	Molecule number in the database (i.e., 1 for H_2O, 2 for CO_2, etc.)
Isotope:	Isotope number (1 = most abundant, 2 = second most abundant, etc.)
ν_0:	Frequency of line center [cm^{-1}]
S:	Linestrength [$cm^{-1}/(molecule\,cm^{-2})$] at the specified temperature
A:	Einstein-A coefficient [s^{-1}] for the transition
γ_{air}:	Air-broadening coefficient [$cm^{-1}\,atm^{-1}$] (HWHM) at $T_0 = 296\,K$ [Note: This is the halfwidth, not the fullwidth → $2\gamma_{air}$ is the fullwidth]
γ_{self}:	Self-broadening coefficient [$cm^{-1}\,atm^{-1}$] (HWHM) at $T_0 = 296\,K$ [Note: This is the halfwidth, not the fullwidth →$2\gamma_{self}$ is the fullwidth]
E'':	Lower-state energy [cm^{-1}] relative to the zero vibrational level
n:	Exponent of temperature dependence of air-broadened halfwidth
δ:	Air-pressure-shift coefficient [$cm^{-1}\,atm^{-1}$] of the transition at $T_0 = 296\,K$
v',v'':	Upper, lower global quanta index (vibrational level)— use this index to look up the vibrational states for v' and v''
Q',Q'':	Upper, lower local quanta index (rotational energy level) [Note: for carbon dioxide, Q' is not specified, and Q'' denotes the branch and lower-state rotational quantum number]

temperatures using the reference temperature, lower-state energy, fundamental vibrational energies, and the reference linestrength [see Eq. (7.75)]; for nonlinear polyatomic molecules, when units of linestrength are $cm^{-1}/molec\,cm^{-2}$,

$$S(T) = S(T_0)\left(\frac{T_0}{T}\right)^{3/2}\left[\frac{1-\exp\left(-\frac{hc\nu_0}{kT}\right)}{1-\exp\left(-\frac{hc\nu_0}{kT_0}\right)}\right]$$
$$\exp\left(-\frac{hcE''}{k}\left(\frac{1}{T}-\frac{1}{T_0}\right)\right)\prod_{i=1}^{n}\left[\frac{1-\exp\left(-\frac{hc\nu_i}{kT}\right)}{1-\exp\left(-\frac{hc\nu_i}{kT_0}\right)}\right], \qquad (D.1)$$

where ν_i are the different fundamental vibrational energies and n is the total number of vibrational modes. The $(T_0/T)^{3/2}$ term results from thc fact that $Q_{rot} \propto T^{3/2}$. Thus, for H_2O, the scaling formula is given as follows:

$$S(T) = S(T_0)\left(\frac{T_0}{T}\right)^{3/2}\left[\frac{1-\exp\left(-\frac{hc\nu_0}{kT}\right)}{1-\exp\left(-\frac{hc\nu_0}{kT_0}\right)}\right]\exp\left(-\frac{hcE''}{k}\left(\frac{1}{T}-\frac{1}{T_0}\right)\right)$$
$$\left[\frac{1-\exp\left(-\frac{hc\nu_1}{kT}\right)}{1-\exp\left(-\frac{hc\nu_1}{kT_0}\right)}\right]\left[\frac{1-\exp\left(-\frac{hc\nu_2}{kT}\right)}{1-\exp\left(-\frac{hc\nu_2}{kT_0}\right)}\right]\left[\frac{1-\exp\left(-\frac{hc\nu_3}{kT}\right)}{1-\exp\left(-\frac{hc\nu_3}{kT_0}\right)}\right], \qquad (D.2)$$

where $\nu_1 = 3657.05\,\text{cm}^{-1}$, $\nu_2 = 1594.75\,\text{cm}^{-1}$, and $\nu_3 = 3755.93\,\text{cm}^{-1}$. Often times it is useful to scale the linestrengths from the [cm^{-1}/molecule-cm^{-2}] units to [cm^{-2} atm^{-1}], for which the conversion is

$$S[\text{cm}^{-2}\,\text{atm}^{-1}] = \frac{7.34 \times 10^{21}}{T_0} \times S[\text{cm}^{-1}/\text{molecule}\,\text{cm}^{-2}]. \tag{D.3}$$

For cases where $T_0 = 296\,\text{K}$, the conversion reduces to

$$S[\text{cm}^{-2}\,\text{atm}^{-1}] = 2.4797 \times 10^{19} S[\text{cm}^{-1}/\text{molecule}\,\text{cm}^{-2}]. \tag{D.4}$$

Note that

$$\frac{S(T)[\text{cm}^{-1}/\text{molecule}\,\text{cm}^{-2}]}{S(T_0)[\text{cm}^{-1}/\text{molecule}\,\text{cm}^{-2}]} = \frac{T}{T_0}\frac{S(T)[\text{cm}^{-2}\,\text{atm}^{-1}]}{S(T_0)[\text{cm}^{-2}\,\text{atm}^{-1}]} \tag{D.5}$$

D.1 Example Calculation of H_2O Absorbance Spectra using the HITRAN Database

To illustrate many of the concepts put forth in this book, this section will present an example calculation of absorbance spectra. The transitions (i.e., lines) of interest here belong to a H_2O doublet near 1392.67 nm that has been used extensively to measure water vapor using TDLAS. Here, we will calculate the absorbance spectrum of this doublet at: T = 1000 K, P = 1 atm, and with 10% H_2O by mole in air over a 10 cm path length. The spectroscopic parameters needed to calculate its absorbance spectrum, taken from HITRAN2012, are given in Table D.3.

D.1.1 Calculation of Linestrength at T

The HITRAN database tabulates linestrengths at 296 K. To calculate the linestrength of these transitions at 1000 K we can use Eq. D.2 or:

$$S(T) = S(T_o)\frac{Q(T_o)}{Q(T)} exp\left[-\frac{hcE''}{k}\left(\frac{1}{T} - \frac{1}{T_o}\right)\right] \times \left[1 - exp\left(-\frac{hc\nu_o}{kT}\right)\right]\left[1 - exp\left(-\frac{hc\nu_o}{kT_o}\right)\right]^{-1} \tag{D.6}$$

for improved accuracy. Eq. D.6 requires an independent evaluation of the partition function Q at T and T_o (available online as supplementary material for the HITRAN database) whereas Eq. D.2 evaluates $Q(T)/Q(T_o)$ assuming a rigid-rotor harmonic oscillator (RRHO). Fig. D.1 compares these two methods, indicating that Eq. D.2

Table D.3 Sample HITRAN2012 output for the H_2O doublet of interest

Line	ν_0	$S(T_o)$	γ_{air}	γ_{self}	E''	n_{air}
1	7185.596571	2.00E-22	0.0342	0.371	1045.0583	0.62
2	7185.596909	5.98E-22	0.0421	0.195	1045.0577	0.62

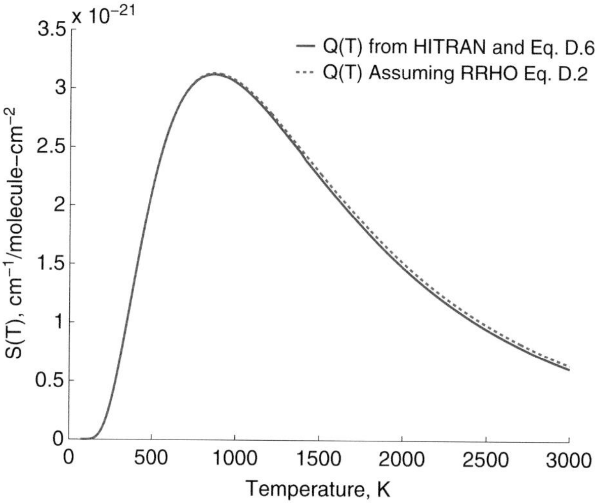

Fig. D.1 Comparison of linestrength as a function of temperature calculated using Eq. D.6 and Eq. D.2 for Lines 1 and 2

is accurate to within 2% of that calculated using Eq. D.6 at 296 to 1500 K, but its accuracy decreases near linearly to within only 5% of that calculated using Eq. D.6 from 1500 to 3000 K.

Using Eq. D.6, the linestrength of Lines 1 and 2 at 1000 K are 1.02×10^{-21} and 3.05×10^{-21} cm^{-1}/molecule-cm^{-2}, respectively.

D.1.2 Calculation of Lineshape Function

Next we will calculate the lineshape function of these transitions using a Voigt profile (to be consistent with HITRAN) and the numerical approximation given in Appendix C. To do so, we must first calculate the collisional FWHM ($\Delta\nu_c$) using Eq. 8.19 (with $\gamma(1000K)$ calculated using Eq. 8.21) and the Doppler FWHM ($\Delta\nu_D$) using Eq. 8.25. To evaluate the self-broadening coefficients ($\gamma_{H_2O-H_2O}$) at 1000 K we will assume a self-broadening temperature exponent of 0.75.

Due to their similar ν_o, $\Delta\nu_D(1000K) = 0.0384$ cm^{-1} for both transitions. For the pressure and mixture of interest, $\Delta\nu_c(1000K) = 0.0587$ and 0.0513 cm^{-1}, for Lines 1 and 2 respectively. Fig. D.2 shows the corresponding Voigt lineshapes of these transitions.

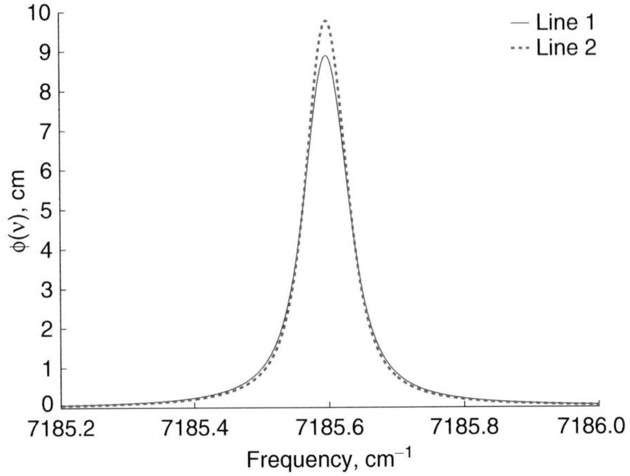

Fig. D.2 Voigt lineshapes of Line 1 and 2 at 1000 K, 1 atm, with 10% H_2O in air

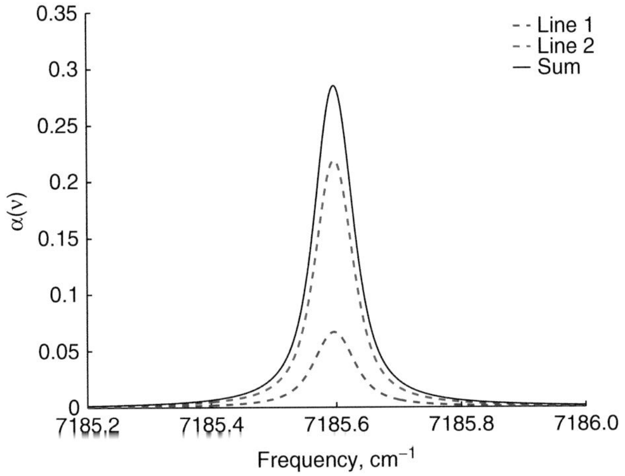

Fig. D.3 Total absorbance spectrum and individual contributions of each line at the conditions of interest

D.1.3 Calculation of Absorbance

With the linestrength and lineshapes of the transitions of interest known, we can now calculate the absorbance spectrum resulting from these transitions. For convenience, we will convert the calculated linestrengths from units of cm^{-1}/molecule-cm^{-2} to

cm^{-2}/atm using Eq. D.3 which yields linestrengths of 7.487×10^{-3} and 2.237×10^{-3} cm^{-2}/atm. Now the resulting absorbance spectrum can be calculated using $\alpha(\nu) = \sum_j S_j(T) P \chi_{H_2O} \phi_j(\nu) L$, where j indicates a given transition and $S(T)$ is in units of cm^{-2}/atm. Note that the absorbance at a given wavelength is given by the sum of that corresponding to each line/transition. Fig. D.3 shows the absorbance of each transition and the sum for this example.

Center of Symmetry

E

The following discussion on the definition of the center of symmetry is taken from Herzberg, *"Molecular Spectra and Molecular Structure, Vol. 2, Infrared and Raman Spectra of Polyatomic Molecules."*

A center of symmetry is usually designated by *i*. By carrying out the corresponding symmetry operation, *reflection at the center (inversion)*, a molecule having such a center is transformed into itself. In other words, if a line is drawn from one atom through the center and continued it will meet an equal atom at the same distance from the center but on the opposite side (if x, y, and z are the coordinates of one atom with respect to the origin, $-x$, $-y$, and $-z$ are the coordinates of the other equal atom). Examples are molecules X_2Y_4, $X_2Y_2Z_2$, XY_2Z_2 if they have the structures indicated in Figs. a–c. A molecule can have only one center of symmetry. There may or may not be an atom at the center of symmetry (see examples $X_2Y_2Z_2$ and XY_2Z_2). All other atoms occur in pairs.

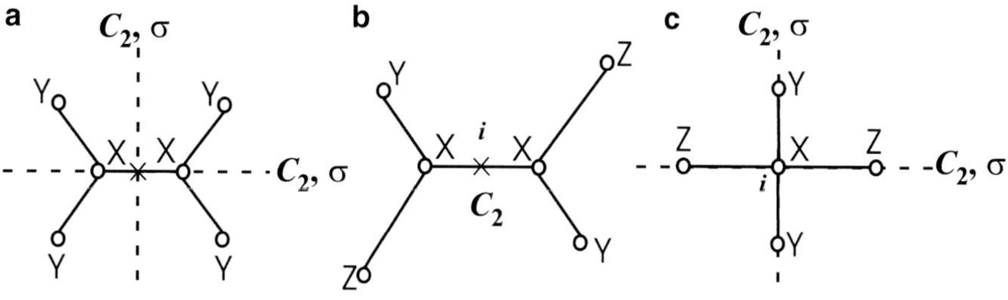

R.K. Hanson et al., *Spectroscopy and Optical Diagnostics for Gases*,
DOI 10.1007/978-3-319-23252-2

Fluorescence Yield: Multi-Level Models

F

Summary: The purpose of this section is twofold:

1. to show that a proper definition of the fluorescence yield requires specification of the spectral response of the light collection system.
2. to show that the simple two-level model also applies (with reasonable or good accuracy) to molecules with multiple coupled energy levels, if the light collection is "broadband" and the relevant A and Q parameters are independent of quantum number.

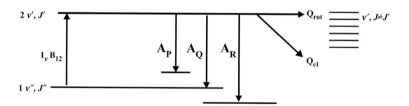

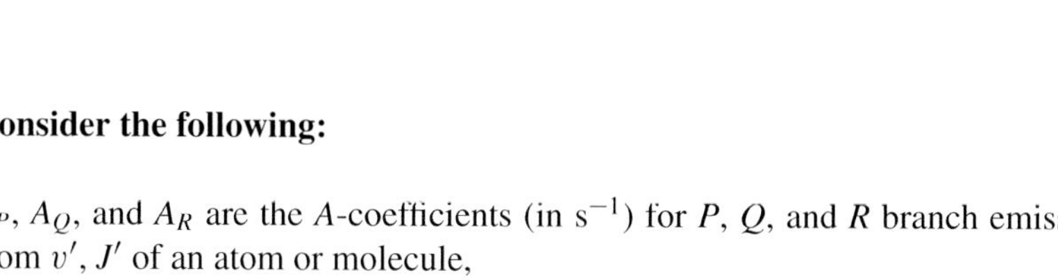

Consider the following:

1. A_P, A_Q, and A_R are the A-coefficients (in s^{-1}) for P, Q, and R branch emission from v', J' of an atom or molecule,

   Q_{el} is the electronic quench rate [s^{-1}]
   Q_{rot} is the rotational transfer rate [s^{-1}]

2. We assume that these rates (probabilities) are the same for all J'. Let's examine two interesting limiting cases:
   (a) "Narrowband collection"—where the collection optics (usually a monochromator) are set to transmit emission from only one line, say the P line.

R.K. Hanson et al., *Spectroscopy and Optical Diagnostics for Gases*,
DOI 10.1007/978-3-319-23252-2

Then the fluorescence yield (FY), which is normally thought of as the fraction of absorbed photons converted into emitted photons, should be more narrowly defined as the *fraction of absorbed photons emitted into the collection bandwidth* i.e.,

$$\boxed{\mathrm{FY} = \frac{\text{probability or rate of desired process}}{\text{sum rate of all processes}}}$$

$$\mathrm{FY} = \frac{A_P}{(A_P + A_Q + A_R) + Q_{\mathrm{rot}} + Q_{\mathrm{el}}} \approx \frac{A_P}{Q_{\mathrm{rot}}}(\text{usually})$$

where Q_{rot} is usually $> Q_{\mathrm{el}} \gg A_P, A_Q, A_R$.

(b) "Broadband collection"—usually just a long pass filter so that fluorescence of all lines is collected. This is the most common case, since it yields a larger number of measured photons. Here we must include the fluorescence from all values of J'. We can view this as a multi-step process, with each J-change constituting a new step (fresh start) subject to the same probabilities.

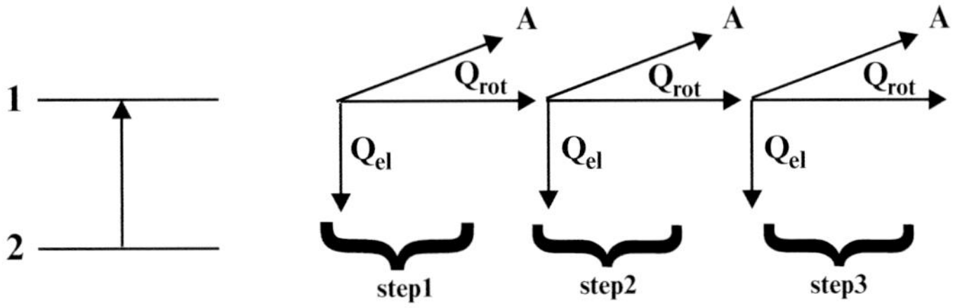

Following absorption, the molecule has three options (A, Q_{el}, Q_{rot})

i. the fluorescence yield for step 1 is

$$\mathrm{FY}_1 = \frac{A}{A + Q_{\mathrm{el}} + Q_{\mathrm{rot}}} \equiv \alpha$$

ii. the probability of going to a new J, and hence having a second chance to fluoresce is

$$\beta = \frac{Q_{\mathrm{rot}}}{A + Q_{\mathrm{el}} + Q_{\mathrm{rot}}}$$

iii. the fluorescence yield for this second step (i.e., its probability) is

$$\mathrm{FY}_2 = \beta\alpha$$

iv. similarly the fluorescence yield for step 3 is

$$FY_3 = \beta\beta\alpha = \beta^2\alpha$$

v. and the total fluorescence yield

$$FY = \sum FY_i = \alpha \sum_{n=0}^{\infty} \beta^n = \frac{\alpha}{1-\beta} = \frac{A}{A + Q_{el}}$$

Thus, we have that important result that the fluorescence yield always involves (in the denominator) the rate (Q) of the primary process which transfers the molecules out of the collection bandwidth!

(c) Note that in the above examples we have not included stimulated emission (with its rate $W_{21} = B_{21}I_\nu$), which implies that we are limiting ourselves to the "weak excitation"or linear LIF regime.

References

1. L.S. Rothman, I.E. Gordon, Y. Babikov, A. Barbe, D. Chris Benner, P.F. Bernath, M. Birk, L. Bizzocchi, V. Boudon, L.R. Brown, A. Campargue, K. Chance, E.A. Cohen, L.H. Coudert, V.M. Devi, B.J. Drouin, A. Fayt, J.-M. Flaud, R.R. Gamache, J.J. Harrison, J.-M. Hartmann, C. Hill, J.T. Hodges, D. Jacquemart, A. Jolly, J. Lamouroux, R.J. Le Roy, G. Li, D.A. Long, O.M. Lyulin, C.J. Mackie, S.T. Massie, S. Mikhailenko, H.S.P. Müller, O.V. Naumenko, A.V. Nikitin, J. Orphal, V. Perevalov, A. Perrin, E.R. Polovtseva, C. Richard, M.A.H. Smith, E. Starikova, K. Sung, S. Tashkun, J. Tennyson, G.C. Toon, V.G. Tyuterev, G. Wagner, The HITRAN2012 molecular spectroscopic database. J. Quant. Spectrosc. Radiat. Transf. **130**, 4–50 (2013)
2. L.S. Rothman, I.E. Gordon, R.J. Barber, H. Dothe, R.R. Gamache, A. Goldman, V.I. Perevalov, S.A. Tashkun, J. Tennyson, HITEMP, the high-temperature molecular spectroscopic database. J. Quant. Spectrosc. Radiat. Transf. **111**(15), 2139–2150 (2010)

Printed in the United States
By Bookmasters